T0074554

The Mathematical Radio

This beautiful art (by an unknown, uncredited artist), with its hint of the coming of color television, appeared on the cover of a 1928 issue of a magazine for radio enthusiasts. The professorial-looking experimenter seated at his workbench bears an uncanny resemblance to how the pure mathematician G. H. Hardy (see the preface) actually looked in 1928, at age 51. If Hardy could have read this book, perhaps he *would* have learned how to solder two wires together without setting himself on fire.

THE MATHEMATICAL RADIO

INSIDE THE MAGIC OF AM, FM, AND SINGLE-SIDEBAND

PAUL J. NAHIN

PRINCETON UNIVERSITY PRESS
Princeton and Oxford

Published by Princeton University Press
41 William Street, Princeton, New Jersey 08540
99 Banbury Road, Oxford OX2 6JX

press.princeton.edu

ISBN 978-0-691-23531-8
ISBN (e-book) 978-0-691-23532-5

British Library Cataloging-in-Publication Data is available

Editorial: Diana Gillooly and Kiran Pandey
Production Editorial: Mark Bellis
Jacket Design: Chris Ferrante
Production: Danielle Amatucci
Publicity: Matthew Taylor and Kathryn Stevens
Copyeditor: Barbara Liguori

This book has been composed in STIX Two Text and Barlow

Printed on acid-free paper. ∞

Printed in the United States of America

10 9 8 7 6 5 4 3 2 1

This book is “dedicated to the one I love”[1]
Patricia Ann

1 I have taken these words from the title of the beautiful 1960s song made famous by both The Shirelles and The Mamas & the Papas. It played endlessly on AM radio 60 years ago, and you can still hear it today on YouTube and Sirius Satellite Radio.

A Note to the Reader

Like any other author, I hope this book appeals to a wide audience, but I *have* written with one particular person in mind: me, 67 years ago, when I was a 16-year-old kid full of excitement about electronics and math but, alas, not so full of technical understanding. I like to think that there are, today, *thousands* (or even more) of 16-year-old boys and girls just like I was in the mid-1950s who, as they read this book, will gain an accelerated understanding of how engineers, physicists, and mathematicians think about technical matters. By intent there is stuff in this book beyond what a 16-year-old almost certainly has so far seen, but better to see the new stuff now, at age 16, than to wait years more, as I did.

At the start of 1922, when broadcast radio was just on the verge of exploding onto the American scene, a member of the Catholic clergy had the foresight to see the enormous social revolution that radio was about to bring about, and he wrote the following: "Jules Verne, in his wildest dreams, never fancied an achievement such as this, and . . . it well-nigh bewilders the imagination."[1] Verne wrote exciting *known engineering* stories, and as you read this book you'll see that the Reverend Coakley was right, that modern radio would have been so utterly unbelievable to Verne (he had died 17 years earlier, almost two years before the first radio transmission of voice and music) that he would have declared it to be pure *fantasy* (a pejorative he reserved to describe the science fiction of H. G. Wells, his great literary competitor).

So, here's the book I wish I could have read all those decades ago. If I had Wells's famous time machine (and knew how to avoid the well-known paradoxes of time travel to the past), I would send a copy back to my young self in 1956 (along, *of course*, with the plans for that wonderful time machine!), to arrive in the mail along with the latest subscription copy of *Popular Electronics* magazine, which I read from cover to cover each month all through my high school years. With the presently known state of physics, I can't do that, but sending this book to you, now, is almost as good.

1 Thomas F. Coakley, "Preaching the Gospel by Wireless," *Catholic World*, January 1922, pp. 516–519.

But what *is* it? I asked.

Ah, if you knew *that* you'd know something nobody knows. It's just It—what we call Electricity.

—an exchange between two characters in Rudyard Kipling's 1902 short story "Wireless," on the mystery of radio

Think of a very long cat. You pull the tail in New York, and it meows in Los Angeles. Radio is the same, only there is no cat.

—"man on the street" idea of radio, showing the mystery continues

Few . . . realize that all through the roar of the big city there are constantly speeding messages between people separated by vast distances, and that, over housetops and even through the walls of buildings and in the very air one breathes, are words written by electricity.

—from a story in the *New York Times* (April 21, 1912) on the sinking of the *Titanic* and on how the "almost magical wireless" had saved hundreds of lives from perishing at sea

[T]here was serious concern among strategists and policy makers that entire segments of the population could be so easily manipulated into thinking that something false was something true. Americans had taken very real, physical actions based on something entirely made up. Pandemonium had ensued. Totalitarian nations were able to manipulate their citizens like this, but in America?

—commentary in Annie Jacobsen's *Area 51: An Uncensored History of America's Top Secret Military Base*, not on the United States Capitol insurrection of January 6, 2021 (the book was published in 2011) but rather on the nationwide hysteria caused in October 1938 by Orson Welles's CBS radio dramatization of a Martian invasion of Earth (based on the 1898 novella *The War of the Worlds* by H. G. Wells).

Contents

Foreword to *The Mathematical Radio*

Andrew J. Simoson

One morning in January 2022, an unusual request appeared in my in-box: "It seems to me that the *book* simply demands an introduction by a mathematician, and I think you're the perfect person to do that."

The *book* refers to the one you might be holding at this moment—which, at my moment of processing the request, consisted of a few first-draft chapters. And the request was by its author, Paul Nahin. Upon scanning through those pages, and encountering Maxwell's equations and extensive allusions to G. H. Hardy's 1940 *A Mathematician's Apology*, my heart missed a few beats. Having been recruited at sporadic semesters over the years by our physics department to teach their Electricity & Magnetism course, yes, I can derive the equations from first principles, but do I know their implications beyond the abstractions? And, yes, Hardy's notions about pure mathematics were familiar to me, as they are to most if not all of you who have read thus far; but how does one say to a mathematical giant, *excuse me, please*, do you really believe what you said?

My response to Paul's request is this threefold-sectioned foreword: (i) Who is Paul Nahin and why among the host of mathematicians did he ask me? (ii) What is the relationship between pure and applied mathematics, and how might we interpret Hardy's claims about the chasm between the two? and (iii) Why is Paul's book intriguing?

Who

Who's your hero?

During my childhood summers, we neighborhood kids were almost always playing sandlot baseball. Each of us, when stepping to home plate, would announce—from names taken from our stash of baseball cards (available from a nearby shop at five for a nickel with a stick of bubble gum as well)—that we were so and so, and point to the fences as did Babe Ruth[1] who thereafter delivered the prophesied homerun. My favorite card was Harmon Killebrew. His name containing both *harm* and *kill* suggested to me that he could crush a ball. Moreover, I vaguely imagined that the person on Killebrew's card wasn't really Killebrew, but that he just pretended to be Killebrew—just as I pretended.

But no, heroes are real.

As I grew, my heroes changed. For example, one of my first submitted journal papers was rejected with an apologetic three-page self-typed personal letter from legendary Paul Halmos (1906–2006). In response to my first published paper came a letter from another mathematical god, analyst Walter Rudin (1921–2010), detailing how my results followed almost immediately from a well-known theorem.[2] Wow! These immortals, as it were, deigned to communicate with mortals such as me.

Then, with increasing years and expanding library, I discovered Pliny the Elder (c. AD 23–79). Like the modern-day one-man Ripley's believe-it-or-not repertoire, Pliny's *Natural History* encyclopedia stretches for a million words, partitioned into 37 volumes.[3] He cataloged both fantastical and actual discoveries. For example, for the former, he said that "when a mountain in Crete was cleft by an earthquake a [human] body 69 feet in height was found." For the latter, of Eratosthenes's calculation that Earth's radius was about 4000 miles, he said, "it was achieved by such subtle reasoning that one is ashamed to be skeptical"; and of Hipparchus's entire body of astronomical feats including solar-eclipse-data-driven calculations that the Moon's distance was about 60 to 70 Earth-radii in the sky, he said, "Hipparchus can never be sufficiently commended." Pliny himself—ever inquisitive—died from breathing residual noxious fumes while investigating the aftermath of the Mount Vesuvius volcanic eruption.

1 On sabbatical at Princeton in 1928, G. H. Hardy—an ardent cricketer from early childhood—wrote, "I read reports in the papers by the hour and worship Babe Ruth and Lou Gehrig [with] batting averages ranging from .387," *The G. H. Hardy Reader*, MAA, 2015, pp. 132–133.

2 For his response to me, see Walter Rudin, Well-distributed measurable sets, *Amer. Math. Monthly* 90 (1983) 41–42.

3 Harvard University Press, 1949. For the three Pliny citations in this paragraph, see vol. 1, book II, pp. 239, 371, and vol. 2, book VII, section xvi, p. 553.

My modern-day Pliny the Elder hero candidate is Professor Emeritus Paul Nahin, a long-time electrical engineer. Well on his way to 37 volumes himself, Paul's expository books are ostensibly written for a literate public who enjoy a blend of story, history, science, and mathematics. At the book fair of the 1999 national Joint Mathematics Meetings (JMM) convention in San Antonio, Princeton University Press's booth featured Paul's hot-off-the-press *The Story of* $\sqrt{-1}$. Beneath this display table was a healthy cache of the same, available for purchase. Curious yet ever frugal, I paged through the book and made a rare decision. I bought my first Nahin book and read it on the flight home.

A few years later, Paul, who just happened to review my first book, characterized it as a "darn good book."[4] Upon publication of a second, I requested that a copy be sent to Paul, perhaps hoping for even heartier commendation. But no, upon receiving the gift, Paul went much further. He wrote a thank you note—and thus began a friendship lasting to this day, in which we share our most recent offerings.[5]

What

Maxwell's equations and Hardy's abyss—these are the two elements that Paul weaves together in *The Mathematical Radio*. To summarize this latter phenomenon, we itemize Hardy's claims from his *Apology*, the first two of which pertain directly to our question, and the remaining ones give insight into the first two items.[6]

1. Engineering is not a useful study for ordinary men.
2. There is the real mathematics of the real mathematicians, and there is what I call the trivial mathematics . . . which includes its practical application, the bridges and steam engines and dynamos.
3. A man's first duty is to be ambitious.

4 In *SIAM Review*, 50:1 (2008) 191–192

5 I have the better of the deal, since his productivity is tenfold that of mine.

6 Hardy's *Apology* with C. P. Snow foreword, Cambridge Univ. Press, 1967, pp. 61, 64, 66, 67, 70, 77, 117, 139.

4. Most people can do nothing at all well.
5. Mathematics is a young man's game.
6. Exposition is for second-rate minds.
7. It is one of the first duties of a professor to exaggerate a little both the importance of his subject and his own importance in it.

Before examining the above list, we go back in time and then leap forward in varying incremental steps to specific episodes until the present date. We ask, *When did we first have the luxury of having both pure and applied mathematics?* Even as far back as the Pleistocene epoch, people have always used a little math so as to pay tribute or taxes, settle property disputes, and build safe structures. But beyond that, when did mathematics bifurcate into quasi-distinct knowledge fields of pure and applied? Somewhat arbitrarily, we submit that a meaningful answer is when we began explaining natural phenomena without reference to mumbo jumbo.

In this context, one of the first philosophers to do so was Thales (620–546 BC). Among Euclid's collected propositions attributed to Thales is Proposition 26: *If two triangles have angle-side-angle or have angle-angle-side in common then the triangles are congruent.* This proposition, as the legend goes, is how Thales determined distances of ships from shore.[7] In Figure 1, imagine that Thales T atop a coastal tower—whose base point A is directly below T—sights ship S (using a larger-than-life compass-like instrument, one arm being a vertical pole and the other free to swivel and lock into position so as to align with his line of sight); whereupon Thales rotates the vertical pole so that the swivel arm now points to marker B (among many markers) along the visible coastline. Since ΔTAS and ΔTAB are congruent, then sea distance AS is coastal distance AB. Thales's trick is both pure and applied geometry; it's a general abstract theorem and untold specific instances of that truism.

Two hundred years later, purportedly emblazoned above the entrance to one of the world's first great schools, Plato's Athens Acad-

7 Euclid's *Elements*, ed. Thomas Heath, vol. I, Dover Press, New York, 1956, pp. 35–36, 304–305.

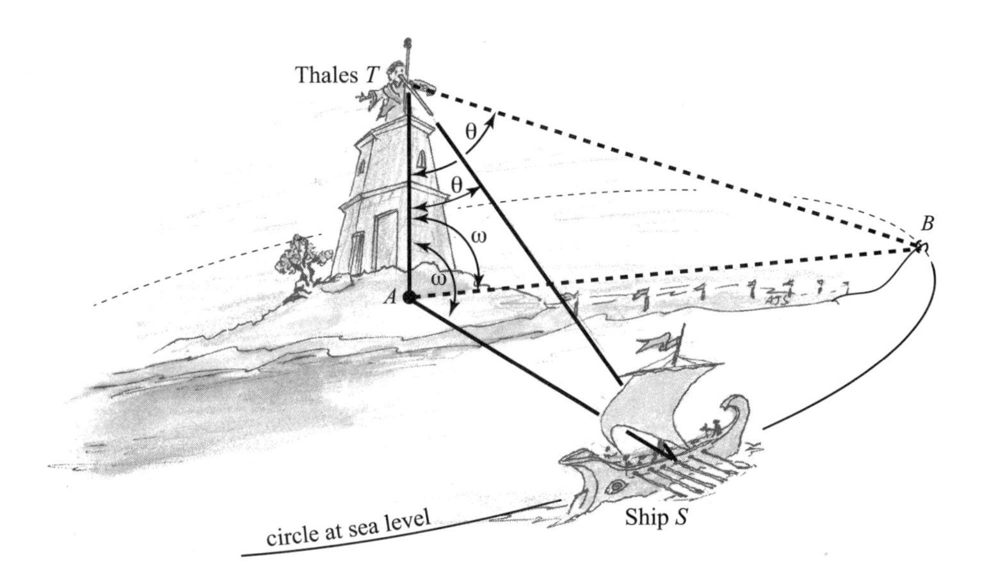

FIGURE 1. Thales measuring a ship's distance at sea, author sketch.

emy, was the motto,[8] "Let no man ignorant of geometry enter here." Plato (c. 429–347 BC) taught that "all the useful arts [such as making shoes] are reckoned mean," and proposed that an ideal education consisted of studying (i) arithmetic, (ii) plane geometry, (iii) solid geometry, (iv) astronomy, (v) music, and (vi) philosophy.[9] A product of this pedagogical tradition was Archimedes (287–212 BC). Most mathematicians believe—among the few opinions which we all might share—that Archimedes was among the greatest mathematicians of all time—perhaps on everyone's top 10 list.[10] Yet, he was also an incredible engineer. Witness Plutarch's description of how Archimedes's boast that, given a place to stand, he could move planet Earth itself, was put to the test by Syracusan King Hiero, who challenged

8 Scholars suspect that this inscription was figurative rather than one cut in stone. A more literal rendition is, "Let no one ignorant of geometry come under my roof," as noted by Bernard Suzanne in "Plato and his dialogues" at https://www.plato-dialogues.org/faq/faq009.htm.

9 See Plato's *The Republic*, ed. B. Jowett, The Modern Library, New York, no date, book VII, pp. 264–278. Plato's last subject includes natural philosophy, or what is now called science.

10 Hardy said, "Archimedes will be remembered when Aeschylus is forgotten," from the *Apology*, p. 81.

Archimedes to launch "a ship of burden which could not be drawn out of the dock without great labour and many men," whereupon,

> [Archimedes, loading the ship] with many passengers and a full freight, sitting himself the while far off, with no great endeavour, but only holding the head of the pulley in his hand and drawing the cords by degrees, he drew the ship in a straight line, as smoothly and evenly as if she had been in the sea.

Yet Plutarch (AD 46–119) went on to say that Archimedes, as Plato believed before him and Hardy would believe long after him, "repudiated as sordid and ignoble the whole trade of engineering, and every sort of art that lends itself to mere use and profit." Instead, Plutarch continued, Archimedes "placed his whole affection and ambition in those purer speculations where there can be no reference to the vulgar needs of life."[11] What are these purer speculations? My favorite is Archimedes' Diophantine riddle of the Sun god's cattle:

> If thou are diligent and wise, O stranger, compute the number of cattle of the Sun, who once upon a time grazed on the fields of the Thrinacian isle of Sicily, divided into four herds of different colors white, black, yellow, and dappled. In each herd were bulls, mighty in number according to these proportions: the white bulls were equal to half and a third of the black together with the whole of the yellow, . . .

and so on, ultimately to give us a total of eight equations. Although Archimedes probably never solved his own riddle, the smallest possible answer is approximately 7.7603×10^{4658} cattle.[12] But just posing this puzzler is the very nature of pure mathematics.

Now, we leap forward to shortly after the fall of Rome. During house arrest prior to execution, Boethius (c. AD 477–524) wrote *The*

11 As evidence that Archimedes held such a view of engineering, Plutarch simply said that Archimedes did "not deign to leave behind him any commentary or writing on such subjects [as engineering]." Nevertheless, Plutarch's explanation suggests that Plutarch himself—and most likely a host of others—held this view. See *Plutarch's Lives*, ed. Arthur Clough, Random House Press, New York, 1932, pp. 376–378.

12 David Bressoud and Stan Wagon, *Computational Number Theory*, Key College Press, Emeryville, CA, 2000, pp. 232–238.

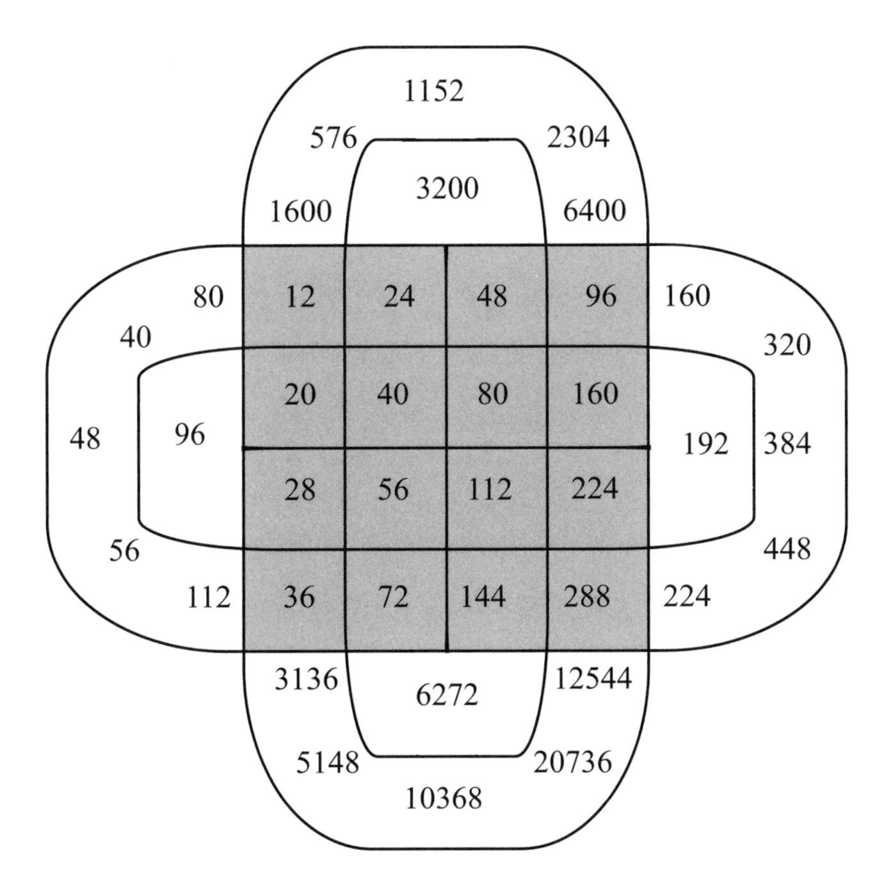

FIGURE 2. An arithmetic puzzle from Boethius' *De Arithmetica.*

Consolation of Philosophy, popularizing Plato's educational slate, now partitioned into the quadrivium and the trivium: arithmetic, geometry, astronomy, music, grammar, rhetoric, and logic. All this material focused on pure rather than applied knowledge, at least in theory.

To illustrate a study item from, say, the pure field of arithmetic, Figure 2 is my adaptation of a puzzle from Boethius's *De Arithmetica*,[13] which was his rough translation of a Greek manuscript of Nicomachus of Gerasa (AD 60–120). Imagine that the gray square

13 Boethius's book was "the source of all arithmetic taught in schools and universities for over a thousand years," Dorothy Schrader, *De Arithmetica*, Book I, of Boethius, *Math. Teacher*, 61:6 (1968) 615–628.

is a 4×4 matrix A, with cell entries a_{ij} in row i and column j, where i and j go from 1 to 4, and that the integers are written in Roman numerals.[14]

Each of the integers outside the gray square can be obtained in two distinct ways by arithmetic operations on the integers inside the gray square. In particular, we allow equality between (i) the sum of two integers, (ii) the product of two integers, (iii) doubling an integer, and (iv) squaring an integer. *Can you find them? Moreover, how are the integers themselves within the gray box related to each other?* For example, from A's row-3, we have $a_{31} \times a_{34} = 6272 = a_{32} \times a_{33}$; and $a_{31} \times a_{33} = 3136 = (a_{32})^2$. *Unimpressed?* Recall that these integers are in Roman numerals,[15] a system never designed for computational purposes. Instead, those computations were performed using an abacus or a counting board with movable tokens. To solve our puzzle, a student would translate, say, two desired numerals from Roman (or Greek) into proper arrangement onto one of these instruments; manipulate the tokens thereon, performing the desired operation; and translate the result into Roman numerals. Ironically, such puzzles as this one develop students' facilities with the abacus—useful skills which a potential employer might prize—but the pure aim of the puzzle was to develop insight into the nature of numbers, so as to discover more (interesting yet arguably useless) relations within number theory.

Next, we leap forward in time to shortly after the bombshell technological innovation of the printing press. My favorite artist of all time is Albrecht Dürer (1471–1528). His 1525 four-volume *The Art of Measurement with Compass and Straightedge* was one of the first printed mathematical texts in German. Therein, he introduced a construction technique for a general family of curves known as the *trochoid*.

Figure 3c is a snippet from Dürer's woodcut "The Circumcision" showing what is now called a *limaçon*—which Dürer constructed

14 To represent the integer ten thousand, linguistic anthropologist Stephen Chrisomalis says that the Romans used the symbol ↂ, explaining that the Roman symbol for one thousand in the time of the Punic Wars was initially written as ↀ, and that over time it "opened up from the bottom to look like an M," from Stephen Chrisomolis, *Reckonings*, MIT Press, Cambridge, MA, 2020, p. 39.

15 At least up until after the days of Fibonacci's 1202 *Liber Abbaci* introducing the fantastic computational capabilities of the Indo-arabic number system.

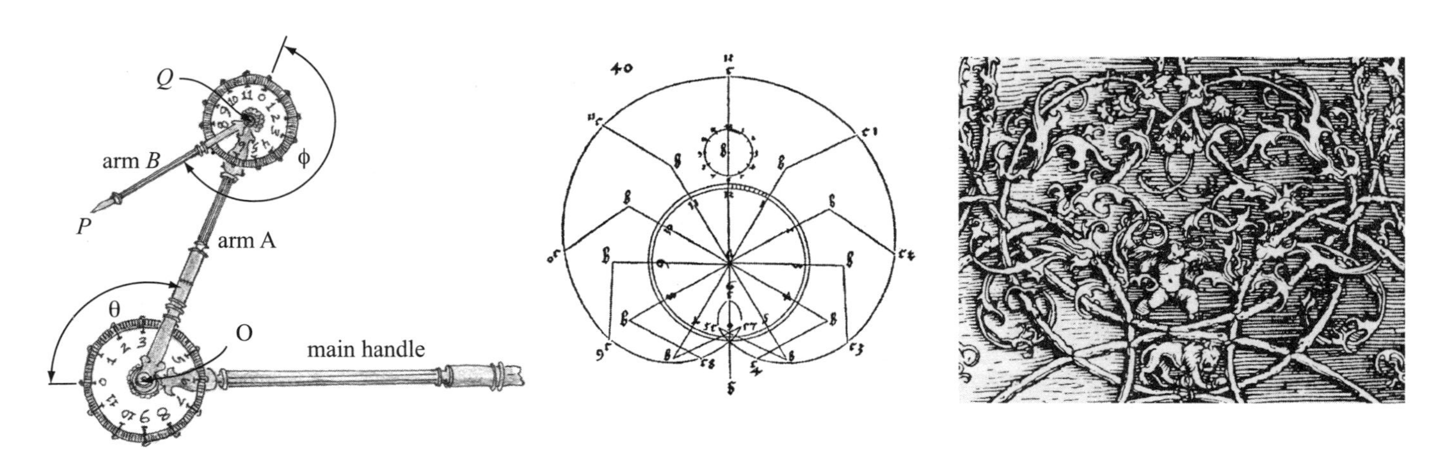

FIGURE 3. Albrecht Dürer's trochoid family of curves.

using a modified compass.[16] To use the instrument, one sets the compass down at point O on a page; arranges arms A and B as desired, so determining angles θ and φ; defines step-size increments for angles θ and φ at the joints O and Q; advances arms A and B by those angle step-sizes and marks the page at point P, and repeats. One hundred years later, geometers such as René Descartes (1596–1650) transformed Dürer's mechanical procedure into algebraic terminology and onto coordinate systems, so much so that the realm of mathematics, which had been classical geometry with a bit of algebra, has become algebra with a bit of classical geometry.

The point of this story is that Dürer applied a mathematical idea to help create and market art. Once this primitive idea percolated into what we might call the mainstream of mathematics, *aha*, the mathematical community delighted in making ever more fanciful curves.[17] This dynamic between pure and applied repeats itself time after after time. Upon which shall we focus? If you are someone like Stephen Wolfram of *Mathematica* fame, why not fine-tune all the world's great mathematical algorithms for universal usage into a convenient computer algebra system (CAS) and make a fortune along the way. However, if you are someone like Grigori Yakovlevich Perelman, who resolved the 1904 Poincaré conjecture and nevertheless turned down the million-dollar Clay Millennium Prize, then you might be like Archimedes, who, in the sack of Syracuse,

> was intent upon working out some problem by a diagram. In [the midst of Archimedes' calculations], a soldier commanded him to [move]. Declining to do so before he had worked out his problem, the soldier, enraged, drew his sword and ran him through.[18]

16 Figure 3a is my sketch of Dürer's modified compass. Figure 3b is figure 40 from book 1 of Dürer's *Underwesung der Messung*, Nuremberg, 1538. Figure 3c is cut 184 from *The Complete WoodCuts of Albrecht Dürer*, ed. Willi Kurth, Crown Publishers, New York, 1941. For further details on how Dürer manipulated his compass to construct the trochoid in the "Circumcision" see Chapter VII, pp. 163–187 of my book, *Voltaire's Riddle*, Mathematical Association of America, 2010.

17 To explore one mathematician's top 10 curves of all time, see Julian Havil, *Curves for the Mathematically Curious*, Princeton University Press, Princeton, NJ, 2019.

18 See *Plutarch*'s, p. 380.

We make just one more leap, this time to the dawn of the twentieth century. At the end of his life, Alfred Nobel (1833–1896), who had made a fortune marketing explosives, was faced with distributing that wealth, ultimately choosing to endow monetary prizes celebrating extraordinary practical discoveries. In 1901, the first prizes[19] were awarded in physics, chemistry, medicine, literature, and peace, but not mathematics. Why? The bottom line is *everyone knows that,* whether we like it or not, *pure mathematics is impractical.* Tracing mathematical development through time, we see that it has continually bifurcated, spawning the disciplines of engineering, accounting, physics, economics, statistics, operations research, computer science, communications technology, and artificial intelligence. Mathematics jettisons, as it were, all the practical aspects of its success, so as not to thwart its grand quixotic quest to understand in full the elusive abstractions of number and space. To personify mathematics, imagine that mathematics is Diogenes, and the world view of what is important is Alexander the Great. One afternoon, approaching from the west, Alexander asked Diogenes, who was drawing figures in the sand, what he could do for him; and Diogenes replied, "Just move out of the sunlight."[20]

Now, I've exaggerated a bit, more or less in the spirit of Hardy's claim (7). Yet, it's hoped that these last few paragraphs have set a foundation upon which to discuss his other claims.

Let's consider points (1) and (2) about engineering. Hardy coauthored about 100 papers with John Littlewood (1885–1977), whose comments often enlighten Hardy's observations, such as this recollection: One day at Trinity College of Cambridge University,

> in 1912, it was proposed that the Engineering students should be taught some *real* mathematics by the *mathematical staff.* I asked F. J. Dykes, the Lecturer in Engineering, what he would like me to select; all he said was: *Give the buggers plenty of slide rule.*[21]

19 In 1968, prizes were also extended to economics.

20 Paintings and sculptures of this classic interaction often depict Diogenes in a drunken stupor, but I like to think he was doodling.

21 J. E. Littlewood, *Littlewood's Miscellany*, Cambridge University Press, 1968, p. 142.

I used the slide rule for all my college physics courses in the early 1970s, but teaching the slide rule in the early twentieth century is like teaching the abacus and the counting board back in the days of Boethius. Yet, Professor Dykes had a valid point, namely, *his students needed help at that fundamental level!* When my older son, with a master's in mechanical engineering, landed his first position at General Electric, he was placed into a multiyear training program in which all the new recruits were taught afresh the ins and outs of the trade. Part of this trade is to realize that the company must be commercially competitive—and while experimentation with new ideas is encouraged—*make sure you all follow the company line*. This same spirit was much in vogue, say, in the days of Johannes Gutenberg (c. 1400–1468) and the printing press; as his team was trying to mass-produce copies of the Bible, Gutenberg continued tinkering with engineering improvements to the press, so much so that his services were terminated shortly after the first few of about 200 complete copies of holy writ rolled out the door. This "it's-good-enough" production-line approach to exploring knowledge, I believe, formed much of the target for Hardy's denigrating remarks about engineering.

When Hardy used the word *useful*[22] in a positive sense, he classified as useful much of college mathematics as well as electricity and magnetism and fluid dynamics. Hardy's main point: "What is useful above all is *technique*, and mathematical technique is taught mainly through pure mathematics." When Hardy used the word pejoratively, he concluded

> If useful knowledge is . . . knowledge which . . . contributes to the material comfort of mankind, then the great bulk of higher mathematics is useless.

Those are blunt and harsh words from a master mathematician who was heartily convinced of both the innate value and ridiculousness of a liberal arts education! In an *Apology* critique, dynamical systems professor Daniel Silver's advice for those reading Hardy's *Apology*, is to remember "that Hardy enjoyed teasing his audience," citing George Pólya who recalled, "Hardy liked to shock people mildly by

22 Hardy's *Apology*, p. 133–135.

stating unconventional views . . . because he liked arguing."[23] For example, listen to Hardy as he closed his 1928 Josiah Gibbs lecture in New York City:

> A month's intelligent instruction in the theory of numbers ought to be twice as instructive, twice as useful and at least ten times as entertaining as the same amount of "calculus for engineers."[24]

Such strident language may charm some, yet it fails to amuse everyone.[25] In a 1941 review,[26] Frederick Soddy, a 1921 Nobel laureate in chemistry, called Hardy's *Apology* "cloistral clowning." Despite this caustic view, consider the robust nature of Soddy and Hardy's relationship: At the end of his review, Soddy mentioned that during a dull faculty meeting in 1931, he passed a note to Hardy asking for the sum on the left of equation (1).

$$\begin{aligned}\sum_{m=1}^{\infty}\sum_{n=1}^{\infty}(2n+1)^{-2m} &= \sum_{n=1}^{\infty}\frac{(2n+1)^{-2}}{1-(2n+1)^{-2}} = \sum_{n=1}^{\infty}\frac{1}{(2n+1)^2-1} \\ &= \sum_{n=1}^{\infty}\frac{1}{4n(n+1)} = \frac{1}{4}\sum_{n=1}^{\infty}\left(\frac{1}{n}-\frac{1}{n+1}\right) = \frac{1}{4}.\end{aligned} \tag{1}$$

"In an incredibly short space of time"—and while the meeting was yet in progress—Hardy passed Soddy the rest of equation (1). As a more thoughtful commentary, here's one by Hardy's friend Norbert Wiener:

> When I returned to Cambridge after working with engineers for many years, Hardy used to claim that the engineering phraseology of much of my mathematical work was a humbug, and that

23 "In defense of pure mathematics," *Sigma Xi*. 103:6 (2015) 418–421.

24 From Hardy's Josiah Gibbs Lecture, New York City, 1928, published as G. H. Hardy, An introduction to the theory of numbers, *Bulletin of the AMS*, 35:6 (1929) 778–818. The lecture is also reprinted in *The G.H. Hardy Reader*.

25 For an exhaustive compendium and commentary on extant reviews of Hardy's *Apology* and for Hardy's allusions therein, see Alan J. Cain's *An Annotated Mathematician's Apology* available at https://ia800904.us.archive.org/9/items/hardy_annotated/hardy_annotated.pdf.

26 See his "Review of G. H. Hardy, *A Mathematician's Apology*," *Nature* 147:3714 (4 Jan. 1941) 3–5.

> I had employed it to curry favor with my engineering friends at MIT. He thought I was really a pure mathematician in disguise, and that these other aspects of my work were superficial. This, in fact, has not been the case.[27]

The physicist turned novelist C. P. Snow, who had read Hardy's original 1940 *Apology* manuscript prior to publication, and who wrote the *Apology*'s foreword appearing in editions since 1967, echoed these thoughts in his 1959 Rede lecture:[28]

> Pure scientists have by and large been dim-witted about engineers and applied science. Their instinct was to take it for granted that applied science was an occupation for second rate minds. I say this sharply because thirty years ago I took precisely that line myself.

Paul is like-minded and—using a clever literary device that may help engage the mathematical community with the mathematics of radio transmission and reception—has written this book as if he were arguing his case before Hardy. I can almost imagine Wiener using Paul's book to rebut points (1) and (2) of the *Apology*. And Paul ably and credibly continues Snow's counterargument as you will discover while reading *The Mathematical Radio*.

Since I've come this far in offering observations about the *Apology*, let me comment on the remaining enumerated Hardy points. Is mathematics only a game for young men and women, point (5)? Hardy said: "I do not know of an instance of a major mathematical advance initiated by a man past fifty."[29] Yet,

> Littlewood remained active in mathematics even at an advanced age: his last paper was published in 1972, when he was 87. One of his most intricate papers, concerning Van der Pol's equation and its generalizations, was written when he was over seventy. 110 pages of hard analysis.[30]

27 *The G. H. Hardy Reader*, p. 158.

28 Available at https://metode.cat/wp-content/uploads/2014/07/Rede-lecture-2-cultures.pdf.

29 Hardy's *Apology*, p. 72.

30 Littlewood's *Miscellany*, p. 15–16.

Paul Erdős (1913–1996) toured the world, visiting one university after another like a Johnny Appleseed planting trees of joint work around the globe, until nearly the end. At a January 2013 Joint Mathematics Meetings in San Diego, Richard Guy (1916–2020) gave a talk on continued fractions. To our packed room of about 500, he introduced himself, "You didn't, most of you, come to hear math, but to see a dinosaur." But really, most of us were thinking, *No, we came to see a dinosaur* and *to hear some math.*

Ambition, is it truly necessary, point (3)? Hardy reminds us that he came to mathematics because "I wanted to beat other boys [at exams]."[31] Littlewood, too, had this competitive element. As Belá Bolobás describes:

> In 1971, a paper was submitted claiming a proof of RH [the Riemann hypothesis]. [Both Littlewood and I examined the proof and] after a few hours of painstaking work he [Littlewood] was relieved to find a mistake.[32]

That is, Littlewood was hoping that he himself would be the one to prove RH! Littlewood, however, acknowledged that not everyone is similarly wired. Of Edmund Landau (1877–1938), Littlewood recalled: "It was said round 1912 that it gave him [Landau] the same pleasure when someone else proved a good theorem as if he had done it himself."[33] If doing mathematics were like playing baseball or cricket, some of us would be all-stars, and others, like me, would be perhaps, at best, umpires. The former require a killer instinct, and the latter, just plain obstinacy.

What about point (6)? With respect to writing, Hardy confessed in a letter to Bertrand Russell:

> I wish you could find some tactful way of stirring up Littlewood to do a little writing. Heaven knows I am conscious of my huge debt to him. But, in our collaboration, he will contribute ideas and ideas only: and that *all* the tedious part of the work has to be done by me. If I don't, nothing would ever get published.[34]

31 Hardy's *Apology*, p. 144.
32 Littlewood's *Miscellany*, p. 16.
33 Littlewood's *Miscellany*, p. 125.
34 *The G. H. Hardy Reader*, p. 155.

That is, Hardy wrote well, and was proud of his ability to put words together. The mathematical community has fully acknowledged that talent. To cite just one example among many: Hardy's 1928 Josiah Gibbs lecture won the 1932 Chauvenet Prize for best expository article in mathematics as published in North America. So Hardy saying that "exposition is for second-rate minds" is like Babe Ruth saying, "Anyone can swing a bat." Yet, stellar writing is a rare gift.

Finally, excellence: is it true that few, if any of us, can do anything well, point (4)? My two-year old grandson can walk upright and utter complete sentences! Wow, I'm impressed. That's my standard for excellence. But suppose you are a Leonardo da Vinci. His deathbed last query to his assistant was, "*Dimmi, dimmi se mai fu fatta cosa alcuna*," that is, "Tell me, tell me if anything ever got done."[35] *Oh my, God bless us all.*

Why

What about *The Mathematical Radio* itself? Why might it appeal to you? Imagine that before Paul begins his "engineer's reply to Hardy," Hardy opens their discussion with this salvo.

> A little chemistry, physics, or physiology has no value at all in ordinary life. We know that the gas will burn without knowing its constitution; when our cars break down we take them to a garage; when our stomach is out of order, we go to a doctor or a drugstore. We live either by rule of thumb or on other people's professional knowledge.[36]

That is, do people—and that includes mathematicians—really want to know how GPS works, how search engines process information, how phone calls reach someone on the other side of the world within seconds, or how the radio works? Everyone has a casual interest, but after seconds into jargon-laden explanations,

35 W. A. Pannapacker, How to procrastinate like Leonardo da Vinci, *Chronicle Review*, 55:24 (2009) B4 at https://www.chronicle.com/article/how-to-procrastinate-like-leonardo-da-vinci/.

36 Hardy's *Apology*, p. 118.

our attention fades. Littlewood found that he himself was prone to this failing:

> I have tried to learn mathematics outside my fields of interest; after any interval I had to begin all over again.[37]

Yes, we all wrestle with new-to-us compressed information and ideas. For example, consider the typical hour-long plenary talk at mathematics conventions: the tacit rule of thumb for such events is that the first twenty minutes is for an interested audience, the next twenty is for the expert, and the last twenty is "hang-on-to-your-hat" territory.

But suppose your interest in the radio is piqued. Paul promises an exposition without engineering jargon—yet sometimes he cannot resist and sneaks in phrases like "magnetic-coupled center-tapped transformers," "carrier signals," "impedance inverters," "superheterodyne receivers," and, my favorite, "Goldilocks frequencies"[38]—perhaps because he is so deeply familiar with those terms. Although I teach undergraduates differential equations along with Maxwell's equations, my insight into applications of electricity and magnetism is limited. The following are two analogies describing this tension or disconnect. Consider first the man blind from birth who was touched by Jesus in Mark 8:24 (KJV); whereupon the man exclaimed, "I see men as trees, walking"; after a second touch, the man "saw every man clearly." In this analogy, I am the blind man—somewhat oblivious to what electrons can do—and Paul is the master. To understand clearly, I need a second touch, as it were, from an engineer with a lifetime of manipulating both abstract equations and concrete circuits of resistors, capacitors, inductors, vacuum tubes, and transistors.

As a second analogy, consider Figure 4, my sketch of Mary Cartwright (1900–1998)—a PhD student of Hardy's and a multiple coauthor with Littlewood—and of a cat,[39] and of a radio broadcasting equations. In this analogy, I am the cat, somewhat befuddled and vaguely alarmed by equations streaming from the radio, and Paul is a Mary figure, clearly explaining the details. In particular, the differ-

37 Littlewood's *Miscellany*, p. 191.

38 Frequencies that are neither too low nor too high.

39 Paul confesses to being a "long-time 'staff person' for multiple cats," expressing the idea that cat owners are servants of the cats.

FIGURE 4. Mary Cartwright and cat with radio.

ential equation to the right of the radio is the Van der Pol equation of radio fame that was studied extensively by both Cartwright and Littlewood and which Paul derives afresh in Chapter 1. To the left of the radio is Maxwell's third equation, which Paul interprets as saying that "a time-varying magnetic field creates an electric field," giving rise to a *displacement current,* which "is what gives life to radio." As he explains, solving this third equation generates the wave equation. That is, electrons involved in oscillating current somehow, and almost magically for someone with my depth of understanding, shed electromagnetic radiation which can be received and induced back into an oscillating current to be transformed into sounds of music, news, and story. Such arc thc epic contents of *The Mathematical Radio.*

One way to read Paul's book is to start on the first page and continue to the last. However, in the spirit of Merlin, a wizard at Camelot, who is imagined to have lived backward in time, another way is to read the appendix first, so giving a warmup to Maxwell's equations and their significance. Then, read Chapter 6, a colorful history of the radio as it affected and transformed our culture. Chapter 1 is a terse review of a typical undergraduate differential equations course involving solution techniques for solving second-order differential equations, including Laplace transforms, Heaviside's step function, and Dirac's impulse function. Chapter 2 follows with Fourier transforms and convolutions. Chapter 3 explains AM radio, wherein waves are modulated in amplitude. In Chapter 4, Paul uses the Hilbert transform—so named by G.H. Hardy!—to explain single-sideband, super-encrypted, ultra-high-frequency radio—which is "the technology behind the ultra-top-secret trans-Atlantic radio/telephone link used" by Churchill and Roosevelt during World War II. In Chapter 5, Paul uses Bessel functions to explain FM radio, wherein waves are modulated in frequency. If you enjoyed the sequence of steps in equation (1), you'll experience, in the same spirit, copious mathematical derivations of why the radio works.

Along the way, you'll learn (i) why radio transmitting antennas must be so high; (ii) why electrons in a current move at snail's pace; (iii) why analyzing the Van der Pol equation shows that *negative resistance* in a circuit can exist and why "the entire development of modern electronics is based on that fact"; (iv) why in a nuclear war, "it is almost certain your car radio with its solid-state chips . . . will no longer work—but old-time radios with vacuum tubes will"; and (v) a host of other fun technical and historical details. As you read you'll discover that the book—at least the first five chapters—reads like a math book. Take your time. Pace yourself. You may find yourself, like Littlewood and me, as well, going back and rereading passages before pressing on. *The Mathematical Radio* unpacks much interesting mathematics. Furthermore, if you are a Nahin fan, you will soon conclude that this book—surpassing by far any of his previous expositions with respect to mathematical sophistication—encompasses a lion's share of what he has been seeking to understand and explain ever since, as he describes, "when I was a 16-year-old kid full of excitement about electrons and math."

March 2022

Preface

This book on radio theory[1] was (odd as it may at first seem) motivated by one of G. H. Hardy's outrageous commentaries on what constitutes "good mathematics." Now, of course, I write that shockingly blunt statement primarily to get the attention of mathematicians, readers who I hope form a significant portion of the book's audience. To such an audience, Hardy needs no introduction, as he was certainly one of the world's great mathematicians of the first half of the twentieth century. For engineers, physicists, and other curious readers who might appreciate just a few additional words, however, the following is a short summary.

Godfrey Harold Hardy (1877–1947) was born into an English family of modest means who nevertheless managed to send him to a good boarding school. He then entered Cambridge University in 1896, where he excelled in mathematics: he was Fourth Wrangler in the much-feared/much-admired-by-all-but-Hardy mathematical Tripos examination (of 1898). Hardy was elected a Fellow of Trinity College in 1900 and became good friends in 1906 with John Edensor Littlewood (1886–1977), who was the 1905 Senior Wrangler. Littlewood and Hardy would, over the next 35 years, coauthor nearly a hundred important papers. In 1913 Hardy famously "discovered" the Indian math genius Ramanujan (1887–1920), an episode that Hardy described as "the one romantic incident in my life." From 1920 to 1931 Hardy was absent from Cambridge, having left to become a professor at Oxford, but then eventually returned to Cambridge, where he remained for the rest of his career.[2]

1 There are a number of scholarly books on radio in print, but all are of a much different nature from this one. They are about either the historical evolution of AM and FM broadcast radio (Tom Lewis's 1991 *Empire of the Air*, Gary Frost's 2010 *Early FM Radio*, or Susan Douglas's 1989 *Inventing American Broadcasting*) or the social impact of radio (Douglas's 1987 *Listening In* or Michele Hilmes's 1997 *Radio Voices*), or they are highly technical (Hugh Aitken's 1985 *Syntony and Spark* and *The Continuous Wave*, David Rutledge's 1999 *The Electronics of Radio*, or my own 1996 *The Science of Radio*). The last two, in particular, as the titles hint, use lots of physics and the jargon of radio engineers (Rutledge and I were both electrical engineering professors in "real life"), but this radio book is, I believe, the first one aimed specifically at *mathematicians*.

2 This absence was the result of the unpleasant atmosphere at Cambridge for all who had opposed England's participation in the First World War (as had Hardy) and the particularly hostile behavior by not just a few colleagues directed toward Hardy for his

FIGURE P1. G. H. Hardy in the mid-1920s, at the height of his powers, in his rooms at Trinity College, Cambridge. Upon seeing this image, one observer famously remarked "To sit that way you have to have been educated in a public [in England, private] school." (By kind permission of the Master and Fellows of Trinity College, Cambridge)

Hardy made many important contributions to mathematics, of which I'll limit myself to mentioning just one. In 1859 the great German mathematician G.F.B. Riemann (1826–1866), in his study of the zeta function $\zeta(s)$ (originally due to the Swiss-born mathematician Leonhard Euler (1707–1783)), defined as $\zeta(s)=\sum_{k=1}^{\infty}\frac{1}{k^s}$, which converges only if the real part of $s>1$, generalized $\zeta(s)$ to have meaning for all values of s in the complex plane. Then, on the basis of quite limited numerical evidence, Riemann conjectured that *all* the complex-valued solutions of $\zeta(s)=0$ (of which there are an infinite number) have a real part of $\frac{1}{2}$. That is, the now famous *Riemann hypothesis* (RH) is that all the complex zeros of the zeta function are on the vertical line $s=\frac{1}{2}+ib$, $i=\sqrt{-1}$ (called the *critical line*), where b varies from minus infinity to plus infinity. The importance of the RH lies in the fact that the proofs of many other theorems are based on the *assumed* truth of the RH. If the RH is eventually shown to be false, new proofs of those other theorems will have to be developed.

Riemann tried (and failed) to prove his conjecture, and so have numerous mathematicians ever since: *all have failed*. Billions upon billions (literally) of complex zeros have been calculated by computer, and every last one of them does, indeed, have a real part of precisely $\frac{1}{2}$. But still, while the results are highly suggestive, that huge pile of computer printouts *proves nothing*, as it would take just one zero off the critical line to invalidate the RH. The RH is generally considered to be the greatest outstanding unsolved problem in mathematics today.

In 1914 Hardy made the first substantial progress toward resolving the RH when he showed that there is an infinity of complex zeros on the critical line. That doesn't prove that *all* the complex zeros are there (Hardy's result, a necessary condition for the RH to be true,

---

defense of the philosopher Bertrand Russell (who had been dismissed from Cambridge as a result of *his* active objection to the war).

isn't sufficient, but it's a start).[3] One doesn't have to know the details of how Hardy proved his result to appreciate that it was a significant advance in understanding the nature of $\zeta(s)$. As an amusing illustration of how Hardy viewed the significance of the RH, on a postcard he sent one year to a friend he listed six New Year's resolutions. The last one was to "murder Mussolini" (which had to be some sort of black humor joke), but number one was to "prove the Riemann hypothesis." Hardy certainly appreciated the difficulty of developing such a proof, and he would have done it if he could. But if Hardy were to have been left alone with Mussolini, and if he had possessed a gun (and didn't accidently use it to shoot off one of his own feet[4]), I can hardly bring myself to believe he would *really* have pulled the trigger.

Now, what did Hardy do that got under my skin to the point that I decided to write this book? It wasn't quite what you might think. Hardy wrote a number of books, but the one that all mathematicians have, with virtual certainty, read is his 1940 *A Mathematician's Apology* (Figure P1 was on the cover of the original publication of the book). A memoir of unusual openness, it contains passages of such melancholy that Hardy's friend the novelist C. P. Snow (who wrote the book's foreword) described it "as a book of haunting sadness," as the cry of a mathematician feeling the irreversible loss of his intellectual power. If one isn't careful, in fact, to read Hardy's *Apology* can be a depressing business, indeed, because, like it or not, it's what's in store for all of us. As Jim Morrison of the 1960s rock band The Doors once famously said about being in this world, "Nobody gets out of here alive" (and as if to make the point with emphatic clarity, Morrison was soon after found dead in his bathtub in 1971, at age 27).[5] Hardy's

3 Littlewood did *not* believe the RH to be true (see my *In Pursuit of Zeta-3*, Princeton University Press, 2021). The Stanford mathematician George Pólya (1887–1985) liked to tell a story that illustrates both the fascination and the difficulty of the RH. When asked by a fellow mathematician to listen to how he was going to prove the RH, Pólya said: "I listened and tried not to interrupt, but at one point I asked for an explanation. He stopped, was silent for a few minutes, then said 'Yes, there is the error.' That was a tragic moment!"

4 Hardy was well-known to be "technologically challenged," and I'll soon elaborate on that point.

5 With that gloomy imagery in mind, perhaps suddenly falling over dead in your tracks at age 27, right after winning the top prize in mathematics (and having just polished off a great dinner), wouldn't actually be such a bad way to go, especially when compared with

late-life depression, caused by the fading of his mathematical powers, almost certainly accounts for this unhappy line in the *Apology*: "Exposition, criticism, appreciation is work for second-rate minds." In a sense, he was denigrating the *Apology* itself.

The claim in the *Apology* that is best remembered by all is Hardy's assertion that his mathematics had never been "useful" and so was, and would be forever, "pure," a virtue I don't think he ever really quite explained. Hardy's claim certainly strikes many engineers and physicists as being overly dramatic, perhaps even borderline insufferably small-minded and snobby, but I long since forgave him for that as I learned more and more of his beautiful mathematics. When, after all, haven't we all said something in haste and then later wished we could take it back?[6]

It was therefore with much curiosity that I started to read a much less well-known essay of his, published earlier the same year as the *Apology*, in a student journal of the undergraduate mathematics society at Cambridge. Titled "Mathematics in War-Time,"[7] I wondered just what Hardy had to say about mathematicians doing "practical stuff." Would it be a foreshadowing of the *Apology*? Well, the last line of the first paragraph was pretty much what I expected, where he wrote that the application of mathematics to war generally filled him "with intellectual contempt and moral disgust." That's pretty strong stuff.[8]

---

just fizzling slowly away like a glass of seltzer water losing its bubbles, as did Hardy in his final, sad years of failing physical health and declining mental agility. He had a heart attack in 1939, at age 62, and early in the year he died he attempted suicide.

6 The Hungarian-born American mathematician Paul Halmos (1916–2006) seemingly echoed Hardy in an essay with the shocking title "Applied Mathematics Is Bad Mathematics" (in *Mathematics Tomorrow*, Springer 1981). But he didn't really mean it. Halmos denied the title's teasing claim, in fact, in the very first sentence! I suspect, however, that Hardy never wished to take back *anything* he wrote in the *Apology*.

7 Reprinted in *The G. H. Hardy Reader*, Cambridge University Press, 2015, pp. 287–290.

8 I think the sort of analysis that was done by the Ballistics Group of the Los Alamos Ordnance Division of the Manhattan Project (the American atomic bomb program in the Second World War) in deriving the escape path for the B-29 bomber *Enola Gay* after it dropped an atomic bomb on Hiroshima, Japan, in August 1945, was the sort of thing Hardy had in mind. He must have been shocked to his core at the news of Hiroshima (and then Nagasaki). See the analysis of the bomb blast escape problem in the corrected paperback edition of my book *Chases and Escapes: The Mathematics of Pursuit and Evasion*, Princeton University Press, 2012. And what, I wonder, would Hardy's reaction be if he could be resurrected only to learn that his lifelong fascination with prime numbers is now central to the mathematics of encryption, a topic of great interest to military intelligence services

But it was the very first line that really stung. There he wrote: "I had better say at once that by 'mathematics' I mean *real* mathematics, the mathematics of Fermat and Euler and Gauss and Abel [who, by the way, died at age 27, just like Jim Morrison], and not the stuff which passes for mathematics in an engineering laboratory." Hardy then brushed aside "ballistics or aerodynamics, or any of the other mathematics which has been specially devised for war," declaring such maths to be "repulsively ugly and intolerably dull; even Littlewood could not make ballistics respectable, and if he couldn't, who can?"[9] All such developments he labeled as being "sinister byproducts" of real mathematics. Sinister? That word strikes me as way over the top, one that should be reserved for denouncing unabashed evil and nothing less.

Hardy's attitude was in keeping with his well-known general unhappiness with technology, preferring (as one joke put it) to use a telephone only if mailing a letter would be too slow. Still, he would use the very technology he thought "ugly" when it was convenient. Hardy's famous ride to visit an ill Ramanujan, for example, was by taxi[10] (powered by an internal combustion engine, which is a mechanical engineering marvel) and not in a horse-drawn carriage. Hardy's odd ability to simultaneously hold such divergent views on the values of "pure" mathematics versus technology and physical science did not go unnoticed. Hardy was elected a Fellow of the Royal Society in 1910, at the young age of 33, but only on his *second* try. As the obituary essay reporting his death put it, perhaps the cause for the initial failure was

---

around the world? And what would Hardy have to say to the mathematicians who broke the German Enigma code and so helped bring down the Nazis?

9 Littlewood served in the British army (Royal Artillery) during the First World War, calculating the trajectories of long-range, high-altitude shells traveling through a realistic model of the atmosphere over a rotating spherical Earth. Hardy, not Littlewood, was the odd one out in this, as other pure mathematicians had also done the same thing. For example, the well-known French mathematician Henri Lebesgue (1875–1941) and the equally famous American mathematician Norbert Wiener (1894–1964)—Hardy knew them both—had during the First World War advised civilian war authorities on ballistics problems. Littlewood's interest in ballistics was lifelong (see his book *Littlewood's Miscellany* with several pages of discussion).

10 In Cab 1729 (a number that after remarking it to be "rather a dull one," Hardy learned from Ramanujan was in fact the smallest number expressible as a sum of two cubes in two different ways: $12^3+1^3=10^3+9^3$). Hardy was also quite willing to travel in the engineering marvels called ocean liners on his visits to friends in Europe and America.

that Hardy "had singularly little appreciation of science."[11]Declaring ballistics and aerodynamics to be "sinister" was pretty obnoxious (to me), but when I got to Hardy's assertion that nothing "could be more soul-destroying than the numerical solution of differential equations," I had had enough. Although its author has now been dead for three-quarters of a century, Hardy's essay clearly demands a response. But how to do that? Hardy wasn't simply a slightly tipsy college don who had overdone the sherry in the faculty club dining room and was now shouting silly stuff to the point his friends were wondering if they should take his car keys away and stuff him into a taxi for a safe ride home. Rather, he was a world-class mathematician who had, yes, gone a bit off the rails but who was, at heart, sincere.

Hardy's dismissive remark about the dreariness of solving differential equations may have been a not-so-subtle swipe at Littlewood, who in 1940 was again doing mathematics in support of the war effort against the Nazis. Indeed, Littlewood had been doing such work for more than two years when Hardy wrote his essay, and so Hardy certainly knew of Littlewood's work, which involved trying to understand the nature of the solutions to the nonlinear, second-order differential equation

$$\ddot{u} - \varepsilon(1 - u^2)\dot{u} + u = 0, \ \varepsilon \text{ an arbitrary constant,}$$

developed no later than 1920 by the Dutch electrical engineer Balthasar van der Pol (1889–1959). Van der Pol had found that

11 "Obituary," *Notices of the Fellows of the Royal Society* 1949, pp. 447–458. It is with no little irony to note that Hardy's professed dislike of applied mathematics didn't prevent him from doing a little of it himself. In 1908 he wrote a letter to the journal *Science* in which, with just some easy algebra and elementary probability, he showed a commonly held belief in genetics was wrong: the end result is that Hardy's name is now forever enshrined in the genetics literature dealing with heredity. I am not the first to grumble about the spirit of Hardy's position on the relative merits of pure and applied mathematics (although, perhaps, I am the first to write a book-length rebuttal in the form of an extended counterexample—I'll elaborate on this in just a bit—a proof technique greatly admired by mathematicians!). You can find a quite interesting essay on others who are/were equally unhappy, by Daniel S. Silver, "In Defense of Pure Mathematics," *American Scientist*, November-December 2015, pp. 418–421.

this equation described the behavior of what are known as *negative resistance oscillators*, which appear in the theory of the intense electrical arcs that were used in the early days of radio (to construct very high-power—as high as a megawatt or even more—Morse-code-keyed frequency-shift transmitters), as well as in the revolutionary new technology of vacuum tube oscillators. In chapter 1 I'll show you where this equation comes from (and how it generates oscillations) once we have transformed most of the electrical engineering jargon of radio engineering into mathematics.

It was immediately clear to me that a rebuttal to Hardy consisting of mere name-calling and/or snide commentary would be equally obnoxious and unworthy. No, to properly respond to Hardy demanded a detailed counterexample showing *mathematics in service to technology*, a technology having obvious wartime service of merit. Thinking back to Littlewood's efforts in trying to unravel the secrets of Van der Pol's differential equation, I then immediately realized what might well have made an impression on Hardy: an exposition on the beautiful mathematics underlying the operation of radio! Such an exposition should certainly discuss AM (amplitude modulation) commercial broadcast radio (the sort of radio Hardy knew about), but also FM (frequency modulation) and single-sideband (SSB) radio, too (both of which were developed in the final years of his life).

It is almost a certainty that a young person today, who has grown up in a world that has literally surrounded him or her with electronics, finds it difficult (if not simply impossible) to imagine the impact the invention of broadcast radio had on preradio society. That happened in the early 1920s,[12] when Hardy was already in his mid-40s, and almost overnight his slow-paced Victorian/Edwardian world

12 Experimental radio had been around since after the turn of the century, and governments were soon making significant use of Morse-code telegraphy via the spark-gap technology of Hertz and Marconi (which was not suitable for voice signal transmission). AM broadcast radio, on the other hand, using vacuum tube technology, produced voice and music radio entertainment for the masses. As Figure P2 shows, the impact of AM

went from one in which the typical evening life in English homes of quiet conversations by the fireplace, or the reading of a book in bed, became that of listening to world news or to a live London concert or to a radio drama, all courtesy of the newly created BBC. *And millions of people were doing the same thing at the same time.* By 1930 it seemed as though everybody had (or at least knew somebody who had) a "wireless."

Radio is perhaps the single most important electronic invention of all, surpassing even the computer in its societal impact (the telephone doesn't depend on electronics for its operation, and television is the natural extension of radio). Even if we drop the "electronic" qualifier, only the automobile can compete with radio in terms of its effect on changing the very structure of society. During the Great Depression the growth of the presence of both the automobile and the telephone in American households faltered, while the spread of radio sets continued unabated even during those disastrous years. The difference was due to the quite different supporting infrastructures required by those quite different technologies (poles and wires, and service stations and highways, respectively, for the telephone and the automobile, while the only thing between a radio station and a radio receiver is air).

Here's how one English physicist recalled the appearance of radio:

> There has never been anything comparable in any other period of history to the impact of radio on the ordinary individual in the 1920s. It was the product of some of the most imaginative developments that have ever occurred in physics, and it was as near magic as anyone could conceive, in that with a few mainly home-made components simply connected together one could conjure speech and music out of the air.[13]

---

broadcast radio was quickly understood (this illustration appeared in a 1922 issue of *Radio Broadcast* magazine).

13 R. V. Jones, writing in his 1978 memoir *Most Secret War*. Reginald Victor Jones (1911–1997) was a key player in British Scientific Intelligence during the Second World War. It was Jones who, in Winston Churchill's words, "broke the bloody [radio] beam" used by the Germans as an electronic bombing aid during the Battle of Britain (see the box at the end of chapter 2). Jones was a pioneer in what is today called *electronic warfare*.

FIGURE P2. "When Uncle Sam Wants to Talk to All the People."

The use of the word *magic* wasn't mere hyperbole. The physicist was using the word in the sense of the famous "Clarke's Third Law" (named after science fiction writer Arthur C. Clarke): "Any sufficiently advanced technology is indistinguishable from magic." The modern AM superheterodyne radio receiver *would* have been magic to the greatest of the Victorian scientists, including the Scottish mathematical physicist James Clerk Maxwell (1831–1879) himself, who first wrote down the differential equations that give life to radio (see the appendix for details). In the Middle Ages such a gadget would have gotten its owner burned at the stake because what else, after all, could a "talking box" be but the work of the devil?

Even present-day quite intelligent people could feel this way. For example, when Supreme Court Chief Justice William Howard Taft was faced with the possibility of hearing arguments about the government regulation of radio, he reportedly said: "I have always dodged this radio question. I have refused to grant writs and have told the other justices that I hope to avoid passing on this subject as long as possible." When asked why he felt this way, he admitted that, for him, "Interpreting the law on this subject is something like trying to interpret the law of the occult. It seems like dealing with something supernatural." And, indeed, during his tenure as Chief Justice from 1921 to 1930, the Supreme Court heard not even a single case dealing with radio.

The "magic" of radio certainly caught the fancy of Einstein, as is clearly obvious from his address that opened the 7th Berlin Radio Exhibition in August 1930. (Hardy was a great admirer of Einstein's, declaring the physicist in his 1940 essay to be "a real mathematician."[14]) After giving a brief summary of the theoretical and experimental foundations of radio (including the work of Maxwell and Hertz), Einstein specifically mentioned the technology of vacuum tubes (!) as "an ideal and simple instrument to generate electric waves." Then, after offering thanks to the large number of unknown engineers who

14 This claim by Hardy does strike me as an odd one because while Einstein was of course quite comfortable with mathematics, he was *not* a creator of new mathematics, which is what I think Hardy meant by a "real" mathematician. When Einstein needed more math than he had learned in school to mathematically express the physical ideas underlying general relativity, he had to be told by a mathematician friend of the tensor calculus, which had been developed years earlier by "real mathematicians." To be fair, however, I should mention that in 1914 Einstein did discover a wonderful mathematical result that ended up being named after a couple of mathematicians! For more on that, see my *Dr. Euler's Fabulous Formula*, Princeton University Press, 2011, pp. 225–226. Einstein visited Oxford in 1931, near the end of Hardy's stay there, and while the mathematician did attend a dinner in honor of the physicist, it isn't known if the two actually had an opportunity to talk.

had reduced all the theory to the physical reality of mass-produced, easily purchased radio sets, Einstein made a critical observation that was obviously directed at anybody who, *like Hardy*, thought technology somehow to be "sinister":

> Everybody who uses the wonders of science and engineering, without understanding it any more than a cow understands the botany of the grass it eats, should be ashamed.

Well, that's pretty direct, but after some thought I have concluded that if Hardy was aware of Einstein's address, he certainly didn't take any of it to heart, given his words a decade later. I see no indication of shame in either his *Apology* or his earlier essay, concerning technology.

But still, maybe a properly written exposition, free as much as possible of electrical engineering and physics jargon, would have held some interest for Hardy. It was clear to all by 1940, a decade after Einstein's address in Berlin, that radio technology would be important to defeating the enormous evil then threatening the entire world. An illustration (Figure P3) that appeared in 1941 in the widely read English publication *Punch* showed resistance fighters listening for the latest news (as well as for the code phrases that initiated guerilla actions against Nazi occupation forces). Hardy was famously an atheist, but he could hardly have missed the near-religious symbolism of the *Punch* illustration, showing a radio bathed in a central light similar to the common image of the baby Jesus in his cradle. The illustration seems to suggest that radio had, in the same way, arrived on Earth just in time to save civilization. The *Punch* illustration may well have been seen by Hardy, but even if so, I do wonder if he fully appreciated the *mathematical* underpinnings of that magical little box.

The technical challenges of writing an exposition on radio for a *mathematician* audience were exciting—but daunting, too. Radio circuits use numerous exotic gadgets, things called resistors, vacuum tubes, transistors, transformers, capacitors, inductors, diodes, and on and on. Radio engineering textbooks are full of talk about magnetic and electric fields, holes, electrons, resonant tank circuits, p-n junctions, Kirchhoff's conservation laws of energy and electric charge,

THE SECRET HOPE

FIGURE P3. Did Hardy understand the math of radio? (*Punch* Cartoon Library/TopFoto).

oscillators, bandpass filters, heterodyning, phase-shifting, and other such matters. To talk of such electrical things to a resurrected Hardy would almost certainly achieve only a bored dismissal with a wave of his hand and a rolling of his eyes. Perhaps he would even have the same reaction as does the elderly gent in Figure P4.

FIGURE P4. How *not* to explain radio to Hardy. ("It's Great to be a Radio Maniac" appeared in *Collier's* magazine in 1924).

No, the only proper response to Hardy, one that would have a chance of success with him, would be a *math* response, with all (or as much as possible) of the electrical gadgetry and exotic tech talk replaced by the language he understood and loved—*mathematics*. And there's *plenty* of math in radio, enough to satisfy even Hardy, I think. When we discuss AM radio, for example, we'll immediately run into complex number algebra and the Laplace and Fourier transforms. With single-sideband (SSB) radio we'll encounter analytic signals and the Hilbert transform (a name due—surprise!—to Hardy). And when we take up FM radio, we will hardly get started before Bessel functions will appear in our analyses.

Could I do all that? I quickly realized it wouldn't be easy, but yes, I almost as quickly thought, I *could* do it. To get started on showing you the level of presentation I have in mind for the rest of this book, imagine that we are seated in the back of a lecture room in the math department on a college campus, a room apparently empty except for the lone figure of Professor Twombly, who is standing in front of a blackboard with a piece of chalk in his hand. I say *apparently* because Twombly doesn't know we are there, and neither does his friend and fellow mathematician, Professor Tweedle, who has just come through an entrance in the front of the lecture room. We are sitting out of sight, but the room acoustics are such that we can easily listen in:

"Well, Twombly," says Tweedle, "are you still trying to patch up your faulty argument from last week that you thought showed the impossibility of the existence of a proof that the Riemann hypothesis is unsolvable?"

Greeting his colleague's words with a forced, half-hearted chuckle (the utter destruction by Tweedle of what he had thought a clever argument was still painful[15]), Twombly replies, "No, no, this is something quite different." He then points at the blackboard, where Tweedle observes Figure P5.

Tweedle stares at what Twombly has drawn, for a full minute, eventually ending up with his nose practically pushed into the board (Tweedle is quite nearsighted). Finally, he steps back, wipes chalk dust off his nose, and says: "Okay, Twombly, I give up. What are you doing?"

Looking a bit embarrassed, Twombly replies: "Tell me, Tweedle, have you ever had students ask you, after you have delivered a great lecture, 'Really interesting stuff, Professor, but . . . what *good* is all of what you just did? How does one use the theory of blah, blah, blah to *do* anything?' And if you ever *have* heard that, what did you say in reply?"

Now looking somewhat embarrassed himself, Tweedle looks down at his feet and then answers: "Well, yes, that does happen to me now and then. I just say, 'Wait until you get a little more into your coursework and *then* you'll see it all come together.' That seems to work."

15 You can read of that earlier encounter between Twombly and Tweedle in my book *In Pursuit of Zeta-3*, Princeton University Press, 2021.

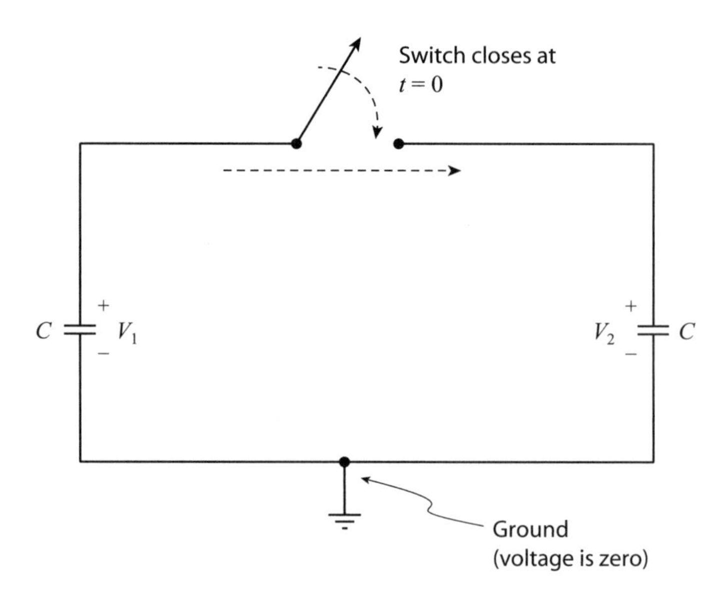

FIGURE P5. Twombly's blackboard drawing.

"Yes," replies Twombly, "but does it, *really*? Students today are pretty sharp and a *lot* more willing to openly label something as BS than we were when we were their age. It's been happening a lot to me lately, particularly in my lower-division classes, and so I've been trying to develop a fast, general-purpose response. That's what *this* is," and at that Twombly points at the blackboard. And then, after a dramatic pause, he adds: "*That's* a mathematical proof that radio waves exist!"

"Really," replies Tweedle, with just a hint of skepticism in his voice, "I thought you had to solve Maxwell's vector-differential electromagnetic field equations to do that. Your drawing just looks like a couple of charged capacitors suddenly switched together. I certainly don't see any electric and magnetic fields, and no divergence or curl operations, either, to derive the Poynting vector describing how electromagnetic energy flows in space."[16]

16 In the appendix I list Maxwell's equations (after the previously mentioned James Clerk Maxwell, who was Second Wrangler in the 1854 Tripos), discuss what those equations *physically* mean, and show how to derive from them traveling-wave solutions for the electric and magnetic fields. Those solutions predict the propagation speed of the fields to be the speed of light, a result that led Maxwell to conclude that light itself is sinusoidally

Hearing that, Twombly stares at Tweedle, open-mouthed. "My Lord, Tweedle, where in the world did you pick all that up? You sound more like one of our colleagues in physics or electrical engineering than a mathematician!"

"Well, Twombly," replies a now really red-faced Tweedle, "perhaps I wasn't quite honest about hearing questions from my students. I've also wanted to show them how math plays a central role in technology, and so I've been auditing some classes in physics and EE. Interesting stuff, as long as you can overlook a disturbing tendency by those guys to reverse the order of integration in double integrals at the drop of a hat. I sometimes wonder if they have ever heard of uniform convergence."

"You continually amaze me, Tweedle, you really do," a now highly impressed Twombly says, "but what you just said *is* in fact the central point of my circuit. With it I've completely sidestepped all the heavy machinery of differential vector calculus, which of course is the classic way to understand how Maxwell's equations make radio possible. With my simple diagram and some elementary math, however, I can show students how to *directly* conclude that radio waves exist. I'm thinking of writing it up for a short paper."

"Okay, Twombly," replies Tweedle, "show me how you do it."

"Right," says Twombly, "but just realize I know I'll be dealing with students who generally know almost nothing about electrical physics or circuits. So, I'll tell them as I go just *what* they need to know, just *when* they need to know it. That makes sense, don't you think?"

"Sure," says Tweedle, "I think the people over in the education school call that the *just-in-time* approach. It gives students less time to forget!"

"Of course," goes on Twombly, who ignores his friend's lame attempt at humor, "I don't have to do that with *you*, now that you're an electrical guru." Twombly says that with a sly grin that Tweedle notices with displeasure but decides to let pass.

"So, Tweedle," continues Twombly, "I'll skip explaining *to you* that electricity comes in tiny little particles called electrons, with each

---

oscillating electromagnetic radiation. Radio waves are, too, but of a vastly lower frequency compared with that of visible light. The appendix also contains the derivation and physical interpretation of the energy-flow Poynting vector and concludes with a simple illustration of how a radio transmitter antenna launches radio frequency energy into space.

having a negative electric charge, and that it's the motion of those electrons in a wire that we call electrical current. That motion is caused by the electric field created by a source of electrons, like a battery, which generates the electrons by chemical reactions. That is, those electrons are the valence electrons in the outermost orbits around atomic nuclei that are so weakly bound to atoms that they can be broken free by the energies of chemical reactions. I think I can do that, since our math kids today are familiar with batteries in their cars and cellphones. You following all this, Tweedle?"

"Oh, sure," says Tweedle, who is getting a bit perturbed by Twombly's superior attitude and so decides to give a little bit of it back: "Don't forget to tell them that charge is *conserved*; that is, charge can't be destroyed. For every negative charge there is a positive charge to balance it. Every time one of those negative valence electrons breaks free of its atomic orbit it leaves behind a positively charged, or *ionized*, atom. So, *all together*, everything is *always* neutral. But what will you say when somebody asks you what charge *is* or what a *field* is?"

That stops Twombly, who apparently hadn't thought of either question. "Well, I'll just say, I'll just say . . . What *will* I say? What would *you* say, Tweedle? You've been hanging out with the physicists and electrical engineers. What do *they* say when *their* students ask that?"

"They're mostly pretty honest about it, Twombly," says Tweedle. "The plain and simple answers are—'I don't know.'"

At that, Twombly looks shocked. "What do you mean, 'don't know'? They've been studying this stuff for nearly two centuries, haven't they?"

"Yeah, they have," agrees Tweedle, "and they don't much like the situation either. The best answer I've heard is simply that charge, q, is a property of some particles that results in a force F on those particles when they're placed in a field, E, a force that makes the particles move. They write that situation mathematically as $F=qE$, and so q is simply a proportionality constant that relates F to E. Or maybe it's E is a proportionality constant that relates F to q. I'm not sure about which way it goes."

"That's it?" says a shocked Twombly, "after 200 years that's all we've got?"

"Well, no, Twombly," replies Tweedle, "there *is* a bit more. If we watch the moving electrical charges that pass by any point in a wire, then the current i at that point is the *rate* at which we see that charge passing. That is, $i = dq/dt$. But as for what E is, well, who knows? The idea of a field being 'lines of force'—whatever *that* means—in space is an old, pre-Maxwell one, dating back to Michael Faraday, but it's pretty much what we have today. An electric field exists, it's thought, in the space between any two points that are at different voltages, like the voltage difference between the two terminals of a battery, but just what's different in that space when the battery isn't there is a mystery. That space looks, smells, feels, whatever, the same whether there is an E in it or not. So, just like what they do with charge, physicists and electrical engineers say that if there is a voltage difference V across any space of length d, then $V = Ed$, and so E is simply a proportionality constant that relates V and d."

"So, all the talk about 'fields' is just games with grammar, is that what you're telling me?" Twombly looks depressed as he asks this.

"No, no," replies Tweedle, "the engineers and physicists definitely think there *is* a physical reality to E because it's thought electric fields can store energy. After all, just stick your hand into any piece of electronics that has recently been powered up, even if now unplugged, and wrap your fingers around a charged capacitor. If it's a big enough capacitor, charged to several hundred volts, you'll need help getting up off the floor. Getting knocked off your feet is clearly not about something that's make-believe!"

"Well, thank the Lord *for that*," says Twombly in a hushed voice. "This electrical terminology business is starting to sound borderline mystical. It seems that to 'explain' each new concept, we just conjure up an even newer one. Like, what's 'voltage'?"

"You're right about that, Twombly," replies Tweedle. "If we keep on just inventing a new theoretical concept to 'explain' the previous one, we're simply playing word games. In the case of voltage, the voltage *difference* between two points is defined as the energy it takes to transport a unit of charge through the electric field produced by that voltage difference. At some point, however, as my friends over in physics say, we have to connect the theoretical stuff to the real, physical world and do an actual *experiment* with actual hardware and measure something."

Listening to that, Twombly seems, just for a moment, to be stunned, but then his face lights up as he thinks he sees a way to turn this state of affairs to his advantage. "Listen, Tweedle, this may *seem* to all be quite grim but, really, it is just a perfect, made-to-order situation for showing students the power of mathematics. Even if I don't have all the terminology buttoned down tight, I can still write *equations*, right? After all, didn't someone really smart once say that math is the study of abstract stuff using symbols whose meaning we don't know? Or something along those lines?"

Tweedle is aghast at that and tells his friend: "I'd be quite sure not to let the dean hear you say that, Twombly. He's always looking for an excuse to cut the department budget, you know, and to tell him we don't understand what we're doing would be asking for real trouble!"

Twombly quickly comes to his senses and agrees. "You're absolutely right, Tweedle. Wash my mouth out and scrub my tongue. But you know what I mean. I mean we can write, besides $i = dq/dt$, that a capacitor C is simply an electrical component that *stores* charge. I was doing a little historical reading a few days ago, in fact, and learned people have been making capacitors for centuries. The first ones were made from metal foil in glass jars, way back in the 1700s. Isn't that amazing? Anyway, a current $i(t)$ transports charge into a capacitor, where it accumulates. As the stored charge builds up, the voltage difference $v(t)$ across the terminals increases in direct proportion to the stored charge, with $1/C$ representing the proportionality constant." With that, Twombly quickly sketches Figure P6 on the blackboard, next to the circuit in his original blackboard drawing.

Tweedle glances at Twombly's new sketch and nods. "Yes, that looks exactly like what one of the EE profs had up on her blackboard a few weeks ago."

"And look at *this*, Tweedle," says a newly energized Twombly. "The instantaneous power $p(t)$ is the *rate* at which energy is delivered to the capacitor, which we can write as $p(t) = v(t)i(t)$. You see that, don't you?"

"Look, Twombly, of course *I* see it, but you're going to be explaining all this to your math students, not to me, says Tweedle. What will you say to *them*?"

"Yes, yes, quite right," replies Twombly. "Well, I'll do it *dimensionally*. That is, I'll first note that power is energy per unit time. Then, as you just pointed out, voltage is energy per unit charge, and current is charge per unit time. Thus, the product vi has the units (energy/charge) times (charge/time) which gives energy/time, the

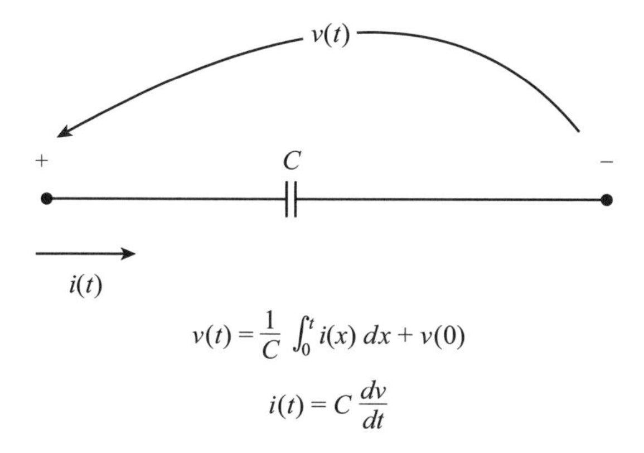

$$v(t) = \frac{1}{C}\int_0^t i(x)\,dx + v(0)$$

$$i(t) = C\frac{dv}{dt}$$

FIGURE P6. The mathematics of a capacitor.

units of power. Now, if we integrate power over an interval of time, call it T, the result is the total energy, W, delivered to the capacitor during that time interval. That is, . . ." and Twombly writes, below his sketch in Figure P5:

$$p = vi = vC\frac{dv}{dt} = \frac{1}{2}C\frac{d(v^2)}{dt},$$

$$W = \int_0^T p(t)\,dt = \frac{1}{2}C\int_0^T \frac{d(v^2)}{dt}\,dt = \frac{1}{2}C\int_0^T d(v^2).$$

"Excellent," says Tweedle, "I see that last line, for W, is the total energy delivered to C during the time interval T. Students should be able to follow that okay, but I do think you should say something about the limits on that last integral. After all, the integration variable is v^2, but the limits are for *time*, not for voltage squared."

"Quite right, Tweedle, quite right, and that's my next step," says Twombly, who amends the energy integral to read

$$W = \frac{1}{2}C\int_{v^2(0)}^{v^2(T)} d(v^2) = \frac{1}{2}C\left[v^2(T) - v^2(0)\right].$$

"Do you see what *this* says, Tweedle?" a now hyperexcited Twombly exclaims. "It says $W = 0$ *if* $v(0) = v(T)$, which is the *math* telling us that any energy stored in the electric field of the capacitor isn't lost but instead is returned from *temporary* storage in the capacitor to the circuitry external to the capacitor *if* the final capacitor voltage equals its initial value. So, we don't need to do an experiment. This is pure math! Tell your buddies in physics and electrical engineering a mathematician did it with nothing but *solid logic*!"

"Please, Twombly, for heaven's sake, *calm down*. This is all very nice, but you've still not explained how any of this has anything to do with radio waves." Tweedle says this with his arms crossed across his chest, in a deliberately challenging tone. Twombly is his friend, yes, but sometimes he's a challenge.

"Okay, Tweedle," replies a completely unchastised Twombly, who hasn't failed to realize it's now put-up or shut-up time. "Prepare to be dazzled!"

"First, a couple of preliminary points. One, the stored energy when the capacitor voltage reaches V is $\frac{1}{2}CV^2$; and, two, the total charge delivered to the capacitor by the current $i(t)$ is

$$Q(T)=\int_0^T i(t)\,dt=\int_0^T C\frac{dv}{dt}\,dt=C\int_0^T dv=C[v(T)-v(0)],$$

or if we take $v(0)=0$, which is a pretty weak assumption,

$$Q(T)=CV."$$

"Mmmm . . . ," murmurs Tweedle, "that's all okay, sure, but I'm *still* waiting for you to pull your radio rabbit out of the hat."

"Right," replies Twombly, who continues without missing a beat. Pointing to his original sketch in Figure P5, he says: "Before the switch is closed at time $t=0$ we have two capacitors, one charged to V_1 volts and the other to V_2 volts, where $V_1 \neq V_2$, and so, for $t<0$, the total stored charge is

$$Q_{t<0}=CV_1+CV_2=C(V_1+V_2),$$

and the total stored energy is

$$W_{t<0}=\frac{1}{2}CV_1^2+\frac{1}{2}CV_2^2=\frac{1}{2}C(V_1^2+V_2^2).$$

But when we close the switch the stored charge will redistribute itself between the two capacitors until both capacitors are at the same voltage. Call that common voltage V. That's because, as you observed earlier, charge is conserved and because, after the switch is closed the two capacitors are connected, as your pals in electrical engineering put it, 'in parallel.'"

Tweedle opens his mouth to say something, but Twombly is really on a roll now and just keeps on going. There's no stopping Twombly when he thinks he is hot on the trail of something amazingly exotic! He says, "For $t > 0$ the total charge is

$$Q_{t>0} = CV + CV = 2CV,$$

and because charge is conserved, we have $Q_{t>0} = Q_{t<0}$ which says

$$2CV = C(V_1 + V_2),$$

or

$$V = \frac{V_1 + V_2}{2}.$$

That means the total stored energy is

$$W_{t>0} = \frac{1}{2}CV^2 + \frac{1}{2}CV^2 = CV^2 = C\left(\frac{V_1 + V_2}{2}\right)^2 = \frac{1}{2}C\frac{(V_1 + V_2)^2}{2}.$$

Now, pay close attention, Tweedle. Here comes the radio rabbit! Notice that

$$\begin{aligned} W_{t<0} - W_{t>0} &= \frac{1}{2}C(V_1^2 + V_2^2) - \frac{1}{2}C\frac{(V_1 + V_2)^2}{2} \\ &= \frac{1}{2}C\left[V_1^2 + V_2^2 - \frac{V_1^2 + 2V_1V_2 + V_2^2}{2}\right] \\ &= \frac{1}{2}C\left[\frac{2V_1^2 + 2V_2^2 - V_1^2 - 2V_1V_2 - V_2^2}{2}\right] \\ &= \frac{1}{4}C[V_1^2 - 2V_1V_2 + V_2^2] = \frac{1}{4}C(V_1 - V_2)^2 > 0 \end{aligned}$$

because any real number squared can't be negative, and $V_1 \neq V_2$."

Tweedle looks at all this for a minute and finally shrugs his shoulders and says: "Okay, Twombly, humor me. Where are your radio waves hiding?"

"My Lord, Tweedle, don't you see it? Charge is conserved in my circuit, yes, but energy isn't! There is *less* energy in the two capacitors for $t>0$ than there is when $t<0$. So, where does the missing energy go? The answer is as close as the chalk dust on your nose—it must be *radiated away*, as radio waves!"

Tweedle stares dumbfounded at his friend, at first undecided on what to do. Should he mumble some sort of amazed grunting sounds of appreciation at Twombly's brilliance, or should he once again—twice in one week!—demolish Twombly's latest "astonishing breakthrough in human thought"? And then he makes up his mind—intellectual honesty must trump all other considerations.

"Well, Twombly," says Tweedle, "I'm afraid I must toss some cold water on your circuit. No, correct that, better make that not water but liquid helium instead. That's the only way your circuit has even a chance of working."

"What are you talking about, Tweedle? What's liquid helium got to do with anything here?'

It has *everything* to do with this," says Tweedle. "Put your circuit in a bath at almost zero absolute temperature, so it becomes superconductive with no resistance, and then what you've done would make at least some sense. But otherwise, you've simply overlooked the real cause for why $W_{t<0} \neq W_{t>0}$: Your missing energy has become *heat* energy, not radio waves. Look here, I'll show you." Tweedle picks up an eraser, rubs out a line in Twombly's circuit of Figure P5, and converts it to Figure P7.

Tweedle then instantly enters his "lecture mode" style of speaking: "The apparent failure of energy conservation in your circuit occurs because you've assumed an impossibly idealized situation. Any circuit you could actually construct would have *some* nonzero resistance in it. I've call it r here, and I'll take it to be arbitrarily small *but not zero*. Now, just like you did for the power and energy delivered to a capacitor by a current $i(t)$, we know that for a resistor the voltage difference $v(t)$ across it is given by $v(t)=i(t)r$."

"Yes, yes, I *know that*, Tweedle," says Twombly, "that's just Ohm's law."

"Right, Twombly," smiles Tweedle, happy to see his friend willing to listen to him. "Therefore, the power is $p(t)=v(t)\,i(t)=\{i(t)r\}\,i(t)=i^2(t)\,r$, and so, in the time interval 0 to T the total energy delivered to the resistor is

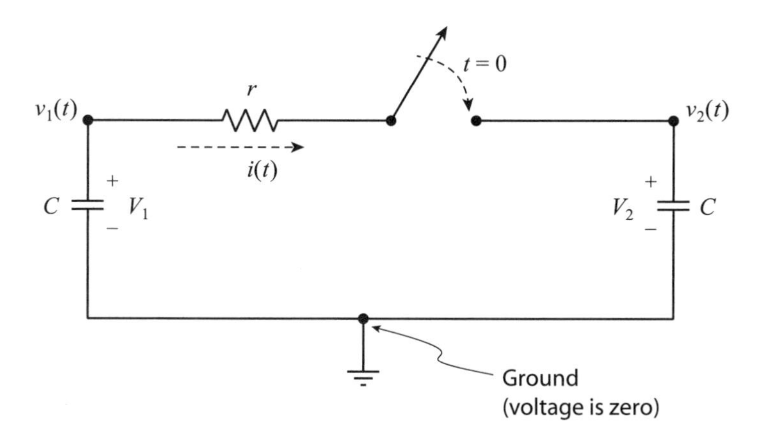

FIGURE P7. Tweedle's modification of Twombly's circuit.

$$w = \int_0^T p(t)dt = \int_0^T i^2(t)r\,dt = r\int_0^T t^2(t)\,dt > 0$$

since the integrand in the last integral is nonnegative. This tells us that W monotonically increases with time, no matter how $i(t)$ varies. That is, the electrical energy delivered to a resistor is permanently 'lost' as heat energy. We can't 'get it back' like you just showed for the energy *temporarily* stored in the electric field of a capacitor."

"Now look here, Twombly," continues Tweedle, "we can work out *exactly* what happens once you close the switch at $t = 0$. I'll assume $V_1 > V_2$, but that's really completely arbitrary, as we could just as well take $V_1 < V_2$. It would all come out the same, either way. Anyway, with $V_1 > V_2$ the current $i(t)$ flows from left to right, from a higher voltage (usually labeled +) to a lower voltage (usually labeled −), and so the voltage on the left capacitor drops and the voltage on the right capacitor rises, and that process continues until the two capacitor voltages meet at V. Specifically,

$$v_1(t) = V_1 - \frac{1}{C}\int_0^t i(x)\,dx,$$

and

$$v_2(t) = V_2 + \frac{1}{C}\int_0^t i(x)\,dx.$$

Since Ohm's law says

$$i(t)=\frac{v_1(t)-v_2(t)}{r}=\frac{V_1-V_2-\frac{2}{C}\int_0^t i(x)dx}{r}=\frac{V_1-V_2}{r}-\frac{2}{rC}\int_0^t i(x)\,dx,$$

then differentiating with respect to t gives us

$$\frac{di}{dt}=-\frac{2}{rC}i.\text{''}$$

Tweedle stops here and looks at Twombly, who appears to be in shock as his thoughts of fame vanish. Still, he is a good enough mathematician to instantly see how to proceed. Stepping up to the blackboard, Twombly says: "This is pretty easy to solve. We could Laplace transform it, of course, but that's getting into second-year stuff. If we use classical differential equation techniques, however, then even a freshman could follow the analysis. So, separating variables we have

$$\frac{di}{i}=-\frac{2}{rC}\,dt,$$

and then, integrating indefinitely and writing K as an arbitrary constant, gives us

$$\ln(i)=-\frac{2}{rC}t+K,$$

or

$$i(t)=e^{-\frac{2}{rC}t+K}=e^Ke^{-\frac{2}{rC}t}=Ae^{-\frac{2}{rC}t},$$

where $A=e^K$. Since

$$i(0)=\frac{V_1-V_2}{r}=A,$$

then

$$i(t)=\frac{V_1-V_2}{r}e^{-\frac{2}{rC}t}.\text{''}$$

Twombly stops after writing this equation for $i(t)$ and stares at it for a few moments. Then, in a voice so low Tweedle has to struggle to hear, Twombly says: "Since, as you said, $r\approx 0$, then this exponentially decaying current will go to zero very fast, and so while the limits on the integral for the total energy delivered to the resistor are technically zero to infinity, essentially all the energy is actually delivered as a very brief burst immediately following the closing of the switch. Am I right with that, Tweedle?"

"Yes, excellent, Twombly," says Tweedle. "The EE guys call $i(t)$ a *transient* because it's there and gone so fast. And if you next compute the W integral for the energy delivered to r, which is converted to heat, you'll get . . ."

Twombly starts writing again as Tweedle lets his prompt hang silently in the air. Twombly mumbles along in step as he writes

$$W = r\int_0^\infty i^2(t)\,dt = r\int_0^\infty \frac{(V_1-V_2)^2}{r^2}e^{-\frac{4}{rC}t}\,dt = \frac{(V_1-V_2)^2}{r}\int_0^\infty e^{-\frac{4}{rC}t}\,dt$$
$$= \frac{(V_1-V_2)^2}{r}\left\{-\frac{rC}{4}e^{-\frac{4}{rC}t}\,\Big|_0^\infty = \frac{1}{4}C(V_1-V_2)^2.\right.$$

Twombly's eyes flick back and forth between this last calculation and his earlier one for the "missing energy" that he thought radiated away as radio waves. Their equality is hard to miss.

"Notice, too," says Tweedle, "how r has disappeared. That is, it makes no difference what it is, *just as long as it's not exactly zero.* That's a pretty neat *mathematical* result, one I don't think is a priori physically obvious. So, there's a terrific 'power of math' illustration for you, one that's hard to top."

Making no reply, but finally convinced that, once again, he has gone slightly off the rails in his enthusiasm, Twombly looks so miserable that Tweedle feels compelled to do something unusual for him. Putting a hand on his friend's shoulder, Tweedle says: "Look, Twombly, this is not a total loss. You've got the makings of a pretty nice lecture here. You could even expand all that we—that is, *you*—have done here to include the third basic circuit component, the inductor, which is basically just a coil of wire. That is, tell your students that if $v(t)$ is the voltage across an inductor L, measured in units of henrys (abbreviated H),[17] and if $i(t)$ is the current in L, then

$$v(t) = L\frac{di}{dt}.$$

17 After the American physicist Joseph Henry (1799–1878).

As the mirror image of how a capacitor temporarily stores energy in an electric field, an inductor L temporarily stores energy $\frac{1}{2}Li^2$ in a magnetic field.[18] If you do that then, as I learned just last week in that EE class I'm auditing, you can show students how to make an oscillator that *does* radiate radio waves![19] So, to celebrate all this, let me buy you lunch at the faculty dining club. I hear they have a great hamburger special today, with an extra big pickle." And with that the two friends depart, leaving us alone in the lecture room and eager to read the appendix, where we'll learn how radio waves *really* come into existence from Maxwell's theory.

Once you've finished taking that first look at the appendix (written specifically with the assumption of a fairly sophisticated reader, say a math, physics, or engineering major who has finished the first two years of college), the book starts its development of radio theory. The style is not the Tweedle/Twombly back and forth,[20] which is used here in the preface mostly for fun, but it is also not a traditional textbook, either. For example, *lots* of historical commentary is included. Further, there is not a lot of discussion about the subtle nuances of electronic circuitry. Most of the mathematical work in this book will be symbolic, and we'll not generally be bothered with the detailed determination of specific circuit component values, which (in radio work) vary over quite wide ranges (resistors go from mere ohms to

18 Note, carefully, that there *is* an important difference between the natures of these two energy storage mechanisms. The energy storage in the electric field of a capacitor continues even if the capacitor is removed from its circuit connections, while the energy storage in the magnetic field of an inductor requires the continued presence of a current. If you remove an inductor from a circuit, its current obviously becomes zero, and so its magnetic field vanishes, along with the stored energy in that field (the collapsing magnetic field creates a transient spark, and the field energy appears as noise, light, and heat).

19 We'll do the mathematics of this in chapter 1, which is the theoretical explanation for the 1887 experimental verification by Hertz of Maxwell's theory.

20 I use the "tutorial conversational" format for Tweedle and Twombly as a tribute to the "Carl & Jerry" stories that appeared in *Popular Electronics* magazine from 1954 to 1964. One of the regular monthly features of that terrific publication were short (2,500 words or so) tales of two teenage boys who got themselves into (and out of) escapades through their mastery of electronics. Carl and Jerry were the Hardy Boys armed with soldering guns! I read all the stories from 1954 to 1958 (until I went off to college), and they were wonderful. The boys have been gone now for more than half a century, but they still live on in my memory. Tweedle and Twombly are, clearly, simply the adult mathematician versions of Carl and Jerry.

millions of ohms, capacitors from picofarads (pF) to millifarads (mF), and inductors from microhenrys (μH) to henrys (H)).

Such detail is definitely important to a radio engineer, of course, who needs to build something that will convince people to part with hard-earned cash, but that's not our concern here. Rather, if I can simply convince you (and Hardy) that there is *some* way to achieve a desired goal, then I'm not going to spend time trying to explain how to achieve that same goal in an even better way. I *am* striving here, instead, to have you (and Hardy) understand how *classic*[21] analog radio "works." What I am *not* striving to do is turn you (*and certainly not Hardy*!) into a radio engineer who knows how to make the best darn radio, *ever*, on planet Earth.[22]

So, you ask, how much "electrical stuff" do you need to know to read this book? Not very much, in fact, as I'll develop that as we go. Here's what I mean by "not very much."

I recall reading, long ago, a funny little bit of dialogue in the classic 1994 Boston crime novel *Cogan's Trade* by George V. Higgins. Two gangsters are discussing plans for breaking into a building protected by an electric alarm system and how that will require bringing on board a new guy "[who knows] bells and stuff." One of the two thugs suggests a candidate who has all the necessary qualifications: "He built one of those quadraphonic things from a kit on his kitchen table[, and he told me] he'd build his own color television [if he ever got a few extra bucks]." To clinch his argument, the thug goes on to

21 There is no discussion of digital radio here and, as much as I might wish otherwise, I am unable to provide instruction on how to make a wonderful science fiction gadget called the *Dirac radio* (a radio that can receive signals from the future). I've also not said anything about the nature of Wi-Fi, a form of radio allowing digital devices to communicate with one another using microwave energy at frequencies far higher than those of AM and FM broadcast radio.

22 Here's an example of what I'm talking about. When we get to FM radio, we'll need a way to turn *frequency* variations in a received radio signal into *amplitude* variations. There are numerous circuits that can do this, but the simplest one (the so-called *slope detector*) is almost trivial to understand. It has some engineering difficulties, however, and so is rarely (if ever) actually used today, but that doesn't matter for a book like this one. As long as I can convince you there is *some* way to do something, then we'll almost always ignore the urge to search for even better engineering solutions. The one exception I make in this approach occurs at the end of the FM discussion, where I take you through a detailed mathematical analysis of the phase quadrature FM detector circuit (including, despite what I wrote just a few sentences ago, the calculation of circuit component values) for a modern FM detector in wide use today.

FIGURE P8. I suspect many of those who saw the cover of the August 1919 issue of *Radio News* on the magazine racks of the day dismissed the artist's vision of what was to come as simply silly—but today we of course know who had his fingers on the pulse of progress.

observe: "It's all the same thing, isn't it? I mean, it's just circuits and stuff." If a Boston mob guy can do "electrical stuff," so can you!

So, with that understanding in mind, off we go (with, I hope, the enthusiasm of the hot-rod driver of Figure P8).

Hardy Asks an Engineer for Help

A mathematician friend, after reading an early draft of this preface, thought I might have been just a bit too hard on Hardy. (But no harder than Hardy himself was in writing an unhappy review of some poor author's book that had committed, in Hardy's eyes, unpardonable sins, as he did in an April 1922 review in *Nature* of a book on the integral calculus: "[This] book may serve to remind us that the early nineteenth century is not yet dead," followed by "it cannot be treated as a serious contribution to analysis," and ending with this fatal stab to the heart—"the book, in short, may be useful to a sufficiently sophisticated teacher, provided he is careful not to allow it to pass into his pupil's hands." How that must have stung!) In any case, my friend suggested I might want to consider including some additional words about Hardy, mathematics, and technology so as not to leave readers with the false belief that, at the end, he was a doddering old goat who sat on park benches talking to pigeons. That was, in fact, definitely *not* the case, and I think my friend correct in her suggestion. It should be acknowledged that in 1943 Hardy coauthored a book on Fourier series, but it should also be noted that it was essentially a rewrite of a book written in 1930 by his coauthor, the Polish-born mathematician Werner Rogosinski (1894–1964). However, I should also note that Hardy *did* publish mathematics up to near the end, with an interesting example being a little analysis he did in 1945 for *The Mathematical Gazette*, titled "A Mathematical Theorem About Golf." The particular details of the problem are of no concern here except to note it is all at the level of an introductory course in probability theory as it is taught to engineers. It's a clever, well-written paper on what, nevertheless, even the admiring editors of *The G. H. Hardy Reader* (note 7) labeled a "rather frivolous" problem. When finished, Hardy was left with a fourth-degree polynomial function $f(x)$, defined over the interval $0 \leq x \leq 0.5$. I was struck, as I read his analysis, by the fact that Hardy was, for some reason, apparently unable to plot $f(x)$. In any case, he wrote: "Mr. A. M. Binnie has plotted $f(x)$ for me." (Alfred Maurice Binnie (1901–1986) was an expert in fluid flow and a lecturer in engineering at both Oxford and Cambridge.) To plot such an $f(x)$ would be, today, a routine exercise in high school algebra, but with the state of computer technology as it was in 1945, perhaps Hardy's need for help is understandable. But still, I can't help but wonder at Hardy's being able to look down on the engineering side of mathematics while at the same time being greatly aided by it.

Chapter 1

Radio Mathematics, Oscillators, and Transmitters

What is the soul of mathematics, and to what wavelength must our souls be tuned to catch its message?

—David Eugene Smith (1860–1944), speaking in 1921 as the retiring president of the American Mathematical Society. Just a few years earlier, this metaphor would have been meaningless to almost everybody.[1]

1.1 Kirchhoff's Laws and FitzGerald's Oscillating Circuit

As you start reading this first chapter (or at any time as you read this book), take a parallel look at the appendix. That will give you an appreciation for the central role high-frequency[2] sinusoidal oscillations play in radio, starting at the transmitter. (Oscillators are in receivers, too, as you'll see in subsequent chapters.) It was understood, right from the moment Maxwell published his *Treatise* (when all that he had written was still pretty much *theory*) that the crucial next step to elevate speculative theory to hard fact was to actually generate the oscillating electromagnetic waves the field equations predict. How to do that?

The key idea for the first (and eventually successful) approach to generating radio frequency (rf) waves came in 1883 from the Irish

1 Smith's address is reprinted under the title "Religio Mathematici" in the October 1921 issue of the *American Mathematical Monthly*.

2 "High-frequency" is dictated by the quarter-wavelength $\left(\frac{1}{4}\lambda\right)$ requirement (discussed in the appendix) for the transmitter antenna. To make $\frac{1}{4}\lambda$ a physically reasonable value, the frequency has to be "high." For example, to build a radio antenna transmitting at the power-line frequency of 60 Hz would be ridiculous, as the wavelength at that frequency is 5 million meters. A $\frac{1}{4}\lambda$ antenna at 60 Hz would be 777 miles high, more than three times the orbital height of the International Space Station!

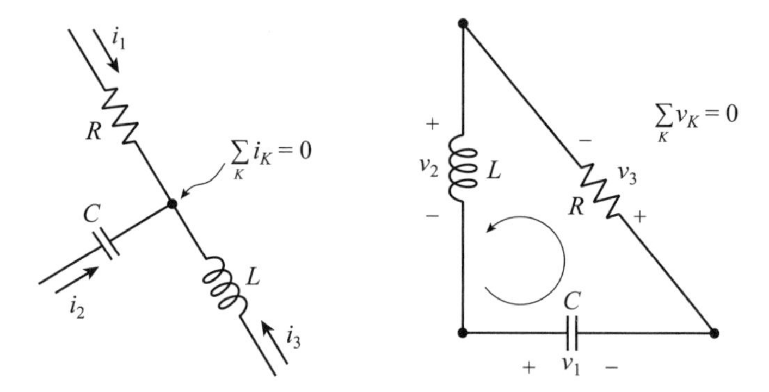

FIGURE 1.1.1. Kirchhoff's two circuit laws.

physicist George Francis FitzGerald (1851–1901). FitzGerald suggested charging a capacitor (with the aid of a static electricity generator) to a high voltage and then letting it discharge through an inductive circuit. (This is actually only slightly more complicated than the circuits of Professors Twombly and Tweedle in the preface.) FitzGerald suggested that oscillations with a wavelength of 10 m (meters) might be achieved (a frequency of 30 MHz (megahertz)—30 million cycles per second. To understand what FitzGerald was talking about requires us first to establish the two fundamental laws obeyed by the electrical circuits you'll find in all radio electronics. These are *Kirchhoff's laws*—after the German physicist Gustav Robert Kirchhoff (1824–1887) who formulated them in 1845—which are simply the laws of conservation of energy and the conservation of electric charge. With reference to Figure 1.1.1, we have

Kirchhoff's current law: The sum of the currents into any node (a point where components are connected together) is zero. This is conservation of electric charge. In other words, charge transported into any node by a current is transported out of the node by another current.

Kirchhoff's voltage law: The sum of the voltage drops (or of the voltage increases) around any closed-loop path in a circuit is zero. This is conservation of energy. You can see this by recalling that voltage is energy per unit charge, and a voltage drop is the energy required to transport a unit charge through the electric field that exists in a component, and so the law says the total energy to go around a closed

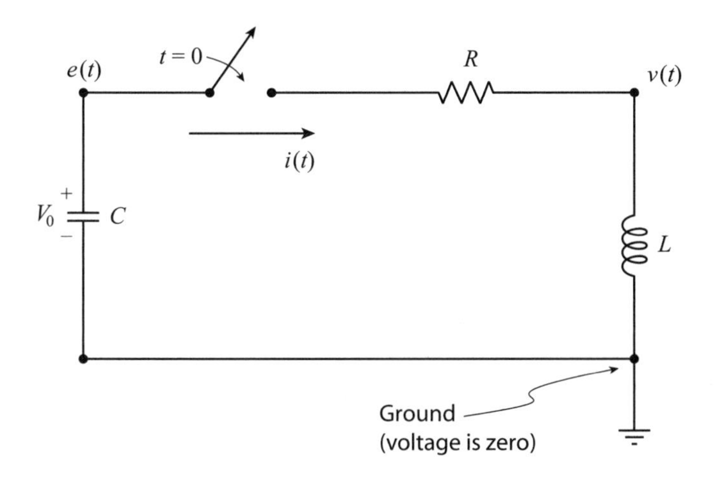

FIGURE 1.1.2. FitzGerald's oscillating circuit.

loop is zero. If it were *not* zero, then we could endlessly transport charge around a closed loop *in the sense for which the energy required is negative* and so become rich selling the gained energy to the local power company! (You'll believe *that* only if you believe in the possibility of a perpetual motion machine.)

Now we can understand what FitzGerald was suggesting. Figure 1.1.2 shows his circuit, with the capacitor C charged to V_0 volts. At $t=0$ we close the switch, and so now there is a path through the resistor R and the inductor L in which the current $i(t)$ can flow.

Just before we close the switch, the stored energy in the circuit is just the energy in the electric field of C ($i(t)=0$ for $t<0$, and so, as Professor Tweedle states at the end of the preface, *initially* there is no stored energy in the magnetic field of L, because there *is* no magnetic field in L for $t<0$).

If $W(t)$ is the total energy in the circuit, then in general we have

$$W(t)=\frac{1}{2}Ce^2(t)+\frac{1}{2}Li^2(t),$$

and if we differentiate with respect to t,

$$\frac{dW}{dt}=Ce\frac{de}{dt}+Li\frac{di}{dt}.$$

But since

$$i(t) = -C\frac{de}{dt},$$

then

$$\frac{dW}{dt} = Ce\left(-\frac{i}{C}\right) + Li\frac{di}{dt} = -i\left(e - L\frac{di}{dt}\right).$$

Now, as the inductor voltage $v(t)$ is

$$v(t) = L\frac{di}{dt},$$

we have

$$\frac{dW}{dt} = -i(e - v).$$

Since Ohm's law says

$$i(t) = \frac{e - v}{R},$$

we have $(e - v) = iR$, and therefore,

$$\frac{dW}{dt} = -i^2R < 0,$$

because no matter what $i(t)$ is, $i^2 \geq 0$. Thus, $\frac{dW}{dt}$ is always negative if we assume $R \geq 0$. This may appear to be a trivial assumption, as *of course R* is positive, right? After all, just go into a store selling electrical parts and ask for a box of negative resistors, and see what the clerk says! *But*, in fact, as we'll get to soon when we discuss how Hardy's friend Littlewood tackled the Van der Pol equation, there *is* such a thing as negative resistance, and, in fact, the entire development of modern electronics is based on that fact.

For now, however, in FitzGerald's preelectronic 1883 circuit R has a positive value, and so we see the initial stored energy in the C continuously decreases once the switch is closed. The central issue raised by FitzGerald, however, was *not* that the energy decreases but rather *how* that decrease occurs. To answer that question, let's look in more detail at $i(t)$. Starting at the ground node (which, by definition is at a voltage of zero) in Figure 1.1.2, let's write Kirchhoff's voltage law as we go around the loop in a clockwise sense (the sum of the voltage drops[3] is zero):

$$-e(t)+i(t)R+L\frac{di}{dt}=0.$$

Differentiating with respect to time,

$$-\frac{de}{dt}+R\frac{di}{dt}+L\frac{d^2i}{dt^2}=0,$$

or, as we observed before, since

$$\frac{de}{dt}=-\frac{i}{C},$$

we have

$$\frac{i}{C}+R\frac{di}{dt}+L\frac{d^2i}{dt^2}=0,$$

or

$$\frac{d^2i}{dt^2}+\frac{R}{L}\frac{di}{dt}+\frac{1}{LC}i=0.$$

The standard method for solving this second-order differential equation is to assume the solution

$$i(t)=Ie^{st},$$

where s is some constant to be determined. Substituting this assumption back into the differential equation, we get

3 As we travel through the C we experience a voltage *rise* from zero to e, which explains why we write $-e$ as the voltage *drop*.

$$Is^2e^{st}+\frac{R}{L}Ise^{st}+\frac{1}{LC}Ie^{st}=0,$$

and so, making the obvious cancellations (which explains why this method works!), we get

$$s^2+\frac{R}{L}s+\frac{1}{LC}=0,$$

a result that lets us solve for what the constant s actually is (in fact, there are *two* such values):

$$s=\frac{1}{2}\left\{-\frac{R}{L}\pm\sqrt{\left(\frac{R}{L}\right)^2-\frac{4}{LC}}\right\}.$$

Now, notice that for given values of L and C, if we have R sufficiently small so that

$$\left(\frac{R}{L}\right)^2<\frac{4}{LC},$$

then with $j=\sqrt{-1}$ we have[4]

$$s=\frac{1}{2}\left\{-\frac{R}{L}\pm j\sqrt{\frac{4}{LC}-\left(\frac{R}{L}\right)^2}\right\},$$

or, more compactly, our two values of s are

$$s_1=-\frac{R}{2L}+j\omega_0,\quad s_2=-\frac{R}{2L}-j\omega_0,\quad \omega_0=\sqrt{\frac{1}{LC}-\left(\frac{R}{2L}\right)^2}.$$

Thus, the most general solution for $i(t)$ is

4 Mathematicians almost always write $i=\sqrt{-1}$ and like to joke that electrical engineers write $j=\sqrt{-1}$ because otherwise they'll confuse $\sqrt{-1}$ with electrical currents in their circuits (which are usually written with the symbol i). This, of course, is nonsense of a near-libelous nature—but, I have to admit, it *is* less confusing not to use i for both concepts. So, if mathematicians will let me write $j=\sqrt{-1}$ and reserve i for currents, I will, in turn, promise not to tell any silly mathematician jokes in this book.

$$i(t) = I_1 e^{s_1 t} + I_2 e^{s_2 t}.$$

To calculate what I_1 and I_2 are, we start with the following important fact about inductors: the current in an inductor cannot change instantly, which follows immediately from its mathematical description. That is, if the current i in inductor could change instantly, then the voltage drop across the inductor would be infinite (because $\frac{di}{dt}$ would be infinite). Engineers and physicists reject the possibility of a physical infinity as nonsense, and so the current in FitzGerald's circuit at $t = 0+$ (immediately after the switch closes) must equal the current at $t = 0-$ (immediately before the switch closes).[5] Since $i(0-) = 0$, then $i(0+) = 0$, too, and we have

$$i(0+) = 0 = I_1 + I_2,$$

and so $I_1 = -I_2 = I$, which gives us

$$i(t) = I(e^{s_1 t} - e^{s_2 t}).$$

To determine what I is, we again use the fact that $i(0+) = 0$, which means (because of Ohm's law) that the voltage drop across R is zero. *That* means, because of Kirchhoff's voltage law, that the initial capacitor voltage V_0 appears across L at $t = 0+$, and so

$$V_0 = L \frac{di}{dt}\Big|_{t=0+},$$

or

$$\frac{di}{dt}\Big|_{t=0+} = \frac{V_0}{L} = I(s_1 e^{s_1 t} - s_2 e^{s_2 t})\Big|_{t=0+} = I(s_1 - s_2),$$

or

$$I = \frac{\frac{V_0}{L}}{s_1 - s_2} = \frac{\frac{V_0}{L}}{j2\omega_0}.$$

5 According to the same sort of argument, it follows that the voltage drop across a capacitor cannot change instantly, as that would require an infinite current.

That is,

$$i(t)=\frac{V_0}{j2\omega_0 L}\left[e^{\left(-\frac{R}{2L}+j\omega_0\right)t}-e^{\left(-\frac{R}{2L}-j\omega_0\right)t}\right]=\frac{V_0 e^{-\frac{R}{2L}t}}{j2\omega_0 L}[e^{j\omega_0 t}-e^{-j\omega_0 t}]$$

$$=\frac{V_0 e^{-\frac{R}{2L}t}}{j2\omega_0 L}j\ 2\sin(\omega_0 t),$$

where I've used *Euler's identity*.[6] Thus,

$$i(t)=\frac{V_0 e^{-\frac{R}{2L}t}}{\omega_0 L}\sin(\omega_0 t),\ \omega_0=\sqrt{\frac{1}{LC}-\left(\frac{R}{2L}\right)^2},$$

assuming that $R<2\sqrt{\frac{L}{C}}$. The current $i(t)$ is said to be an *alternating current*, popularly known as "ac."

So, FitzGerald was correct in saying his circuit will, if R is sufficiently small, oscillate sinusoidally at a particular frequency determined by the values of the circuit components.[7] Does that, however, mean the circuit will generate rf waves? We can explore that question by re-doing the calculation done by Professors Tweedle and Twombly in the Preface: the evaluation of the heat energy integral

$$\int_0^\infty i^2 R dt=\left(\frac{V_0}{\omega_0 L}\right)^2 R\int_0^\infty e^{-\frac{R}{L}t}\sin^2(\omega_0 t)dt.$$

This is a straightforward (if *slightly* messy) freshman calculus calculation, and I'll let you confirm that its value is $\frac{1}{2}CV_0^2$, precisely the

6 Euler's identity, $e^{jx}=\cos(x)+j\sin(x)$, (due to the Swiss-born mathematician Leonhard Euler (1707–1783), is at the very heart of AM, FM, and SSB radio theory, and we will use it repeatedly in this book.

7 The oscillations are a manifestation of the circuit's stored energy sloshing back and forth between the electric field of the C and the magnetic field of the L. Electrical engineers demonstrate the poetic nature of their souls by picturesquely calling the LC combination a *tank circuit*, a reference to the sloshing of water waves back and forth in a disturbed water tank.

value of the initial stored energy in the capacitor. FitzGerald's circuit, therefore, *as it stands*, is no better than Twombly's in generating radio waves. But, unlike Twombly's, all FitzGerald's circuit needs is one final touch—the addition of an antenna! (This is where you really need to read the appendix, particularly the end of it.)

From the oscillating current in FitzGerald's circuit, the resulting oscillating magnetic field of the L can be coupled via Faraday's electromagnetic induction (as shown in Figure 1.1.3) into the antenna, to serve as the oscillating voltage that drives the conduction electrons in the antenna back and forth. That motion, as explained in the appendix, creates kinks in the electric field in the space around the antenna, kinks which in turn give rise to a Poynting energy-flow vector always directed away from the antenna.

As it stands in Figure 1.1.3, FitzGerald's circuit won't transmit for long, because the initial energy in the C is quickly dissipated as heat in the R and as rf waves from the antenna. The early radio experimenters attempted to keep the oscillations going by periodically injecting new energy into the circuit, by incorporating a repeatedly operating spark gap, reaching speeds of up to 20,000 sparks per second. With each new spark a pulse of energy was injected, and such radio transmitters sounded like machine guns! This was okay for Morse code wireless telegraphy but totally inadequate for use in what would become modern voice-and-music radio, and I'll not pursue that approach to radio in this book.[8]

A much different approach was to introduce a *negative resistance* into the oscillator circuit, to counter the energy loss caused by the positive R and the rf radiation. This was achieved, most importantly, with the invention in 1906–1907 of the triode electronic vacuum tube that so captured Einstein's imagination, but it was preceded in the nineteenth century by the *electric arc*. We'll briefly discuss the arc once we have established more mathematical results in the next section.

8 You can find a detailed mathematical discussion of spark-gap radio in my book *The Science of Radio*, Springer 2001. Such radios are now mostly of historical interest, as spark-gap radio has been illegal since 1923, for reasons based on the mathematics (which itself remains quite interesting).

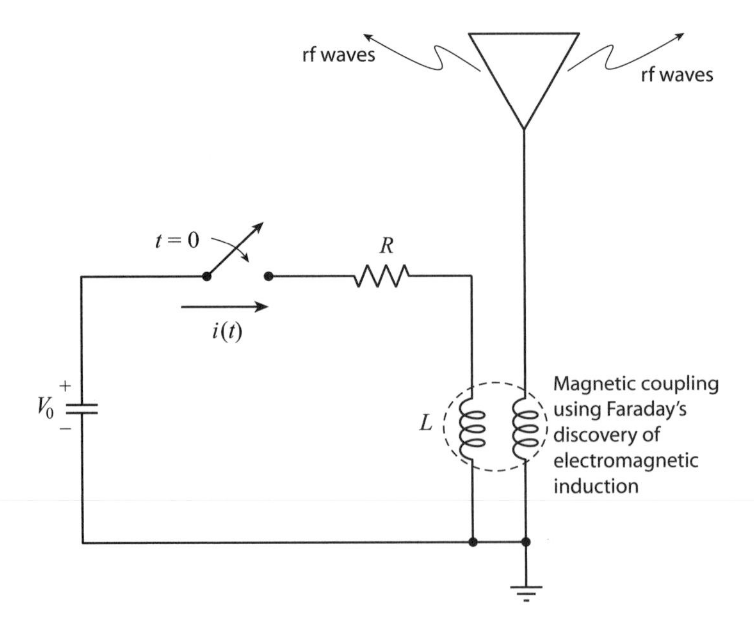

FIGURE 1.1.3. FitzGerald's circuit as a transmitter (the oscillating current $i(t)$ in L creates an oscillating magnetic field that, in turn, creates an oscillating voltage in the antenna which drives conduction electrons in the antenna back-and-forth, creating electric field kinks.

1.2 Laplace Transforms, AC Impedance, and Transfer Functions

We will be concerned in all our discussions of radio with electrical signals that vary *sinusoidally* with time, that is, with signals like $\cos(\omega_0 t)$ and $\sin(\omega_0 t)$, where ω_0 is the *angular frequency* (in radians per second). (As used here, ω_0 is an *arbitrary* frequency and is not the particular ω_0 of the previous section.) If f_0 is the frequency in hertz (what used to be called cycles per second), then $\omega_0 = 2\pi f_0$. AM radio frequencies are in the interval 540 to 1600 kHz (kilohertz),[9] while FM radio operates in the interval 88 to 108 MHz (megahertz). We will find that the differential equations that describe how numerous radio circuits work are *linear*, which means that the sum of two solu-

9 The first radio program Hardy heard was almost certainly broadcast by the BBC London-based station 2LO, which began operating in 1922 at 842 kHz (that is, at 842 kilocycles).

tions to the differential equations is also a solution. Thus, rather than studying the behavior of a circuit in response to, say, a voltage signal like $\cos(\omega_0 t)$ or $\sin(\omega_0 t)$, we can do both problems simultaneously by studying the solution to the complex voltage signal $e^{j\omega_0 t}$, because Euler's identity says $e^{j\omega_0 t} = \cos(\omega_0 t) + j\sin(\omega_0 t)$.

This is because the solution for the signal $e^{j\omega_0 t}$ is the sum of the solution to the signal $\cos(\omega_0 t)$ and the solution to the signal $j\sin(\omega_0 t)$. The solution to the signal $j\sin(\omega_0 t)$ will be the solution to $\sin(\omega_0 t)$ multiplied by the constant j (again, by linearity), and so the solution to the signal $\cos(\omega_0 t)$ will be the real part of the solution to $e^{j\omega_0 t}$, and the solution to the signal $\sin(\omega_0 t)$ will be the imaginary part of the solution to $e^{j\omega_0 t}$. This simple idea leads to the enormously useful concept of *ac impedance*, which allows us (for *sinusoidal* time functions) to treat capacitors and inductors as obeying Ohm's law, which up to now has been limited to resistors.

Since $e^{j\omega_0 t} = \cos(\omega_0 t) + j\, sin(\omega_0 t)$, it follows that $e^{-j\omega_0 t} = \cos(-\omega_0 t) + j\sin(-\omega_0 t) = \cos(\omega_0 t) - j\sin(\omega_0 t)$. Thus, $\cos(\omega_0 t) = \frac{1}{2}[e^{j\omega_0 t} + e^{-j\omega_0 t}]$, and $\sin(\omega_0 t) = \frac{1}{j2}[e^{j\omega_0 t} - e^{-j\omega_0 t}]$, and both of these expressions have simple physical interpretations. In the complex plane, $e^{j\omega_0 t}$ and $e^{-j\omega_0 t}$ are vectors of unit length (because $\cos^2(\omega_0 t) + \sin^2(\omega_0 t) = 1$), making angles $\omega_0 t$ and $-\omega_0 t$ with the real axis, respectively, as shown in Figure 1.2.1a. Indeed, since these two angles increase as t (time) increases, $e^{j\omega_0 t}$ and $e^{-j\omega_0 t}$ are counterrotating vectors, both with real part $\cos(\omega_0 t)$ and with imaginary parts $\sin(\omega_0 t)$ and $-\sin(\omega_0 t)$, respectively. If we sum these two vectors as they rotate, it is obvious their imaginary parts cancel and their real parts add, to give an oscillating result that always lies along the real axis. If, however, we subtract $e^{-j\omega_0 t}$ from $e^{j\omega_0 t}$, we simply multiply $e^{-j\omega_0 t}$ by -1 (which reflects $e^{-j\omega_0 t}$ through the origin) and add, as shown in Figure 1.2.1b. This addition results in the real parts cancelling and the imaginary parts adding, to give us an oscillating result that always lies along the imaginary axis.

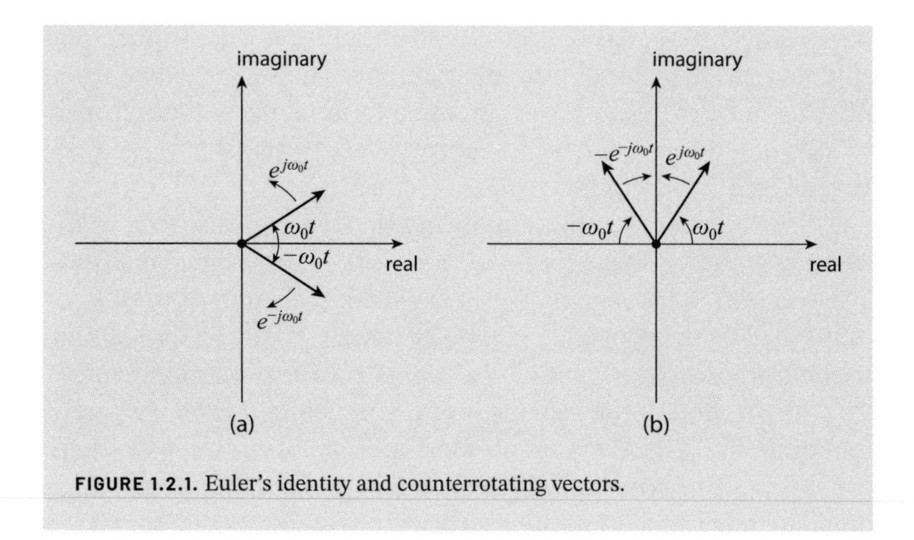

FIGURE 1.2.1. Euler's identity and counterrotating vectors.

To start our development of the impedance concept, consider Fitz-Gerald's series circuit again, but now powered by a complex-valued voltage source, as shown in Figure 1.2.2.

We assume there is, most generally, an initial current $i(0+)$ in the circuit, as well as an initial charge $q(0+)$ in the capacitor. Thus, with $i(t)$ the current for $t \geq 0$, the differential equation that describes the circuit is, using Kirchhoff's voltage loop law (starting at the negative terminal of the voltage source and going clockwise around the loop),

$$-v(t) + iR + \frac{1}{C}\left[\int_0^t i(x)dx + q(0+)\right] + L\frac{di}{dt} = 0,$$

or

$$Ee^{j\omega_0 t} = iR + \frac{1}{C}\left[\int_0^t i(x)dx + q(0+)\right] + L\frac{di}{dt}.$$

Taking the Laplace transform (see the following box) with $a = -j\omega_0$ for the term on the left, we have[10]

10 Notice that on the right we have $\mathcal{L}\{q(0+)\} = \int_0^\infty q(0+)e^{-st}dt = -\frac{q(0+)}{s}e^{-st}\,|_0^\infty = \frac{q(0+)}{s}$, where, to evaluate the upper limit, $\lim_{s\to\infty} e^{-st} = 0$, as s is defined to have a *positive* real part.

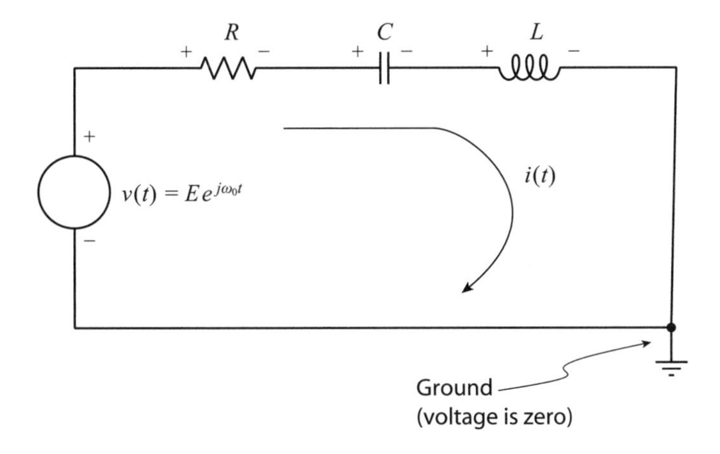

FIGURE 1.2.2. FitzGerald's circuit with a complex-valued voltage source.

$$\frac{E}{s-j\omega_0} = I(s)R + \frac{1}{C}\left[\frac{I(s)}{s} + \frac{q(0+)}{s}\right] + L\left[sI(s) - i(0+)\right]$$

and it is an easy matter to solve for $I(s)$:

$$I(s) = \frac{1}{sL + R + \frac{1}{sC}}\left[\frac{E}{s-j\omega_0} + Li(0+) - \frac{q(0+)}{sC}\right].$$

In radio theory we will be interested in time functions $f(t)$ that vanish for $t<0$. Physically, we interpret the instant $t=0$ as when we "turn $f(t)$ on." The Laplace transform of $f(t)$ is $\mathcal{L}\{f(t)\} = F(s) = \int_0^\infty f(t)e^{-st}dt$; the variable s is a complex variable with a positive real part to ensure convergence of the integral, but we can often formally work with the Laplace integral as if s is real. The value of the transform lies in its conversion of certain "complicated" operations in the time domain (like differentiation and definite integration) into "simple" algebraic ones. Specifically, $\mathcal{L}\left\{\frac{d}{dt}f(t)\right\} = sF(s) - f(0+)$, and $\mathcal{L}\left\{\int_0^t f(x)dx = \frac{1}{s}F(s)\right\}$. Tables of transforms have been created over the decades for a vast number of time functions, but the one most useful in radio analyses is $\mathcal{L}\{e^{-at}\}=\frac{1}{s+a}$, where a is a constant. When $a=0$, this says $\mathcal{L}\{1\}=\frac{1}{s}$. More precisely, in the $a=0$ case we are dealing with the function $f(t)=\begin{cases}1, & t>0\\ 0, & t<0\end{cases}$, which is called the *Heaviside step function*, often written as $H(t)$, in honor of Oliver Heaviside (see the appendix), who made extensive use of it. Of course, there are also other important *transform pairs* of $f(t)\leftrightarrow F(s)$, but the exponential time function transform will do 95% of the work for us here.

The factor

$$\frac{1}{sL+R+\frac{1}{sC}}=\frac{sC}{s^2LC+RCs+1}=\frac{s}{s^2L+Rs+\frac{1}{C}}=\frac{s}{L\left(s^2+\frac{R}{L}s+\frac{1}{LC}\right)}$$

can be written in the form

$$\frac{s}{L(s-s_1)(s-s_2)},$$

where s_1 and s_2 are each a function of R, L, and C.[11] Thus, the Laplace transform of the current is

$$I(s)=\frac{s}{L(s-s_1)(s-s_2)}\left[\frac{E}{s-j\omega_0}+Li(0+)-\frac{q(0+)}{sC}\right]$$

$$=\frac{\frac{E}{L}s}{(s-s_1)(s-s_2)(s-j\omega_0)}+\frac{si(0+)}{(s-s_1)(s-s_2)}-\frac{q(0+)/LC}{(s-s_1)(s-s_2)}$$

$$=\frac{\frac{E}{L}s+(s-j\omega_0)si(0+)-(s-j\omega_0)\frac{q(0+)}{LC}}{(s-s_1)(s-s_2)(s-j\omega_0)},$$

11 I'll leave it for you to confirm (it's easy!) that s_1 and s_2 are both either *real and negative* or both complex with *negative real parts* for any choice of positive values for R, L, and C. That's all we'll need to know about s_1 and s_2, as you'll soon see.

or

$$I(s)=\frac{\frac{E}{L}s+(s-j\omega_0)\left[si(0+)-\frac{q(0+)}{LC}\right]}{(s-s_1)(s-s_2)(s-j\omega_0)}.$$

If you examine $I(s)$, you see it has the form of a fraction with a numerator that is quadratic in s divided by a denominator that is cubic in s. It is therefore clear that we can write $I(s)$ as the partial-fraction expansion

$$I(s)=\frac{N_1}{s-s_1}+\frac{N_2}{s-s_2}+\frac{N_3}{s-j\omega_0},$$

where N_1, N_2, and N_3 are constants. If we now return to the time domain (using the exponential transform pair), this says

$$i(t)=N_1e^{s_1t}+N_2e^{s_2t}+N_3e^{j\omega_0t}.$$

Since s_1 and s_2 are either both negative or are both complex with negative real parts (see note 11), we see that the first two terms go to zero as $t\to\infty$. These two terms, which disappear with increasing time, represent *transient currents*. The third term, however, does *not* vanish as $t\to\infty$ but endlessly oscillates (because of Euler's identity). This persistent term is called a *steady-state current*. We can calculate N_3 by multiplying through $I(s)$ by the factor $s-j\omega_0$ and then taking the limit $s\to j\omega_0$. That is,

$$N_3=\lim_{s\to j\omega_0}(s-j\omega_0)I(s)=\lim_{s\to j\omega_0}\frac{\frac{E}{L}s}{(s-s_1)(s-s_2)}.$$

Since *by definition*

$$(s-s_1)(s-s_2)=s^2+\frac{R}{L}s+\frac{1}{LC},$$

then

$$N_3=\frac{\frac{E}{L}j\omega_0}{(j\omega_0)^2+\frac{R}{L}j\omega_0+\frac{1}{LC}}=\frac{\frac{E}{L}}{j\omega_0+\frac{R}{L}+\frac{1}{j\omega_0LC}}=\frac{E}{j\omega_0L+R+\frac{1}{j\omega_0C}}.$$

So, the steady-state current (the current after all transients have become insignificant) is

$$i(t)=\frac{Ee^{j\omega_0 t}}{R+j\omega_0 L+\frac{1}{j\omega_0 C}},$$

and this current is in response to the voltage $v(t)=Ee^{j\omega_0 t}$. Thus, *for the special case of sinusoids*, we see that we have a result that "looks like" Ohm's law; that is, if we write

$$Z(j\omega_0)=R+j\omega_0 L+\frac{1}{j\omega_0 C}=R+j\left(\omega_0 L-\frac{1}{\omega_0 C}\right)$$

as a sort of "resistance" (radio engineers call the frequency-dependent $Z(j\omega_0)$ the *ac impedance* at frequency ω_0), then for the steady state we have (where the symbols for voltage V and current I are written in uppercase to emphasize we are considering *only sinusoidal* time functions)

$$V(j\omega_0)=Z(j\omega_0)I(j\omega_0).$$

The unit of impedance is ohms, but unlike a resistance, which is purely real, an impedance is generally complex (the imaginary part of Z is called the *reactance*). This result, you'll notice, holds for *any* $i(0+)$ and *any* $q(0+)$; that is, while the initial conditions affect the transient terms, *they play no role in the steady-state term*. If $\omega_0=\frac{1}{\sqrt{LC}}$, then $|Z|$ is minimized (equal to R, with zero reactance), and ω_0 is called the *resonant frequency*.

We have the further observation that, at *any* frequency, the ac impedance of a resistor is R, the ac impedance of L at frequency ω is $j\omega L$, and the ac impedance of C at frequency ω is $\frac{1}{j\omega C}$. (Notice that I'm now writing ω, not ω_0, since the frequency of the input $v(t)$ is arbitrary, and a subscript is not necessary.) When working with ac impedances we can treat inductors and capacitors, *mathematically*, just like we treat resistors. So, when impedances are in series (as

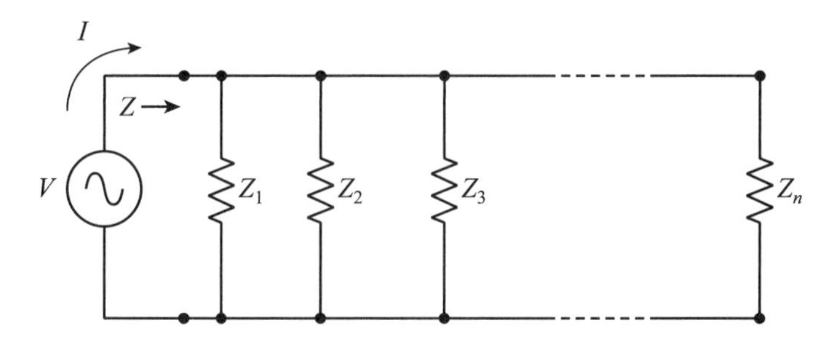

FIGURE 1.2.3. Z is the equivalent impedance of n parallel impedances.

they are in FitzGerald's circuit) they add. When impedances are in parallel their *reciprocals* add, a slightly nonobvious result we can see as follows with reference to Figure 1.2.3.

We have the impedance "seen" by the voltage source V as

$$Z = \frac{V}{I},$$

while from Kirchhoff's current law we have

$$I = \frac{V}{Z_1} + \frac{V}{Z_2} + \frac{V}{Z_3} + \ldots + \frac{V}{Z_n},$$

and so

$$\frac{I}{V} = \frac{1}{Z} = \frac{1}{Z_1} + \frac{1}{Z_2} + \frac{1}{Z_3} + \ldots + \frac{1}{Z_n}.$$

The special case of $n = 2$ leads to the very useful rule that two impedances in parallel are equivalent to their product divided by their sum. When analyzing radio circuits it is helpful to notice that the impedance of a capacitor is very large at low frequencies (infinite at zero frequency, or direct current (dc)) but tends to zero as the frequency increases, while the opposite is true for an inductor. (In the next section we'll use the fact that at $\omega = 0$ the dc resistance of an ideal inductor is zero, while the ac impedance can be quite large for any high-frequency energy that may also be present.)

The frequency behaviors of inductors and capacitors can be used to construct circuits that are of central importance in radio. As an example, consider the circuit of Figure 1.2.4. To emphasize that we are assuming sinusoidal voltages and currents *only*, I've written the input and output voltages in uppercase letters showing explicit dependence on the frequency variable ω (and not in lowercase as arbitrary functions of the time variable, t). You'll see this circuit again, later in this chapter, where I'll show you how it can be used to build an oscillator. For now, to support that discussion we'll need to know what electrical engineers call the *transfer function* $H(j\omega)$ of the circuit; that is, we'll now calculate

$$H(j\omega) = \frac{V_o(j\omega)}{V_i(j\omega)}.$$

A systematic way of calculating $H(j\omega)$ is based on the clever idea of *loop currents*, labeled as $I_1(j\omega)$ and $I_2(j\omega)$ in Figure 1.2.4. The loop-current approach to writing Kirchhoff's voltage loop law was introduced into circuit theory by Maxwell in his 1873 *Treatise*, and it is now a routine part of electrical engineering. The loop currents, *individually*, are fictitious, but they combine to give the actual currents in each component. For example, the current in the left C is $I_1 - I_2$ downward (or $I_2 - I_1$ upward), while the current in the right C is I_2 (to the right). The physical significance of I_1 is that it's the current that must be supplied from whatever is the source of the input voltage V_i. Writing Kirchhoff's voltage loop equations for the two loops in Figure 1.2.4, we have

$$-V_i + I_1 R + \frac{1}{j\omega C}(I_1 - I_2) = 0,$$

and

$$\frac{1}{j\omega C} I_2 + I_2 R + \frac{1}{j\omega C}(I_2 - I_1) = 0,$$

which can bc written in the form demanded by *Cramer's rule*[12] for solving these two simultaneous algebraic equations for I_1 and I_2:

12 After the Swiss mathematician Gabriel Cramer (1704–1752). Cramer published the rule in 1750, but in fact it had appeared two years earlier in a posthumously published work by the Scottish mathematician Colin MacLaurin (1698–1746).

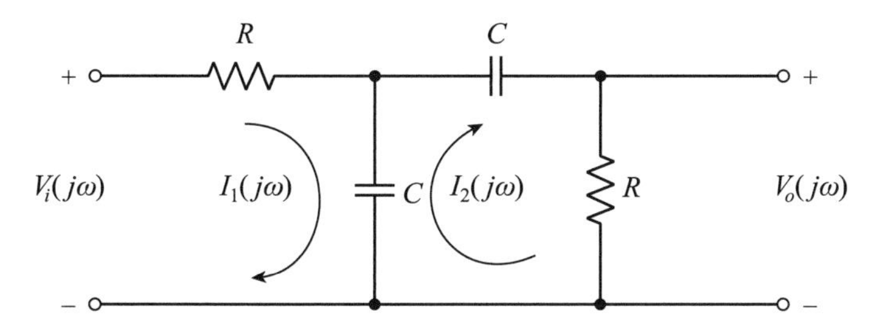

FIGURE 1.2.4. What is $\frac{V_o(j\omega)}{V_i(j\omega)}$?

$$I_1\left(R+\frac{1}{j\omega C}\right)+I_2\left(-\frac{1}{j\omega C}\right)=V_i$$

$$I_1\left(-\frac{1}{j\omega C}\right)+I_2\left(R+\frac{2}{j\omega C}\right)=0.$$

With the 2×2 system determinant D defined as

$$D=\begin{vmatrix}\left(\mathrm{R}+\frac{1}{j\omega C}\right) & \left(-\frac{1}{j\omega C}\right)\\ \left(-\frac{1}{j\omega C}\right) & \left(R+\frac{2}{j\omega C}\right)\end{vmatrix}=R^2-\left(\frac{1}{\omega C}\right)^2-j\frac{3R}{\omega C},$$

Cramer's rule says that

$$I_1=\frac{\begin{vmatrix}V_i & \left(-\frac{1}{j\omega C}\right)\\ 0 & \left(R+\frac{2}{j\omega C}\right)\end{vmatrix}}{D}=V_i(j\omega)\frac{R+\frac{2}{j\omega C}}{R^2-\left(\frac{1}{\omega C}\right)^2-j\frac{3R}{\omega C}},$$

and

$$I_2 = \frac{\begin{vmatrix} \left(R + \frac{1}{j\omega C}\right) & V_i \\ \left(-\frac{1}{j\omega C}\right) & 0 \end{vmatrix}}{D} = V_i(j\omega)\frac{-j\frac{1}{\omega C}}{R^2 - \left(\frac{1}{\omega C}\right)^2 - j\frac{3R}{\omega C}}.$$

Finally, observing that $V_o(j\omega) = I_2(j\omega)R$, we see that the transfer function is

$$\frac{V_o(j\omega)}{V_i(j\omega)} = H(j\omega) = \frac{-j\frac{R}{\omega C}}{R^2 - \left(\frac{1}{\omega C}\right)^2 - j\frac{3R}{\omega C}}.$$

We can draw one immediate, quite interesting conclusion from this result: $H(j\omega)$ is purely real when the real part of the denominator vanishes. That is, if $\omega = \frac{1}{RC}$, then we have $H(j\omega) = +\frac{1}{3}$. We won't pursue the implications of this (which are profound) until later in this chapter except to note for now that this property can be used to construct a sinusoidal oscillator. Oscillators are obviously important in radio *transmitters*, but less obvious at this point is that oscillators are also used in radio *receivers*. I'll remind you of Figure 1.2.4 again later in this chapter.

The transfer functions of all but the simplest circuits (for example, all resistors) used in radio will be complex. That is, $H(j\omega)$ will, in general, consist of both amplitude *and* phase response functions, and so

$$H(j\omega) = |H(j\omega)|e^{j\theta(\omega)},$$

where $\theta(\omega)$ is the phase shift that occurs from input to output for a sinusoid at frequency ω. I'll say more about $\theta(\omega)$ later in this chapter.

You'll notice that we did not need $I_1(j\omega)$ to find $H(j\omega)$. Knowledge of $I_1(j\omega)$ nevertheless provides important information. Knowing $I_1(j\omega)$ in terms of $V_i(j\omega)$ allows us to calculate $\frac{V_i(j\omega)}{I_1(j\omega)} = Z_i(j\omega)$, the ac impedance "seen" by the signal source that generates $V_i(j\omega)$. This is important to know because that impedance determines the current the signal source has to be able to provide. So,

$$Z_i(j\omega)=\frac{R^2-\left(\frac{1}{\omega C}\right)^2-j\frac{3R}{\omega C}}{R+\frac{2}{j\omega C}},$$

which, at the frequency $\omega=\frac{1}{RC}$ reduces to

$$Z_i(j\omega)=\frac{-j\frac{3R}{\frac{1}{RC}C}}{R+\frac{2}{j\frac{1}{RC}C}}=\frac{-j3R^2}{R-j2R}=-j\frac{3R}{1-j2}=R\left[-j\frac{3(1+j2)}{(1-j2)(1+j2)}\right]$$
$$=R\left[-j\frac{3+j6}{1+4}\right]=R\left(\frac{6-j3}{5}\right)=R(1.2-j0.6).$$

Thus, while the transfer function of the circuit of Figure 1.2.4 is purely real at $\omega=\frac{1}{RC}$, the input impedance is complex (the negative imaginary part of $Z_i(j\omega)$ means the input impedance "acts like" a capacitor (which, given the components in the circuit, should be no surprise!).

Be particularly careful to notice this important conclusion from our result for $Z_i(j\omega)$: while we can vary either R or C (actually either *both* of the matched R's together or *both* of the matched C's together, because the two resistors are assumed to be equal, and the two capacitors are assumed to be equal[13]) to vary the frequency at which $H(j\omega)$ is purely real, if we choose to vary the two R's we will also vary $Z_i(j\omega)$. If we choose to vary the two C's, however, then we can vary the frequency at which $H(j\omega)$ is purely real while keeping the input impedance fixed. In that case, the V_i signal source "sees" an

13 To simultaneously vary multiple matched-value components, radio engineers use a "ganged shaft" that allows turning a single control-panel knob to rotate a shaft on which all the variable components are mechanically mounted.

unchanging current demand, a property of great importance in building the variable-frequency oscillator circuits we will later encounter in radio receivers.

To end this discussion of transfer functions, let me show you one more thing we can do with them. Suppose we apply a sinusoid at frequency α as the input. That is, suppose $v_i(t) = \sin(\alpha t)$. What is the resulting output $v_o(t)$? From Euler's identity we have

$$v_i(t) = \frac{1}{j2}\left(e^{j\alpha t} - e^{-j\alpha t}\right).$$

The output of a circuit is simply the sum of each complex exponential term of $v_i(t)$ multiplied by the transfer function of the circuit *evaluated at the frequency of the input term.*[14] So, for the circuit of Figure 1.2.4,

$$\begin{aligned} v_o(t) &= \frac{1}{j2}\left[e^{j\alpha t}H(j\alpha) - e^{-j\alpha t}H(-j\alpha)\right] \\ &= \frac{1}{j2}\left[e^{j\alpha t}\frac{-j\dfrac{R}{\alpha C}}{R^2 - \left(\dfrac{1}{\alpha C}\right)^2 - j\dfrac{3R}{\alpha C}} - e^{-j\alpha t}\frac{-j\dfrac{R}{-\alpha C}}{R^2 - \left(\dfrac{1}{-\alpha C}\right)^2 - j\dfrac{3R}{-\alpha C}}\right]. \end{aligned}$$

There are a lot of j's in this expression, but since the input $v_i(t)$ is real-valued, and since the circuit of Figure 1.2.4 is made from real hardware, we know that $v_o(t)$ *has* to be real, too. Is it? Yes, and you can see that *by inspection* if you write

$$v_o(t) = \frac{1}{j2}\left[e^{j\alpha t}\frac{-j\dfrac{R}{\alpha C}}{R^2 - \left(\dfrac{1}{\alpha C}\right)^2 - j\dfrac{3R}{\alpha C}} - e^{-j\alpha t}\frac{j\dfrac{R}{\alpha C}}{R^2 - \left(\dfrac{1}{\alpha C}\right)^2 + j\dfrac{3R}{\alpha C}}\right].$$

14 Note, carefully, that there are *two* frequencies here: $+\alpha$ and $-\alpha$. The concept of a negative frequency might seem a bit "science fictiony," but we have already encountered a simple physical interpretation, namely, the *clockwise*-rotating vector in the box that opens this section. Later in the book, when we get to the sidebands of a modulated rf carrier wave, in both AM and FM radio, you'll see that "negative" frequencies possess an undeniable physical reality.

The expression inside the square brackets is the difference of conjugates and so is equal to the imaginary part of the first term times $j2$ (which is cancelled by the $\frac{1}{j2}$ in front of the brackets). Specifically,

$$v_o(t) = \mathrm{Im}\left[e^{j\alpha t} \frac{-j\frac{R}{\alpha C}}{R^2 - \left(\frac{1}{\alpha C}\right)^2 - j\frac{3R}{\alpha C}} \right]$$

$$= \mathrm{Im}\left[e^{j\alpha t} \frac{-j\frac{R}{\alpha C}\left\{R^2 - \left(\frac{1}{\alpha C}\right)^2 + j\frac{3R}{\alpha C}\right\}}{\left\{R^2 - \left(\frac{1}{\alpha C}\right)^2\right\}^2 + \left(\frac{3R}{\alpha C}\right)^2} \right],$$

or

$$v_o(t) = \mathrm{Im}\left[\frac{\{\cos(\alpha t) + j\sin(\alpha t)\}\left\{-j\frac{R}{\alpha C}\left[R^2 - \left(\frac{1}{\alpha C}\right)^2 + j\frac{3R}{\alpha C}\right]\right\}}{\left\{R^2 - \left(\frac{1}{\alpha C}\right)^2\right\}^2 + \left(\frac{3R}{\alpha C}\right)^2} \right].$$

Thus,

$$v_o(t) = \frac{\frac{3R^2}{(\alpha C)^2}\sin(\alpha t) - \frac{R}{\alpha C}\left\{R^2 - \left(\frac{1}{\alpha C}\right)^2\right\}\cos(\alpha t)}{\left\{R^2 - \left(\frac{1}{\alpha C}\right)^2\right\}^2 + \left(\frac{3R}{\alpha C}\right)^2}.$$

As a partial check on our calculations, notice that if $\alpha = \frac{1}{RC}$, then this expression reduces to $v_o(t) = \frac{1}{3}\sin\left(\frac{t}{RC}\right)$ when $v_i(t) = \sin\left(\frac{t}{RC}\right)$, giving $H\left(j\frac{1}{RC}\right) = +\frac{1}{3}$, which we found earlier.

1.3 Van der Pol's Negative Resistance Oscillator Equation

Almost from the invention of the electric battery in 1799–1800, by the Italian physical chemist Alessandro Volta (1745–1827), it was known that a low-voltage source (a few tens of volts) able to supply a continuous large current (hundreds of amperes) could generate an electrical arc of intense brilliance.[15] That is, if two electrodes in contact, carrying this current, are slowly pulled apart to form a gap, the electrode current can continue to flow across the gap, appearing as a *flame* of ionized atmospheric gases and vaporized electrode material. By the 1890s it was known that a plot of the gap current versus the gap voltage drop had the surprising behavior shown in Figure 1.3.1.

The surprising feature of Figure 1.3.1 is, of course, that the current-voltage curve of the electric arc has a *kink*, that is, an interval where a *decrease* in gap current is associated with an *increase* in gap voltage drop, behavior certainly not at all like the Ohm's law linear behavior of a resistor. The total voltage drop divided by the total gap current is always positive, but *in the kink* the *dynamic* ratio $\frac{di}{dv}$ is negative (the slope of the kink is negative), and for that reason the electric arc was said to have a negative ac resistance. With some very clever engineering, this feature of the electric arc was used to neutralize the energy-dissipating positive R of FitzGerald's oscillating circuit, allowing the construction of very powerful radio transmitters. Arc radio[16] itself is of only historical interest today, but the mathematical theory of negative resistance is of continuing interest, as it also appears in the electronics of modern radio.

15 The Cornish chemist Humphry Davy (1778–1829), mentor to the young Faraday, invented the arc lamp in 1809.

16 Not to be confused with the biblical Ark of the Covenant, said (in the first Indiana Jones movie, the 1981 *Raiders of the Lost Ark*) to be "a radio for speaking to God." The Ark, built by Moses according to detailed instructions from God (Exodus 25) to hold the stone tablets of the Ten Commandments, is described in various ancient Jewish legends as being surrounded by sparks and so was perhaps electrical in nature. Further, when Uzzah touched the Ark (2 Samuel 6:67) he instantly died (electrocuted?). In Exodus 25:22 the Lord tells Moses he will speak to him from the Ark, and this was the motivation for the movie line claiming the Ark to be a radio. Well, I have to admit it's a thought-provoking assertion, but we'll find *our* inspiration for radio in this book to be more from Maxwell's *Treatise* than we will from the Bible.

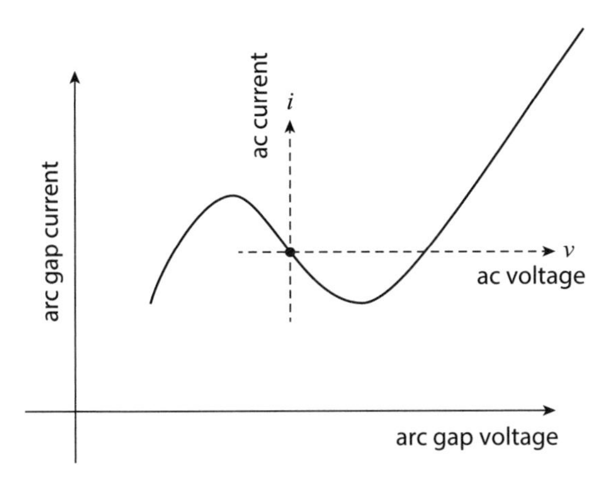

FIGURE 1.3.1. Current versus voltage for an electric arc.

In Figure 1.3.2 we see FitzGerald's circuit connected in parallel with an arc which is powered by a low-voltage, high-current dc energy source (which is itself in series with an inductor called the *choke coil*, a name that will be explained in just a moment). To understand what is happening in this enhanced FitzGerald circuit, you have to visualize two distinct current loops. First, there is the dc loop formed by the energy source, the choke coil (with a small ohmic resistance important at dc but presenting a relatively high ac impedance at the frequency at which the circuit oscillates), and the arc. Second, there is an ac current loop formed by the R, L, C, and again, the arc. The total ac resistance in this second loop is the sum of R and the dynamic ac resistance of the arc (which, being negative, can result in a net ac resistance of zero). Because of the choke coil, the oscillations in the ac loop cannot "leak back" through the dc source, which typically has a very low resistance. Such leakage would result in energy loss via heating of the dc source.

The arc current consists, then, of two components: a large, steady dc current, on top of which is superimposed an oscillating (that is, an ac) component. This is indicated in Figure 1.3.1 by the dashed axes centered on the midpoint of the negative resistance kink. If we imagine the arc operates at that midpoint ($i = v = 0$) when there are no oscillations, then when we say the circuit of Figure 1.3.2 oscillates, we mean we are interested in the current/voltage *deviations* around

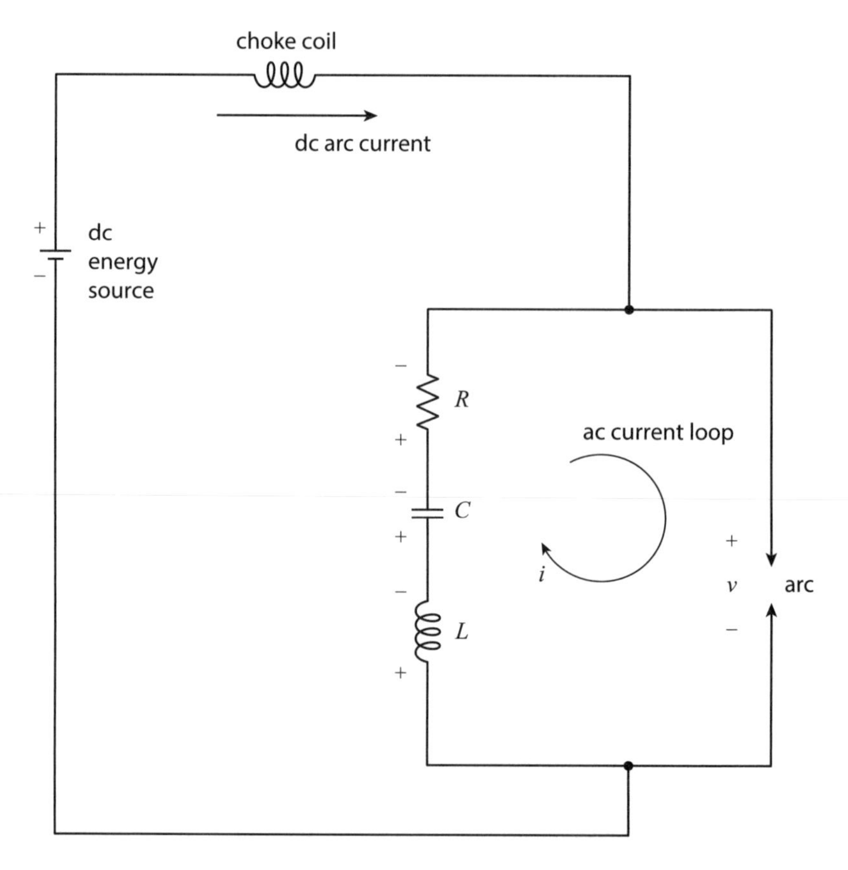

FIGURE 1.3.2. The ac current and voltage (i, v) in FitzGerald's series circuit, in parallel with an electric arc.

that dc midpoint. Now, be sure to understand that all this business about the electric arc *itself* is not the central issue. What is important is the *negative resistance* of the arc; I use the arc here simply to give you a physical model.to envision. When we get to the superregenerative radio receiver in chapter 3, you'll again see negative resistance mentioned in connection with oscillatory behavior.

In any case, such arc-enhanced versions of FitzGerald's circuit were unable to oscillate at frequencies beyond about 60 kHz or so,[17]

17 Because of engineering difficulties that are discussed in *The Science of Radio* (note 8), none of which I'll pursue here because Hardy couldn't have cared less about such things. If I really wanted to drive Hardy into a coma, I could next tell him that these same difficulties (and their solutions) appear in the physics of circuit breakers with superfast tripping times. But I wouldn't actually want to do that, and so I won't do it here, either.

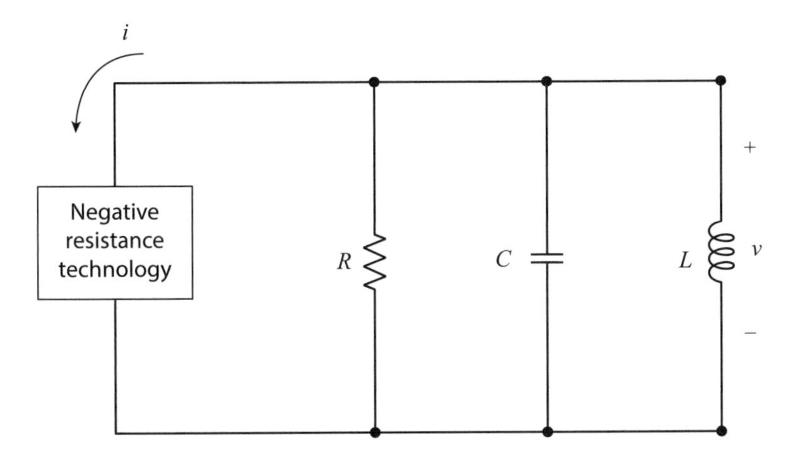

FIGURE 1.3.3. Van der Pol's parallel version of FitzGerald's oscillating series circuit.

and by the early 1920s the use of arc transmitters in radio was commercially dead. Van der Pol, however (see the box in the preface) studied a mathematically equivalent[18] parallel version (as shown in Figure 1.3.3) of FitzGerald's series circuit while retaining the idea of a negative resistance kink in the ac voltage/current behavior of whatever nonlinear technology (present or future) was under study.

As stated in note 18, from Kirchhoff's current law Van der Pol immediately wrote

$$i+\frac{v}{R}+C\frac{dv}{dt}+\frac{1}{L}\int v\,dt=0,$$

and he modeled the kink in the negative resistance technology box at the far left of Figure 1.3.3 with the equation

$$i=-av+bv^3,$$

18 Here's what *mathematically equivalent* means. If we write Kirchhoff's voltage law for the ac loop in Figure 1.3.2, we have $v+L\frac{di}{dt}+\frac{1}{C}\int i\,dt+iR=0$.

If we write Kirchhoff's current law for Figure 1.3.3, we have $i+\frac{v}{R}+C\frac{dv}{dt}+\frac{1}{L}\int v\,dt=0$, which is the voltage loop equation with v and i swapped, with R and $1/R$ swapped, and with L and C swapped. With those trivial symbol changes, FitzGerald's original *series* ac circuit becomes Van der Pol's *parallel* ac circuit. Electrical engineers say that each circuit is the *dual* of the other; the behavior of i in Figure 1.3.2 is the behavior of v in Figure 1.3.3.

where a and b are positive constants. Differentiating both equations, we get

$$\frac{di}{dt}+\frac{1}{R}\frac{dv}{dt}+C\frac{d^2v}{dt^2}+\frac{1}{L}v=0,$$

and

$$\frac{di}{dt}=-a\frac{dv}{dt}+3bv^2\frac{dv}{dt}.$$

I'll leave it to you to confirm that substituting the second equation into the first one and doing a little algebra gives the result

$$LC\frac{d^2v}{dt^2}+\left[L\left(\frac{1}{R}-a\right)+3bLv^2\right]\frac{dv}{dt}+v=0.$$

Next, we change the variable to $x=\omega_0 t$, where $\omega_0=\frac{1}{\sqrt{LC}}$. Then, $dx=\omega_0 dt$, or $dt=\frac{dx}{\omega_0}$, and so

$$\frac{dv}{dt}=\frac{dv}{\frac{dx}{\omega_0}}=\omega_0\frac{dv}{dx}.$$

Thus,

$$\frac{d^2v}{dt^2}=\frac{d}{dt}\left(\frac{dv}{dt}\right)=\frac{d}{dt}\left(\omega_0\frac{dv}{dx}\right)=\frac{d}{\frac{dx}{\omega_0}}\left(\omega_0\frac{dv}{dx}\right)=\omega_0\frac{d}{dx}\left(\omega_0\frac{dv}{dx}\right)$$

$$=\omega_0{}^2\frac{d^2v}{dx^2}=\frac{1}{LC}\frac{d^2v}{dx^2}.$$

Substituting these two results for $\frac{dv}{dt}$ and $\frac{d^2v}{dt^2}$ into the equation just before we change the variable to x, we have

$$\frac{d^2v}{dx^2}+L\left[\left(\frac{1}{R}-a\right)+3bv^2\right]\omega_0\frac{dv}{dx}+v=0,$$

or after just two or three more easy steps, we arrive at

$$\frac{d^2v}{dx^2} - \left[\sqrt{\frac{L}{C}}\left(\frac{1}{R} - a\right) - 3b\sqrt{\frac{L}{C}} + v^2\right]\frac{dv}{dx} + v = 0.$$

Now, to finish, we write $\varepsilon = \sqrt{\frac{L}{C}}\left(\frac{1}{R} - a\right)$ and make the change of variable $v = hu$, where h is the constant such that $h^2 = \frac{\varepsilon}{3b\sqrt{\frac{L}{C}}}$. In just a couple more easy steps of algebra we arrive at Van der Pol's equation:

$$\frac{d^2u}{dx^2} - \varepsilon(1 - u^2)\frac{du}{dx} + u = 0.$$

In this equation u is a normalized v as a function of x (which, in turn, is a normalized time). The parameter ε has absorbed the values of a, R, L, and C, while h (which is an amplitude-scaling parameter relating u and v) has absorbed b. Van der Pol's nonlinear differential equation is not "easy" to solve, and he was able to find analytical solutions only for the case of $\varepsilon \ll 1$, for which he found the remarkable result that the solutions are periodic with a normalized amplitude of 2. In *The Science of Radio* (note 8) I work through, in detail, how Van der Pol did this. It's elementary, but pretty tricky. Van der Pol was a very clever engineering analyst (and I think even Hardy would have concluded that).

Hardy's friend Littlewood came to his study of Van der Pol's equation in response to a January 1938 memorandum from the British Radio Research Board asking for "really expert guidance" from pure mathematicians in helping engineers understand the behavior of "certain types of non-linear differential equations involved in radio engineering." A copy of the memorandum was sent to the London Mathematical Society, where it caught the eye of Mary Cartwright (1900–1998), an English mathematician who started her doctoral studies at Oxford under Hardy (but finished with a different thesis advisor when Hardy left Oxford for a sabbatical leave at Princeton University and Caltech, 1928–1929). Cartwright knew Littlewood (who had, in June 1930, traveled to Oxford to supervise her doctoral examination), she got him interested in Van der Pol's equation, and the two of them decided to jointly respond to the 1938 memorandum.[19]

Their starting point was a 1934 paper published by Van der Pol that included "graphically integrated" solutions for various values of ε. Littlewood was able to show that the oscillation amplitude was *not* exactly 2 for ε "small," but later Cartwright and Littlewood further showed that as $\varepsilon \to \infty$ the oscillation amplitude *did* approach 2 from above. With a modern home computer and powerful software (I use MATLAB), it is today easy to confirm these results. Figure 1.3.4, for example, shows a computer solution for one of the values of ε in Van der Pol's paper,[20] and it is virtually identical with Van der Pol's graphical solution (see the following box for how this figure was created). To quote from Van der Pol's paper, the solution "represents the slow building up of an approximately sinusoidal oscillation." The final amplitude of those oscillations does appear to be pretty close to 2. For $\varepsilon \gg 1$ the oscillations are decidedly *not* sinusoidal.

The computer-generated solution to Van der Pol's differential equation was obtained by the standard method of writing an *n*th-order differential equation as a system of *n* first-order differential equations. That is, we start by defining

$$u_1(x) = u(x), \quad u_2(x) = \frac{du_1}{dx}.$$

Then,

$$\frac{du_1}{dx} = \frac{du}{dx} = u_2, \quad \frac{du_2}{dx} = \frac{d^2u_1}{dx^2} = \frac{d^2u}{dx^2} = (1 - u_1^2)u_2 - u_1.$$

19 See Shawnee L. McMurran and James J. Tattersall, "The Mathematical Collaboration of M. L. Cartwright and J. E. Littlewood," *American Mathematical Monthly*, December 1996, pp. 837–845.

20 B. van der Pol, "The Nonlinear Theory of Electric Oscillations," *Proceedings of the IRE* (Institute of Radio Engineers), September 1934, pp. 1051–1086.

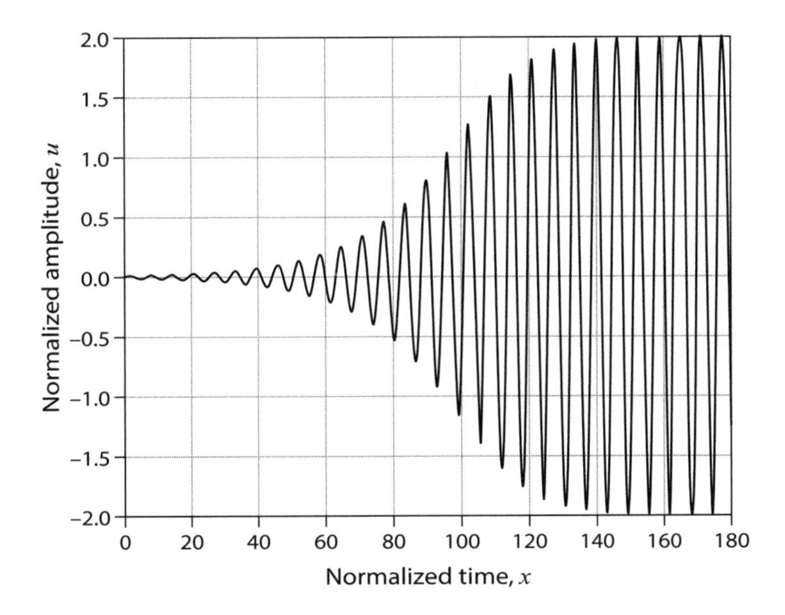

FIGURE 1.3.4. Computer solution of Van der Pol's equation for $\varepsilon = 0.1$.

Once the initial conditions of $u(0) = u_1(0)$ and $\frac{du}{dx}|_{x=0} = u_2(0)$ are given along with the value of ε, a solution is uniquely determined. It is clear that Van der Pol defined $u(0) = 0$, but it is not so clear what value he used for the first derivative of u at time $x = 0$. By trial and error, a value of 0.01 reproduced his solution, and that is the value used in the vector *uinit* (for *u initial*). The MATLAB code **vdpol.m** (which calls the subroutine **vdpolsub.m**) does the job, including generating a plot.

```
%vdpol.m
xinterval=[0 180];
uinit=[0;0.01];
[x,u]=ode45(@vdpolsub,xinterval,uinit);
plot(x,u(:,1),'-k')
grid
xlabel('normalized time, x')
ylabel('normalized amplitude, u')
```

```
%vdpolsub.m
function uvector=vdpolsub(x,u);
epsilon=0.1;
uvector=[u(2);epsilon*(1-u(1)^2)*u(2)-u(1)];
```

I include the computer code here not to turn anyone into a MATLAB programmer but simply to illustrate how easy it is to do today what might be what Hardy had in mind in 1940 when he claimed (in the preface) that numerically solving differential equations was "soul-destroying." The code took a mere minute or two to write, and then, with the push of a button, it produced and plotted the solution. I suspect that even Hardy would have had a difficult time resisting playing with the code!

1.4 Filters and the Wiener-Paley Theorem

Filters are a class of circuits of great importance in radio work. They allow energy to pass through them (or not to pass) as a function of frequency. The ones we'll be concerned with are of three general types: low-pass filters (LPF), high-pass filters (HPF), and band-pass filters (BPF), names that are self-descriptive. Elementary examples of the HPF and LPF are shown in Figure 1.4.1, and they are easily understandable based on the frequency behavior of capacitors.

At low frequencies C has a high impedance, and so low-frequency energy is essentially blocked from getting to the output of the HPF circuit on the left of Figure 1.4.1. But since the ac impedance of C decreases with increasing frequency, energy at high frequency is not blocked. Things work in reverse in the LPF circuit on the right of Figure 1.4.1. (Can you see how to make an HPF and an LPF using an L instead of a C?)

The *ideal* BPF, an important theoretical[21] circuit in radio work, allows energy located in an interval of intermediate frequencies (the

21 I say *theoretical* because, as you'll see by the end of this section, the *ideal* BPF is impossible to actually construct (but radio engineers can get pretty close to ideal). For a bit more on this, see Challenge Problem 1.6.

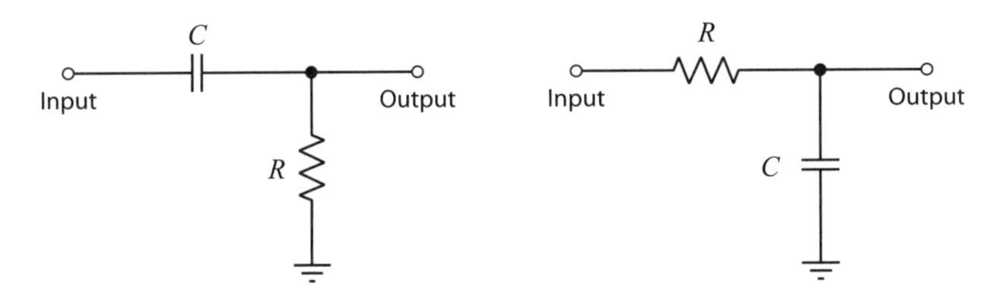

FIGURE 1.4.1. A simple high-pass (left) and a low-pass filters (right).

filter's *passband*) to move through it without attenuation, while completely blocking from the output all the input energy that lies outside the passband. A sketch of the magnitude of the transfer function of the ideal BPF is shown in Figure 1.4.2; it said to be *ideal* because $|H(j\omega)|=0$ when it isn't equal to 1.[22] Such a magnitude response is said to have *vertical skirts* (a term motivated by the resemblance to nineteenth-century hoop skirts). The *bandwidth* of the ideal BPF is $2\,\Delta\omega$ where, if $\omega_c=2\pi f_c$ is the center frequency of the passband, the passband is the interval $\omega_c-\Delta\omega<|\omega|<\omega_c+\Delta\omega$.

These elementary ideas give us a hint on how to make a BPF: simply have an LPF feeding its output into the input of an HPF (only "Goldilocks frequencies" can survive passage through both filters, that is, frequencies that are "not too low" but also "not too high"). If we do that, we get the circuit of Figure 1.2.4, which we've already analyzed. That circuit uses four components, while in Figure 1.4.3 a different BPF uses just three components. Again, it is easy to see how this new BPF works just by remembering the frequency behavior of inductors and capacitors. At very low frequencies the L provides a low-impedance electrical path to ground, effectively bypassing the high-impedance path through the C. At high frequencies the L and C exchange roles, with the C providing a low-impedance electrical path to ground that bypasses the high-impedance path through the L. At "middle" Goldilocks frequencies, however, the impedances of the L and C are both "high," and so the output voltage can be significant.

22 Notice that in Figure 1.4.2 $|H(j\omega)|=1$ for both positive and negative ω. That's because of Euler's identity, which says a real sinusoidal signal is made from both $e^{j\omega t}$ and $e-^{j\omega t}=e^{j(-\omega)t}$. So, to let sin (ωt) or cos(ω) "through," the ideal BPF must let both $+\omega$ and $-\omega$ through.

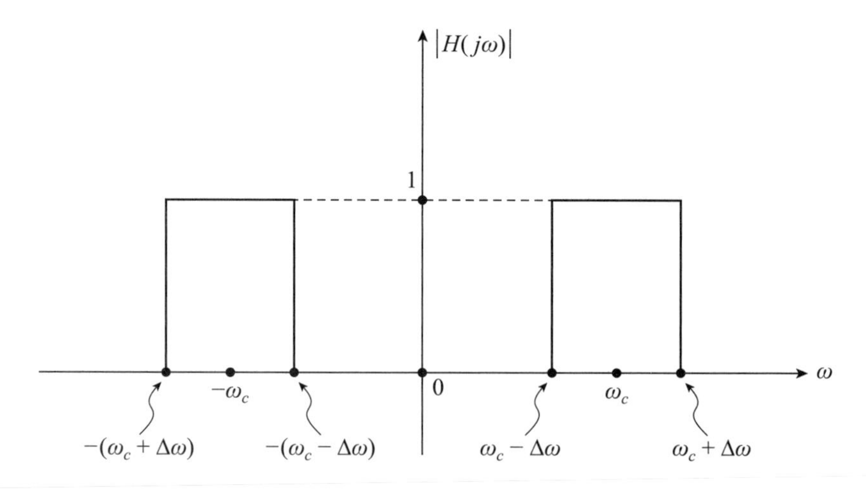

FIGURE 1.4.2. Amplitude response of the ideal BPF transfer function.

If you apply Kirchhoff's voltage-loop law to Figure 1.4.3, you should be able to see that

$$H(j\omega)=\frac{Z_2}{Z_1+Z_2},$$

where $Z_1=R$, and Z_2 is the impedance of L and C in parallel; that is,

$$Z_2=\frac{j\omega L\frac{1}{j\omega C}}{j\omega L+\frac{1}{j\omega C}}=\frac{\frac{L}{C}}{j\left(\omega L-\frac{1}{\omega C}\right)}.$$

With just a bit of algebra it then follows that

$$H(j\omega)=\frac{1}{1+jR\left(\omega C-\frac{1}{\omega L}\right)},$$

and

$$\theta(\omega)=-\tan^{-1}\left\{R\left(\omega C-\frac{1}{\omega L}\right)\right\},$$

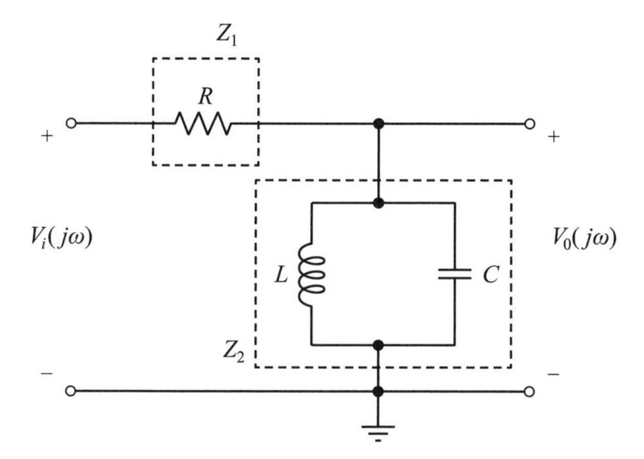

FIGURE 1.4.3. A simple band-pass filter.

where you'll recall from section 1.2 that $\theta(\omega)$ is the phase response of the transfer function; that is,

$$H(j\omega) = |H(j\omega)|e^{j\theta(\omega)}.$$

Now, we define the *center frequency* of the filter as

$$\omega_c = \frac{1}{\sqrt{LC}},$$

and

$$\omega_c = \frac{1}{RC}.$$

Thus, once one of the values of R, L, and C is assigned, the values of the other two components are determined for a given ω_c.[23] With these two definitions for ω_c, it is then just a few more easy steps of algebra to arrive at

23 For example, suppose we wish to build our BPF to have a center frequency of $f_c = 455$ kHz (a particularly interesting frequency that, as you'll learn later in the book, appears in AM radio receivers). Further, suppose we want to use $L = 100$ μH. Using our two definitions for ω_c then tells us that $C = 0.0012$ μF, and $R = 350$ ohms.

$$\left|H(j\omega)\right| = \frac{1}{\sqrt{1+\left(\frac{\omega}{\omega_c}-\frac{\omega_c}{\omega}\right)^2}},$$

and

$$\theta(\omega) = -tan^{-1}\left(\frac{\omega}{\omega_c}-\frac{\omega_c}{\omega}\right).$$

Plots of these two response functions as functions of the normalized frequency $\frac{\omega}{\omega_c}$ are shown in Figure 1.4.4. The amplitude response plot shows that the circuit of Figure 1.4.3 is a *very* approximate BPF.

Knowledge of $|H(j\omega)|$ is not enough to completely describe the ideal BPF because it doesn't say anything about the phase response of the filter. To determine the phase response of the ideal BPF we'll impose the physical requirement that there be zero *phase distortion*. Phase distortion occurs if energies at different frequencies take different times to transit the filter. So, a signal's *shape* will be unaltered by passage through the filter if all the energy of the signal lies in the filter passband. (The signal *amplitude*, however, may be altered.) So, suppose that all the energy propagating through the filter experiences the same time delay $t_0 > 0$. That is, for $\omega > 0$ the input signal $e^{j\omega t}$ emerges at the output as $e^{j\omega(t-t_0)} = e^{j\omega t}e^{-j\omega t_0}$. But, by the very definition of the transfer function $H(j\omega)$, the output is $H(j\omega)e^{j\omega t}$. Thus,

$$H(j\omega)e^{j\omega t} = e^{j\omega t}e^{-j\omega t_0},$$

and so

$$H(j\omega) = e^{-j\omega t_0}.$$

In the $\omega > 0$ passband the transfer function is

$$H(j\omega) = \left|H(j\omega)\right|e^{j\theta(\omega)} = e^{-j\omega t_0},$$

and we see that $|H(j\omega)| = 1$ in the passband (as it should for an ideal BPF) and that

$$\theta(\omega) = -\omega t_0.$$

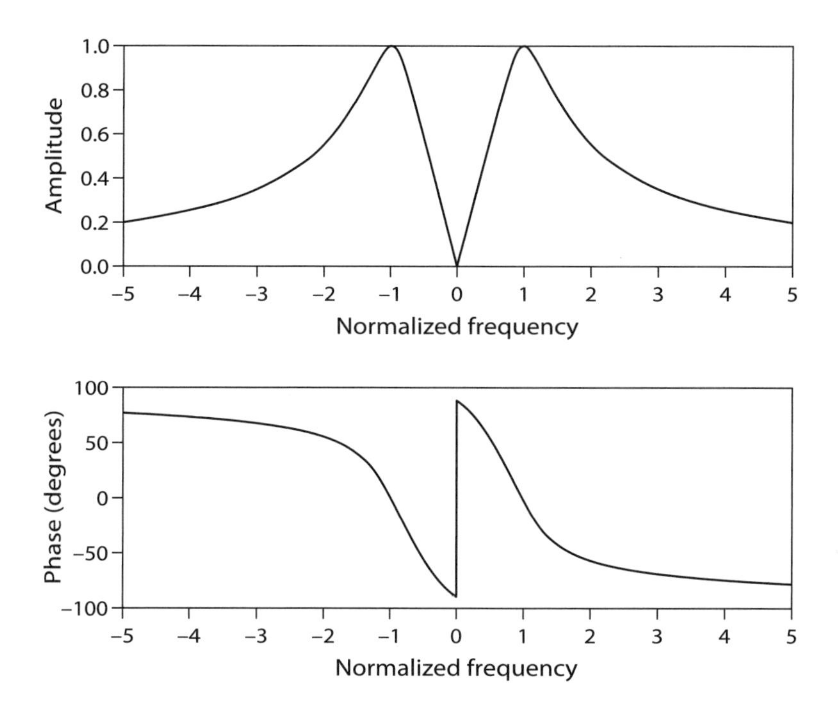

FIGURE 1.4.4. Amplitude (top) and phase (bottom) responses of the filter in Figure 1.4.3.

Thus,

$$\frac{d\theta}{d\omega} = -t_0 < 0$$

because $t_0 > 0$. That is, the phase response of the ideal BPF varies linearly (with negative slope) with frequency in the $\omega > 0$ passband.[24] The second plot in Figure 1.4.4 shows that the circuit of Figure 1.4.3 exhibits just this sort of behavior, especially at frequencies near ω_c (that is, at normalized frequencies around ±1).

Radio engineers have learned how to make filter circuits that closely *approach* the ideal of vertical skirts and zero phase distortion but at the price of increasing complexity. I emphasize the word *approach* because, in 1933, it was proven that it is impossible to actually achieve the ideal. Called the Wiener-Payley theorem—after

24 You'll notice I've stated that all this is for the case of $\omega > 0$. What about the $\omega < 0$ case? Well, it all comes out the same, as you can (and should) verify by simply repeating the argument in the text, arriving at $\theta(-\omega) = \omega t_0$ for $\omega < 0$. That is, $\theta(\omega) = -\omega t_0$, just as before.

the American mathematician Norbert Wiener (1894–1964) and the English mathematician Raymond Payley (1907–1933)[25]—this result says that for the transfer function of a filter to be physically realizable a certain integral must exist (be finite). Specifically, a necessary and sufficient condition for $|H(j\omega)|$ to be the magnitude of the transfer function of a causal filter[26] is

$$\int_{-\infty}^{\infty} \frac{|\ln\{H(j\omega)\}|}{1+\omega^2} d\omega < \infty.$$

For the ideal BPF, with $|H(j\omega)| = 0$ for almost all ω (all ω not in the passband), it is obvious this integral is not finite, because the log function blows up when its argument is zero). And so, just like that, Wiener and Payley showed that the ideal BPF is impossible (and for the same reason, so are the ideal LPF and HPF). The Wiener-Payley integral doesn't say anything about the phase of the transfer function; all it tells us is that *if* $H(j\omega)$ passes the integral test, *then* there exists some phase function (not necessarily one with zero phase distortion) that when combined with $|H(j\omega)|$ results in the transfer function of a causal filter.

In the second volume of his autobiography,[27] Wiener wrote of this result:

> Payley attacked [the problem] with vigor, but what helped me and did not help Payley was that it is essentially a problem in electrical engineering [much of Wiener's work in mathematics was inspired by his interactions with colleagues in the MIT electrical engineering department, much like Professor Tweedle's in the preface]. It had been known for many years that there is a

25 Payley's math teachers at Cambridge were Littlewood and Hardy. He was just 26 when he was killed in an avalanche while skiing in Canada.

26 A causal filter is one that no matter what its input signal has zero output response at all times before the input signal is applied. With what is known to modern science, this causality requirement for physical realizability seems plausible. If one could build a filter that violated causality, then time machines would surely soon follow! (For more on that issue, see my *Time Machine Tales*, Springer, 2017.) An outline of how to derive the Wiener-Payley integral appears in A. Papoulis, *The Fourier Integral and Its Applications*, McGraw-Hill, 1962, pp. 215–217.

27 *I Am a Mathematician: The Later Life of a Prodigy*, MIT Press, 1956, pp. 168–169.

> certain limitation on the sharpness with which an electric wave filter cuts a frequency band off, but the physicists and engineers had been quite unaware of the deep mathematical grounds for those limitations. In solving what was for Payley a beautiful and difficult chess problem, completely contained within itself, I showed at the same time that the limitations under which the electrical engineers were working were precisely those which prevent the future from influencing the past. [It is this last sentence, which must have perplexed many of Wiener's 1950s readers, that prompted my comment on time machines in note 26.]

Wiener was a pure theoretician, but like Littlewood (and unlike Hardy), he was willing to apply math to military service. During the First World War he, like Littlewood did for the British Army, performed ballistic shell trajectory calculations for the US Army.

Well, no matter that ideal LPF, HPF, and BPF circuits are impossible—their properties are *so* convenient that in all that follows we will imagine all our filter circuits in radio to be ideal. We can get away with this because, as I mentioned earlier, radio engineers have learned how to construct circuits pretty close to the ideal. I won't get into the details of how they do that (that subject is, alone, an entire course in an electrical engineering major) but will instead be content to know radio engineers can build excellent approximations to the ideal.[28]

1.5 An Electronic Phase-Shift Oscillator

As all of our discussions so far have indicated, radio of any sort depends on the existence of electrical oscillations. One very intuitive approach to building a circuit that can generate *sustained* oscillations uses the idea of an electrical signal endlessly circulating, without attenuation, around a closed loop at some specified frequency ω_0 and

28 In sharp contrast, zookeepers and aquarium operators can't argue that unicorns and mermaids would be so convenient to attracting paying visitors that they are going to simply imagine they actually have a unicorn or a mermaid in residence. They can't get away with that because there are no "excellent approximations" to unicorns and mermaids. Unicorns, mermaids, and ideal filters don't exist, but one is "less impossible" than are the other two.

no other. This idea is shown in block-diagram form in Figure 1.5.1 as a negative-gain (a 180° phase-shift) amplifier with voltage gain G whose output signal Ge_0 is the input signal to a frequency-dependent circuit that at the particular frequency ω_0 introduces another reversal of sign (a second 180° phase shift) to produce the input signal e_0 to the amplifier. $H(j\omega)$ is the transfer function of the phase-shifting network, and so, for oscillation at frequency ω_0, we have

$$Ge_0H(j\omega_0)=e_0,$$

or

$$GH(j\omega_0)=1.$$

This last result is called the *Barkhausen criterion* for oscillation, after the German physicist Heinrich Barkhausen (1881–1956), who stated it in 1921.[29]

The negative gain of the amplifier means it has introduced a 180° phase shift from its input to its output, and the frequency-dependent phase shifter is imagined to introduce another 180° phase shift from its input (the amplifier output) to its output (the amplifier input). The total loop phase shift is then 360°. This creates a self-sustaining situation at a unique frequency determined by the phase shifter (the frequency where it achieves a 180° phase shift). If the phase shifter is constructed from just resistors and capacitors, then $|H(j\omega)|<1$, and so the voltage gain of the amplifier will have to be greater than 1 in magnitude for the circuit to oscillate (see the following box). It is also possible to use a positive gain amplifier (0°phase shift) if the phase-shifter network also introduces 0° phase shift at some frequency, giving a total loop phase shift, at that frequency, of 0°.

If you look back at the circuit in Figure 1.2.4, where we showed its $|H(j\omega)|=+\frac{1}{3}$ at $\omega=\frac{1}{RC}$, we can use that circuit to make an

29 As I said, this is intuitive, but there is *lot* more that can be said, as it is only a necessary (but not sufficient) condition for oscillation. For our purposes here, however, we'll not need to be concerned with this issue.

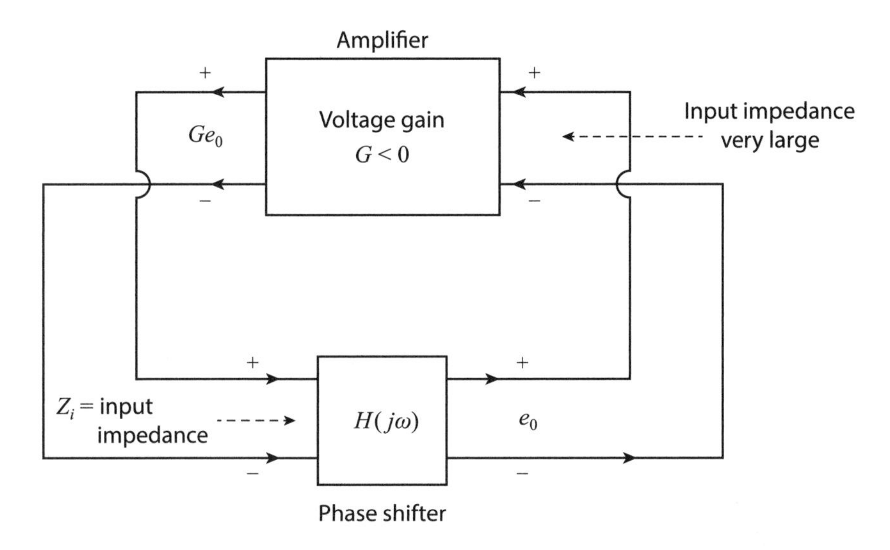

FIGURE 1.5.1. The closed-loop phase-shift oscillator.

oscillator if the amplifier voltage gain is at least +3, to satisfy the Barkhausen criterion. If we want to use a negative-gain amplifier, as we argued in the opening of this section, the phase-shift network needs to introduce a second 180°phase shift. One commonly used circuit that does this is shown in Figure 1.5.2, using three identical resistors R and three identical capacitors C.

If you calculate the transfer function $H(j\omega)=\dfrac{V_o(j\omega)}{V_i(j\omega)}$, you should be able to show (start by defining the three obvious loop currents shown in Figure 1.5.2 and then write the Kirchhoff voltage loop equations, just as we did for Figure 1.2.4) that at the particular frequency $\omega_0=\dfrac{1}{RC\sqrt{6}}$, $H(j\omega_0)=-\dfrac{1}{29}$. This means the amplifier in Figure 1.5.1 must now be able to achieve a voltage gain of at least −29 to satisfy the Barkhausen criterion. When writing the loop equations, use the assumption that the amplifier has infinite input impedance, and so there

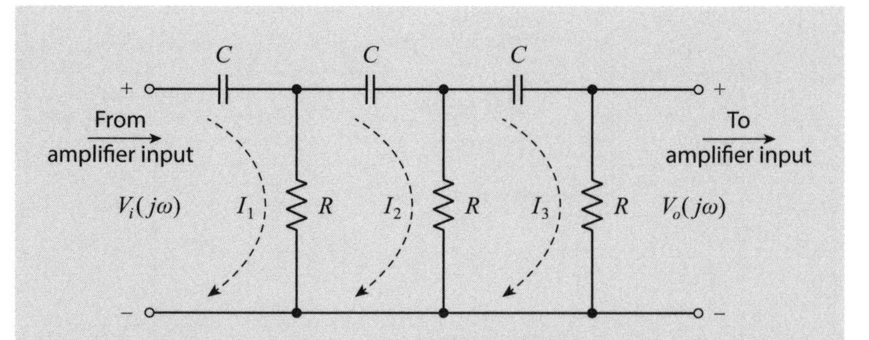

FIGURE 1.5.2. A phase-shift network for an oscillator using a negative gain amplifier.

is zero amplifier input current. Thus, $V_o(j\omega)=I_3R$. With this assumption, show that the input impedance to the circuit at $\omega=\omega_0$ is $Z_i=R(0.83-j2.7)$. Thus, if it is desired to keep Z_i the same no matter what oscillation frequency ω_0 is desired—thereby keeping fixed the current the amplifier has to supply to the input of the phase-shifting network—one should vary the value of C to vary ω_0 and *not* vary the value of R.

The results we've developed up to now all flow from the simple expressions we derived earlier for the ac impedances of resistors, capacitors, and inductors. It sometimes seems almost a miracle that such elementary impedance expressions can have such profound implications, but here's one more such result for you to try your hand at. Figure 1.5.3 appeared as part of a larger circuit for an FM oscillator, and in a paper describing how the oscillator works, the inventor made the following claim:[30] "[T}he phase shifter . . . used in this oscillator . . . consists of L in series with R, and this series combination is shunted by another consisting of C in series with [another resistor of equal value]. If the elements are so chosen that $R=\sqrt{\frac{L}{C}}$, the impedance looking into the phase shifter [Z in the figure] is purely resistive [that is, has zero imaginary part] *and has the value R at all*

30 O. E. De Lange, "A Variable Phase-Shift Frequency-Modulated Oscillator," *Proceedings of the IRE*, November 1949, pp. 1328–1330.

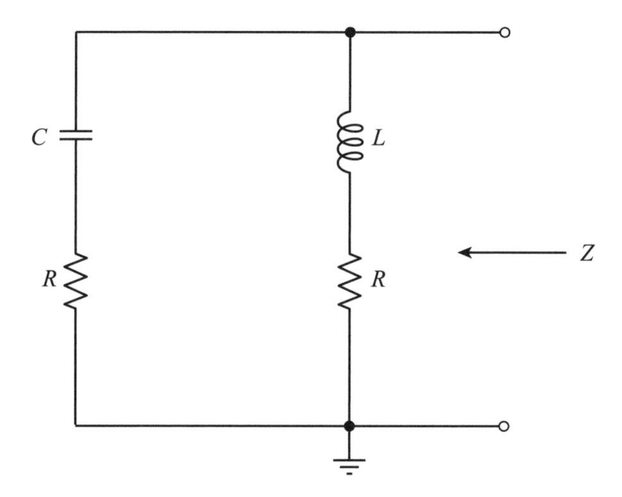

FIGURE 1.5.3. A circuit with a remarkable property.

frequencies [my emphasis added]." See if you can show this is, in fact, true (a conclusion that, I think, can be understood *only* with the aid of mathematics). (*Hint*: You'll find it helpful to remember the result we found in section 1.2: two impedances in parallel are equivalent to their product divided by their sum).

SPECIAL NOTE: Up to now I've written quantities like $H(j\omega)$ and $Z(j\omega)$ with an explicit j, to emphasize they are complex valued. From this point on, I'll drop the j, and simply write $H(\omega)$, $Z(\omega)$, and so on.

1.6 Fourier Series, Transforms, and Signal Spectrums

In this section we'll review some mathematics that we'll find invaluable in understanding radio. This will be a fairly fast presentation because mathematical readers have almost surely already seen all of what we'll discuss; what might be new for such readers, however, may be the intimate connection of this theoretical math with practical radio. To start, suppose $x(t)$ is a periodic, real-valued time function with period T. That is, for T some positive constant $x(t)=x(t+T)$, $-\infty<t<\infty$. We will impose few other restrictions on

$x(t)$, one of which is that $x(t)$ has finite *energy*, W, in a period.[31] That is, we assume

$$W = \int_{\text{period}} x^2(t)\, dt < \infty.$$

For functions in the real world of working radios, this is a very weak restriction. This definition for the energy of a signal is motivated by an analogy from electrical engineering (a claim that should catch the attention of our resurrected Hardy!). If $x(t)$ is a periodic voltage, then if it is applied across a resistor R resulting in the current $i(t)$, the instantaneous power is

$$p(t) = x(t)i(t) = x(t)\frac{x(t)}{R} = \frac{x^2(t)}{R} = x^2(t)$$

if $R = 1$ ohm. Since the energy W is the integral of power, we have

$$W = \int_{\text{period}} p(t)\, dt,$$

and our definition of W as the integral of $x^2(t)$ follows. This definition is used (by radio engineers and mathematicians alike) even if $x(t)$ is not a voltage and even if there isn't an R (of *any* value) in sight.

The *Fourier series*[32] of a periodic $x(t)$ is the complex exponential series

$$x(t) = \sum_{k=-\infty}^{\infty} c_k e^{jk\omega_0 t},\ \omega_0 = \frac{2\pi}{T},$$

where the c_k are constants (generally, they are complex constants). How do we know we can express $x(t)$ in this way? Suppose we write $x_N(t)$ as a truncated Fourier series with a finite number of terms; that is, for N some finite positive integer

$$x_N(t) = \sum_{k=-N}^{N} c_k e^{jk\omega_0 t},\ \omega_0 = \frac{2\pi}{T}.$$

31 We will, however, find that one particular time function will be *so* useful that we will admit it into our discussion of radio even though it has infinite energy in a finite time interval. This is the *Dirac impulse function*, and I'll say a lot more about it in the next section.

32 After the French mathematician Joseph Fourier (1768–1830).

Further, suppose we write E_N as the *integrated squared error* that $x_N(t)$ makes as an approximation to $x(t)$. That is,

$$E_N = \int_{\text{period}} \{x(t) - x_N(t)\}^2 \, dt = \int_{\text{period}} \left\{ x(t) - \sum_{k=-N}^{N} c_k e^{jk\omega_0 t} \right\}^2 dt.$$

E_N is the energy of the error. It can be shown[33] that if we pick the c_k to be such that E_N is minimized, then

(a) For any value of N, the resulting c_k are independent of the value of N, and
(b) $\lim_{N \to \infty} (\min E_N) = 0$.

That is, the complex exponential Fourier series of $x(t)$ tends toward having zero energy in the error made by $x_N(t)$ as an approximation to $x(t)$ as we let N increase without bound. This result gives physical meaning to the complex exponential series that, at first glance, seems to be an abstract mathematical definition.

As part of showing that the truncated Fourier series of $x(t)$ converges in energy to the actual $x(t)$ as we include more and more terms, we also find that

$$c_k = \frac{1}{T} \int_{\text{period}} x(t) e^{-jk\omega_0 t} \, dt.$$

Notice that while the c_k are, in general, complex, since $x(t)$ is real-valued, then the special case of $k = 0$ says

$$c_0 = \frac{1}{T} \int_{\text{period}} x(t) \, dt$$

is always real-valued. The value of c_0 is, physically, the average value of $x(t)$ over a period and is commonly called the *dc level* of $x(t)$. Notice, too, that the conjugate of c_k, c_k^*, is (remember, $x(t)$ is real)

33 See *The Science of Radio* (note 8), pp. 136–140, for the mathematical details.

$$c_k^* = \left\{\frac{1}{T}\int_{\text{period}} x(t)e^{-jk\omega_0 t}\,dt\right\}^* = \frac{1}{T}\int_{\text{period}} x^*(t)(e^{-jk\omega_0 t})^*\,dt$$
$$= \frac{1}{T}\int_{\text{period}} x(t)e^{jk\omega_0 t}\,dt = c_{-k}.$$

We can use this result to show that, like c_0, all the c_k for $k \neq 0$ also have important physical significance.

We start by writing the integrand of the energy integral W as

$$x^2(t) = \left\{\sum_{k=-\infty}^{\infty} c_k e^{jk\omega_0 t}\right\}^2 = \sum_{m=-\infty}^{\infty}\sum_{n=-\infty}^{\infty} c_m c_n e^{j(m+n)\omega_0 t},$$

and so

$$W = \int_{\text{period}} \sum_{m=-\infty}^{\infty}\sum_{n=-\infty}^{\infty} c_m c_n e^{j(m+n)\omega_0 t}\,dt$$
$$= \sum_{m=-\infty}^{\infty}\sum_{n=-\infty}^{\infty} c_m c_n \int_{\text{period}} e^{j(m+n)\omega_0 t}\,dt.$$

This last integral is particularly easy to evaluate. For any integration interval of one period duration, starting at the arbitrary time t', we have

$$\int_{t'}^{t'+T} e^{j(m+n)\omega_0 t}\,dt = \frac{e^{j(m+n)\omega_0 t}}{j(m+n)\omega_0}\Big|_{t'}^{t'+T} = \frac{e^{j(m+n)\omega_0(t'+T)} - e^{j(m+n)\omega_0 t'}}{j(m+n)\omega_0}$$
$$= \frac{e^{j(m+n)\omega_0 t'}\{e^{j(m+n)\omega_0 T} - 1\}}{j(m+n)\omega_0}.$$

Now, recall that $\omega_0 T = 2\pi$, and that $e^{j(m+n)2\pi} = 1$ for all integer m and n. Thus, the integral is *almost* always equal to zero but not *always*, because for the special cases where $n = -m$ the last expression becomes the indeterminate $\frac{0}{0}$. For those special cases, *first* set $n = -m$ in the energy integral and *then* do the integral:

$$\int_{\text{period}} e^0\,dt = \int_{t'}^{t'+T} dt = T.$$

Thus,

$$W=\sum_{m=-\infty}^{\infty} c_m c_{-m} T = T\sum_{m=-\infty}^{\infty} c_m c_m^* = T\sum_{m=-\infty}^{\infty} |c_m|^2 = T\left[c_0^2 + 2\sum_{m=-\infty}^{\infty} |c_m|^2\right],$$

or dividing through by T to get the *average power* of $x(t)$ over a period, we have (changing the dummy summation index back to k)

$$\frac{W}{T} = c_0^2 + 2\sum_{k=1}^{\infty} |c_k|^2,$$

a result called *Parseval's theorem.*[34]

Parseval's theorem has an elegant physical interpretation. As a preliminary comment to developing that interpretation, the reciprocal of the period, $\frac{1}{T}$, is called the *fundamental frequency* of $x(t)$. Combining the c_{-k} and c_k terms of the Fourier series of $x(t)$ using Euler's identity, we see that

$$\begin{aligned} c_{-k}e^{-jk\omega_0 t} + c_k e^{jk\omega_0 t} &= c_k^*(e^{jk\omega_0 t})^* + c_k e^{jk\omega_0 t} = (c_k e^{jk\omega_0 t})^* + c_k e^{jk\omega_0 t} \\ &= 2\ \mathrm{Re}\{c_k e^{jk\omega_0 t}\} = 2\left[a_k \cos(k\omega_0 t) - b_k \sin(k\omega_0 t)\right], \end{aligned}$$

where a_k and b_k are some real-valued constants such that $c_k = a_k + jb_k$. By elementary trigonometry, this represents a sinusoid with amplitude $\sqrt{a_k^2 + b_k^2}$ at frequency $k\omega_0$. This sinusoid is called the kth *harmonic* of $x(t)$. So, the average power of $x(t)$ is the sum of the dc power (c_0^2) and all the individual powers of the harmonics of $x(t)$; the average power of the kth *harmonic* is $2|c_k|^2$. This physical interpretation means that all our math is not simply abstract symbolism but instead has a strong connection to the real world of energetic radio signals.

So far, we've thought of $x(t)$as periodic. What if it isn't? We then can't, *it would seem*, write $x(t)$ as a Fourier series, right? Well, maybe we *can* if we extend what we mean by the "period" of a periodic signal. What if we say a nonperiodic $x(t)$ *is* periodic, with an *infinite*

34 After the French mathematician Marc Antoine Parseval des Chenes (1755–1836). This result (in remarkably different form from what we've developed here) dates from 1799, a *long* time before radio appeared.

period? We just happen to be observing $x(t)$ in its "current period"! This is, of course, nothing more or less than a sneaky trick, but it leads to something of enormous value. If we let $T \to \infty$ in the Fourier series math, it can be shown[35] that we get what is called the *Fourier transform pair*:

$$X(\omega) = F\{x(t)\} = \int_{-\infty}^{\infty} x(t)e^{-j\omega t}dt,$$

$$x(t) = F^{-1}\{X(\omega)\} = \frac{1}{2\pi}\int_{-\infty}^{\infty} X(\omega)e^{j\omega t}d\omega.$$

In these equations $X(\omega)$ is the Fourier transform of $x(t)$, and the second equation shows that we can recover $x(t)$ from $X(\omega)$, a claim written as $x(t) \leftrightarrow X(\omega)$. The second equation of the pair is called the inverse Fourier transform. As was the case for the Fourier series, these equations may at first look abstractly mysterious, but there is a real, physical significance to the Fourier transform.

With what we've already done as inspiration, let's define the energy of $x(t)$ as

$$W = \int_{-\infty}^{\infty} x^2(t)\, dt,$$

and so we have

$$\begin{aligned} W &= \int_{-\infty}^{\infty} x(t)x(t)\, dt = \int_{-\infty}^{\infty} x(t)\left[\frac{1}{2\pi}\int_{-\infty}^{\infty} X(\omega)e^{j\omega t}\, d\omega\right]dt \\ &= \int_{-\infty}^{\infty} \frac{1}{2\pi}X(\omega)\left[\int_{-\infty}^{\infty} x(t)e^{j\omega t}dt\right]d\omega \\ &= \int_{-\infty}^{\infty} \frac{1}{2\pi}X(\omega)X^*(\omega)\, d\omega = \int_{-\infty}^{\infty} \frac{1}{2\pi}|X(\omega)|^2\, d\omega. \end{aligned}$$

That is, we can calculate the energy of $x(t)$ either in the time domain or in the frequency domain (getting the same answer, of course!). This result,

$$W = \int_{-\infty}^{\infty} x^2(t)\, dt = \int_{-\infty}^{\infty} \frac{1}{2\pi}|X(\omega)|^2\, d\omega,$$

35 *The Science of Radio* (note 8), pp. 168–170.

is called the *Rayleigh/Plancherel energy theorem*;[36] it is the nonperiodic equivalent of Parseval's theorem for periodic signals. (The following box shows a purely mathematical use of this theorem as an example I am certain Hardy would have loved.)

The integral on the right in the Rayleigh/Plancherel energy theorem is interpreted as being a description of how the energy of $x(t)$ is distributed in frequency. The integrand is called the *energy spectral density* (ESD), with the idea that

$$\left.\begin{array}{c}\text{energy of } x(t) \text{ in the} \\ \text{frequency interval} \\ \omega_1 \le \omega \le \omega_2\end{array}\right] = \int_{\omega_1}^{\omega_2} \frac{1}{2\pi} |X(\omega)|^2 \, d\omega.$$

In radio, the signal $x(t)$ that we start with is the output of a microphone into which someone speaks or sings or plays a musical instrument. The energy of such a signal is typically distributed over the finite frequency interval $-\omega_m \le \omega \le \omega_m$, and that distribution (centered on $\omega = 0$) is called the *baseband spectrum*. The baseband spectrum $|X(\omega)|$ is said to be *bandlimited*, because the spectrum is of finite width along the ω-axis (in AM radio, ω_m is, at most, a few kilohertz). The spectrum $|X(\omega)|$ is depicted as shown in Figure 1.6.1; the actual distribution may, in fact, *not* be a triangle, which is simply a generic way to indicate there is energy in that frequency interval, whatever the actual distribution might be.

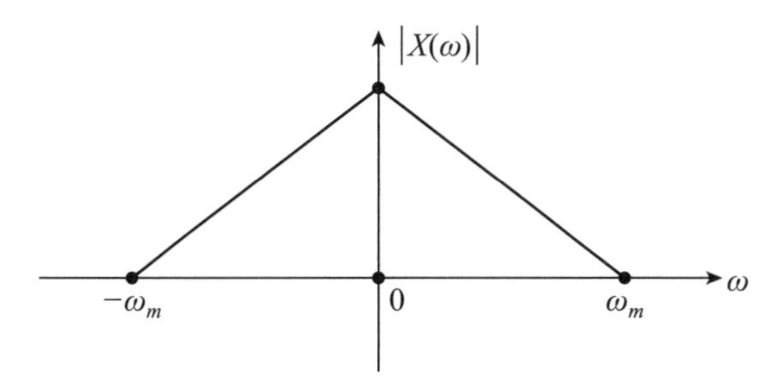

FIGURE 1.6.1. A bandlimited baseband spectrum.

36 After the winner of the 1904 Nobel Prize in Physics, Lord Rayleigh—the English mathematical physicist John William Strutt (1842–1919)—who stated the theorem in 1889, and the Swiss mathematician Michel Plancherel (1885–1967) who, in 1910, rigorously established the conditions under which the theorem is true.

Suppose $x(t)=\begin{cases} e^{-t}, & 0\leq t\leq 1 \\ 0, & \text{otherwise} \end{cases}$. The energy of $x(t)$ is $W=\int_0^1 e^{-2t}\,dt=\left(\frac{e^{-2t}}{-2}\right)\Big|_0^1=\frac{1-e^{-2}}{2}=\frac{e^2-1}{2e^2}$.

The Fourier transform of $x(t)$ is

$X(\omega)=\int_0^1 e^{-t}e^{-j\omega t}\,dt=\int_0^1 e^{-(1+j\omega)t}\,dt=\left\{-\frac{e^{-(1+j\omega)t}}{(1+j\omega)}\right\}\Big|_0^1$, which, with a little algebra, becomes $X(\omega)=\frac{1-\frac{1}{e}\cos(\omega)+j\frac{1}{e}\,sin(\omega)}{1+j\omega}$.

So, $|X(\omega)|^2=\frac{\left\{1-\frac{1}{e}\cos(\omega)\right\}^2+\left\{\frac{1}{e}sin(\omega)\right\}^2}{1+\omega^2}$, or, again with just a bit of algebra, $|X(\omega)|^2=\frac{1+\frac{1}{e^2}-\frac{2}{e}\,cos(\omega)}{1+\omega^2}$. Now, from the energy theorem we can therefore

write $\frac{1}{2\pi}\int_{-\infty}^{\infty}\frac{1+\frac{1}{e^2}-\frac{2}{e}\,cos(\omega)}{1+\omega^2}\,d\omega=\frac{e^2-1}{2e^2}$, or,

with just a bit more algebra, we arrive at

$\int_{-\infty}^{\infty}\frac{cos(\omega)}{1+\omega^2}\,d\omega=\frac{e^2+1}{2e}\int_{-\infty}^{\infty}\frac{d\omega}{1+\omega^2}-\frac{\pi(e^2-1)}{2e}$. Since

$\int_{-\infty}^{\infty}\frac{d\omega}{1+\omega^2}=tan^{-1}(\omega)\Big|_{-\infty}^{\infty}=\frac{\pi}{2}-\left(-\frac{\pi}{2}\right)=\pi$, then we have the beautiful conclusion (involving two of the most important constants in mathematics)

$$\int_{-\infty}^{\infty}\frac{\cos(\omega)}{1+\omega^2}\,d\omega=\frac{e^2+1}{2e}\pi-\frac{\pi(e^2-1)}{2e}=\frac{\pi}{e}.$$

This is an integral that often appears in textbooks on the theory of complex variables as an illustration of doing a contour integral in the complex plane. (See, for example, my *Inside Interesting Integrals* (2d ed.), Springer 2020, problem 8.3(b).) The Rayleigh/Plancherel energy theorem reduces that calculation to one of freshman calculus, and I think Hardy would have liked this little bit of "engineering math." You can find a generalization of this integral, again using the Rayleigh/Plancherel energy theorem, in my book *Dr. Euler's Fabulous Formula*, Princeton University Press, 2011, pp. 211–213, where it is shown that, for any real m,

$$\int_{-\infty}^{\infty} \frac{\cos(m\omega)}{a^2+\omega^2}\,d\omega = \frac{\pi}{a}e^{-|m|a}.$$

One property of the spectrum of any real-valued signal that the generic triangular shape in Figure 1.6.1 *does* possess, however, is that of evenness (symmetry around $\omega = 0$). This is easy to show and quite important to know in theoretical discussions of radio. The Fourier transform of $x(t)$ is the complex-valued

$$X(\omega) = \int_{-\infty}^{\infty} x(t)e^{-j\omega t}\,dt$$
$$= \int_{-\infty}^{\infty} x(t)\cos(\omega t)\,dt - j\int_{-\infty}^{\infty} x(t)\sin(\omega t)\,dt.$$

Writing

$$X(\omega) = \mathrm{Re}\{X(\omega)\} + j\,\mathrm{Im}\{X(\omega)\},$$

the real and imaginary parts of $X(\omega)$ are, since $x(t)$ is real,

$$\mathrm{Re}\{X(\omega)\} = \int_{-\infty}^{\infty} x(t)\cos(\omega t)\,dt, \quad \mathrm{Im}\{X(\omega)\} = -\int_{-\infty}^{\infty} x(t)\sin(\omega t)\,dt.$$

Since $\cos(\omega t)$ is even and $\sin(\omega t)$ is odd, we see that

$$\mathrm{Re}\{X(-\omega)\} = \int_{-\infty}^{\infty} x(t)\cos(-\omega t)\,dt = \int_{-\infty}^{\infty} x(t)\cos(\omega t)\,dt = \mathrm{Re}\{X(\omega)\}$$

and

$$\mathrm{Im}(X) = -\int_{-\infty}^{\infty} x(t)\sin(-\omega t)dt = \int_{-\infty}^{\infty} x(t)\sin(\omega t)dt = -\mathrm{Im}\{X(\omega)\}$$

That is, $\text{Re}(\omega)$ is even and $\text{Im}(\omega)$ is odd. Since

$$|X(j\omega)|^2 = \left[\text{Re}\{X(\omega)\}\right]^2 + \left[\text{Im}\{X(\omega)\}\right]^2,$$

and since $[\text{Re}\{X(\omega)\}]^2$ and $[\text{Im}\{X(\omega)\}]^2$ are each even, then $|X(j\omega)|^2$ will be even, and so, too, of course, the spectrum $|X(\omega)|$ will be even, since $|X(\omega)|$ is never negative.[37]

The significance of this result is this: if you need to generate a time signal that has a noneven spectrum, then you're in trouble, because you're trying to generate a non-real-time signal (that is, a complex-valued signal). What would such a thing look like? Nevertheless, we'll actually encounter such spectrums in chapter 4, in our discussion of single-sideband radio, and the genius of the early SSB radio engineers is illustrated by how they got around the "difficulty" of generating complex-valued time signals.

1.7 Impulses in Time and Frequency

A most useful mathematical concept in electrical physics in general, and certainly in radio theory, is that of the *impulse*. Impulses occur in analytical radio analyses whenever a physical quantity (voltage, current, charge, energy) is encountered in concentrated form. That is, an impulse describes the situation where "everything occurs at a single point" (a point in time or in space or in frequency). This is, in fact, the defining property of an impulse—it has vanishing *duration*. In radio theory, we'll mostly run into impulses in the form of energy concentrated at some specific frequency. The mathematical description of the impulse, written as $\delta(x)$, that we will use is the one popularized by the English mathematical physicist Paul Dirac (1902–1983) during his work in quantum mechanics:[38]

37 If, in addition to being real, $x(t)$ is even (odd), then $X(\omega)$ is real (imaginary). See if you can show this (it's not difficult).

38 Dirac was a "man for all seasons." He had an undergraduate degree in electrical engineering and a doctorate in mathematics, and he won a share of the 1933 Nobel Prize in Physics. The impulse had actually been around long before Dirac: he learned of it from his undergraduate reading of the mathematics of Oliver Heaviside (see the appendix), but it was Dirac who used impulses to resolve long-outstanding puzzles in quantum physics.

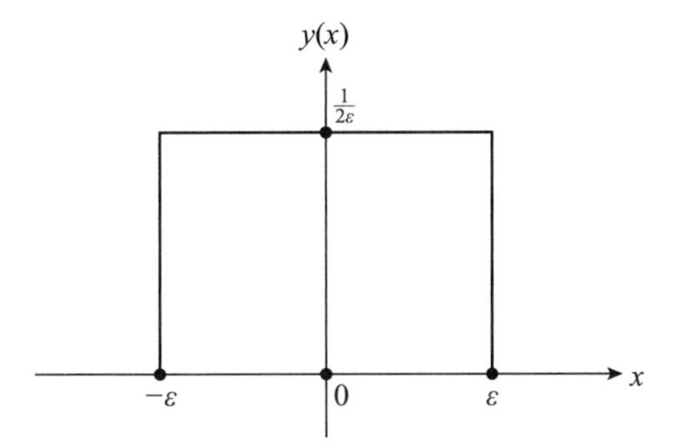

FIGURE 1.7.1. $\delta(x) = \lim_{\varepsilon \to 0} y(x)$.

$$\delta(x) = \begin{cases} \infty, & x = 0 \\ 0, & x \neq 0 \end{cases},$$

with the property $\int_{-\infty}^{\infty} \delta(x)dx = 1$.

In this case, we say the impulse has *unit strength* (we'll later encounter impulses of different strengths but *always* of vanishing duration).

A simple way to visualize the admittedly "unusual" $\delta(x)$ is as the limit of the simple pulse function $y(x)$ shown in Figure 1.7.1 as $\varepsilon \to 0$. The figure shows $y(x)$ is even for any $\varepsilon > 0$, and that property carries through as $\varepsilon \to 0$. Thus, the impulse is even; that is, $\delta(-x) = \delta(x)$.

We can use Figure 1.7.1 as support in deriving one of the most important properties of the impulse. Imagine we write $\varphi(x)$ as a *continuous* (this requirement is important!) but otherwise arbitrary function of x. It then follows that

$$\int_{-\infty}^{\infty} \delta(x)\varphi(x)\, dx = \varphi(0).$$

We can see this by writing the integral as

$$\int_{-\infty}^{\infty} \lim_{\varepsilon \to 0} y(x)\varphi(x)\, dx = \lim_{\varepsilon \to 0} \frac{1}{2\varepsilon} \int_{-\varepsilon}^{\varepsilon} \varphi(x)\, dx.$$

Now, as we let $\varepsilon \to 0$ we argue that over the ever-shrinking integration interval $\varphi(x)$ is little changed from its value at $x = 0$, that is, $\varphi(x) \approx \varphi(0)$

in the interval $-\varepsilon \leq x \leq \varepsilon$, and this is because $\varphi(x)$ is continuous. So, since $y(x) \to \delta(x)$ as $\varepsilon \to 0$, we have

$$\int_{-\infty}^{\infty} \delta(x)\varphi(x)\, dx = \frac{1}{2\varepsilon}\varphi(0)2\varepsilon = \varphi(0).$$

This is, of course, a bit of engineering derring-do that would have driven Hardy nuts![39]

When we pulled the $\lim_{\varepsilon \to 0} \frac{1}{2\varepsilon}$ outside the integral we were arguing that it's okay to reverse the order of two limiting operations (recall that the integral itself is defined as a limiting process, too), and that reversal needs to be justified in the minds of pure mathematicians. I won't do that here and will instead simply tell you that all I've done here has been justified.[40] As a last comment on this point, impulses need not occur only at $x=0$. Suppose, for example, that we have a unit-strength impulse at $x=x_0$, where x_0 is not necessarily zero. Then, our last result generalizes to (an impulse is located where its argument vanishes)

$$\int_{-\infty}^{\infty} \delta(x - x_0)\varphi(x)\, dx = \varphi(x_0),$$

a result called the *sampling property* of the impulse.

Two properties of the impulse are quite useful:

$$x\delta(x) = 0,$$

and

$$\delta(ax) = \frac{1}{|a|}\delta(x), \text{ for any constant } a \neq 0.$$

39 You'll recall, from the preface, those mathematicians that Hardy named as "real mathematicians." Dirac was not among the honored, perhaps because he was always quite open about his appreciation of the engineering mathematics he learned as an undergraduate electrical engineering student. Hardy almost certainly would not have been amused by Dirac's praise of "engineering math."

40 In the early 1950s, by the French mathematician Laurent Schwartz (1915–2002). The mathematicians are not being pedantic by asking for formal proofs. An example of a physical problem where the reversal of two limiting processes is *not* valid can be found in my *Mrs. Perkins's Electric Quilt*, Princeton University Press, 2009, pp. 24–36.

These statements may look a bit odd, but they are quite easy to establish once we keep in mind a point that Dirac particularly emphasized concerning expressions involving impulses: *they have mathematical meaning only when inside an integral.* So, to say $x\delta(x)=0$ simply means that if placed inside an integral, $x\delta(x)$ and 0 produce the same result. To see this, notice that for any finite but otherwise arbitrary function $\varphi(x)$ we have

$$\int_{-\infty}^{\infty} 0\varphi(x)\, dx = 0,$$

and

$$\int_{-\infty}^{\infty} x\delta(x)\varphi(x)\, dx = \int_{-\infty}^{\infty} \delta(x)\{x\varphi(x)\}\, dx = x\varphi(x)\,|_{x=0} = 0\varphi(0) = 0.$$

In the same way, if you write

$$\int_{-\infty}^{\infty} \delta(ax)\varphi(x)\, dx,$$

change the variable to $u=ax$, and treat the two cases of $a>0$ and $a<0$ separately, you'll see you get precisely the result that $\frac{1}{|a|}\delta(x)$ would produce.[41]

You may be thinking at this point that all this looks pretty far removed from radio, so let me next show you the intimate connection between radio and impulses. We start by writing the impulse in time at $t=t_0$ as $v(t)=\delta(t-t_0)$ and then ask, What is its Fourier transform? To answer that, we just write

$$V(\omega) = \int_{-\infty}^{\infty} v(t)e^{-j\omega t} dt = \int_{-\infty}^{\infty} \delta(t-t_0)e^{-j\omega t} dt,$$

or associating $\varphi(t)$ with $e^{-j\omega t}$, we have

$$V(\omega) = e^{-j\omega t_0}.$$

41 Dirac's papers are full of similar impulse identities, and here's another: $t\delta'(t) = -\delta(t)$, where $\delta'(t)$ is the time derivative of $\delta(t)$. (If you think $\delta(t)$ is strange, what do you think of its derivative?) For fun, see if you can establish this identity. *Hint*: For $\varphi(t)$ an arbitrary *differentiable* function, integrate $\int_{-\infty}^{\infty} t\delta'(t)\varphi(t)\, dt$ by parts, and compare the result with the value of $\int_{-\infty}^{\infty} -\delta(t)\varphi(t)\, dt$.

In particular, the unit-strength impulse located at $t_0=0$ has the very simple Fourier transform of 1. Notice, too, that since $\left|e^{-j\omega t_0}\right|=1$ for any real t_0, then the energy spectral density of the impulse is a *constant* over *all* frequencies ω, from minus infinity to plus infinity. For that reason the impulse is said to have a *flat* spectrum. A flat spectrum is also called a *white* spectrum, in analogy with white light (which is a uniform mix of all visible frequencies).[42] From the Rayleigh/Plancherel energy theorem we immediately see that the impulse has infinite energy, yet another indication (as if we actually needed one) that the impulse is, indeed, a rather "interesting" function.

With the Fourier transform pair that we've just derived ($\delta(t) \leftrightarrow 1$) we can now use the second half of the transform pair equations (the inverse transform) to write

$$\delta(t)=\frac{1}{2\pi}\int_{-\infty}^{\infty} e^{j\omega t}\, d\omega.$$

This is an astonishing statement, because the integral simply doesn't make any sense if we attempt to actually evaluate it (because $e^{j\omega t}$ doesn't approach a limit, as $\omega \to \pm\infty$). The only way we can make any sense of the statement (at the level of this book) is that the right-hand side denotes the same concept (an impulse) that the ink squiggles on the left do. That is, any time we encounter the right-hand squiggles (the integral) we'll simply replace them with the squiggles $\delta(t)$. Notice, too, that if we interchange the variables ω and t on both sides of the statement in all their occurrences (thus retaining the truth of the equality[43]) we arrive at

$$\delta(\omega)=\frac{1}{2\pi}\int_{-\infty}^{\infty} e^{j\omega t}\, dt,$$

an impulse in the *frequency* domain.

42 Extending the analogy with light even further, time signals that have nonflat (nonwhite) spectrums are said to have *colored* (or *pink*) spectrums. Who says radio engineers aren't romantic souls?

43 This trick is based on the observation that the particular ink squiggles we use in our equations are all historical accidents. The *only* constraint we have to obey is to be consistent. I'll show you this trick again in the final example of this section.

To find the time function that pairs with $\delta(\omega)$ we simply put $\delta(\omega)$ into the inverse transform equation and use the sampling property of the impulse to get

$$\frac{1}{2\pi}\int_{-\infty}^{\infty}\delta(\omega)e^{j\omega t}\,d\omega=\frac{1}{2\pi}(e^{j\omega t})\big|_{\omega=0}=\frac{1}{2\pi},$$

a constant. That makes physical sense, too, because a constant has all its energy (obviously infinite) at dc, that is, at $\omega=0$, which is just where $\delta(\omega)$ is located. This gives us the transform pair $1\leftrightarrow 2\pi\delta(\omega)$. These two Fourier transform pairs,

$$\delta(t)\leftrightarrow 1$$
$$1\leftrightarrow 2\pi\delta(\omega)$$

occur repeatedly in radio theory.

Here's an example of that, building on my comments in the paragraph that opens this section. When a radio station goes "on the air" transmitting just its *carrier signal*[44] at frequency ω_c, it is broadcasting a sinusoidal signal we can model as $v(t)=A\sin(\omega_c t)$, where A is the carrier amplitude (since A is simply a scale factor, we lose no generality by taking $A=1$). A natural question to ask is, Where is the energy of the carrier? This is not a mystery of the ages, of course—the energy is at the carrier frequency. The mathematics of this claim is not quite so obvious, as you'll soon see, but Fourier theory will show us the way. From Euler's identity we have

$$v(t)=\sin(\omega_c t)=\frac{e^{j\omega_c t}-e^{-j\omega_c t}}{j2},$$

and so the Fourier transform of $v(t)$ is

$$V(\omega)=\int_{-\infty}^{\infty}\frac{e^{j\omega_c t}-e^{-j\omega_c t}}{j2}e^{-j\omega t}dt$$
$$=\frac{1}{j2}\left[\int_{-\infty}^{\infty}e^{j(\omega_c-\omega)t}dt-\int_{-\infty}^{\infty}e^{j[-(\omega_c+\omega)]t}dt\right].$$

44 The carrier is so named because the voice/music information broadcast by a radio station rides "piggyback" on the carrier via a process called *modulation*. The why and how of modulation (which is just a fancy name for what amounts to multiplying) is the topic of the next chapter.

Since we've previously established that

$$\delta(\omega)=\frac{1}{2\pi}\int_{-\infty}^{\infty} e^{j\omega t}dt,$$

and since the impulse is even, we have

$$\int_{-\infty}^{\infty} e^{j(\omega_c-\omega)t}dt=2\pi\delta(\omega_c-\omega)=2\pi\delta(\omega-\omega_c),$$

and

$$\int_{-\infty}^{\infty} e^{j[-(\omega_c+\omega)]t}dt=2\pi\delta[-(\omega_c+\omega)]=2\pi\delta(\omega+\omega_c).$$

Thus,

$$\begin{aligned}V(\omega)&=\frac{1}{j2}[2\pi\delta(\omega-\omega_c)+2\pi\delta(\omega+\omega_c)]\\&=-\pi j[\delta(\omega-\omega_c)+\delta(\omega+\omega_c)].\end{aligned}$$

Notice that $V(\omega)$ is purely imaginary, as it should be, since $v(t)$ is odd (see note 37 again).

The formal answer to the question, Where is the energy of the carrier? is given by calculating the energy spectral density, which involves $|V(\omega)|^2$. But how do we square impulses? Since the impulses in $V(\omega)$ are located at $\pm\omega_c$, it seems plausible, whatever "squared impulses" might mean, that the carrier energy is also located at $\pm\omega_c$. But what *is* the energy spectral density? If it has physical meaning, we should be able to plot it, right? Well, we *can* if we're just a bit sneaky about things.

The impulses in the carrier transform occur because our model for the carrier signal extends in time from minus infinity to plus infinity. To "sneak up" on those impulses, to get something we can square, let's model the carrier as a sinusoid that exists for just N complete cycles, find its transform (which will be nonimpulsive), and for that transform calculate the energy spectral density. Such a signal is called a *sinusoidal burst*, which we'll denote by $v_b(t)$. Then, we'll let $N\rightarrow\infty$ and see what happens to the energy spectral density as the lengthening burst comes ever-closer to approximating the carrier, using the idea that $\lim_{N\rightarrow\infty} v_b(t)=v(t)$.

We'll start our analysis by observing that the duration of each complete cycle of the burst is $\frac{2\pi}{\omega_c}$ and so, centering the burst on $t=0$, we have

$$v_b(t)=\begin{cases} sin(\omega_c t), & -\frac{2\pi}{\omega_c}\bullet\frac{N}{2}\leq t\leq\frac{2\pi}{\omega_c}\bullet\frac{N}{2} \\ 0, & \text{otherwise} \end{cases},$$

or

$$v_b(t)=\begin{cases} sin(\omega_c t), & -\frac{N\pi}{\omega_c}\leq t\leq\frac{N\pi}{\omega_c} \\ 0, & \text{otherwise} \end{cases}.$$

If we now calculate $V(\omega)$, it is a routine (if moderately messy) exercise to show that the energy spectral density of $v_b(t)$ is

$$\frac{1}{2\pi}|V_b(\omega)|^2=\frac{2}{\omega_c^2\pi}\bullet\frac{\sin^2\left(N\pi\frac{\omega}{\omega_c}\right)}{\left[\left(\frac{\omega}{\omega_c}\right)^2-1\right]^2},$$

and this result is independent of whether the integer N is even or odd.[45] Figure 1.7.2 shows four plots of this result (without the constant factor of $\frac{2}{\omega_c^2\pi}$) for several values of N, over the normalized frequency interval $-2\leq\frac{\omega}{\omega_c}\leq 2$. It is clear from those plots that as N increases, the energy in $v_b(t)$ is concentrated in an ever-narrowing interval of frequencies centered on the carrier and that this occurs very quickly with increasing N.[46]

45 See, for details, my *Dr. Euler's Fabulous Formula*, Princeton University Press, 2011, pp. 242–243.

46 As a little calculation for you to do, see if you can show that the peak value of the plots, at the carrier frequency, is $N^2\frac{\pi^2}{4}$, and confirm that the plots are consistent with that value. *Hint*: You may find using L'Hôpital's rule, *twice*, to be helpful.

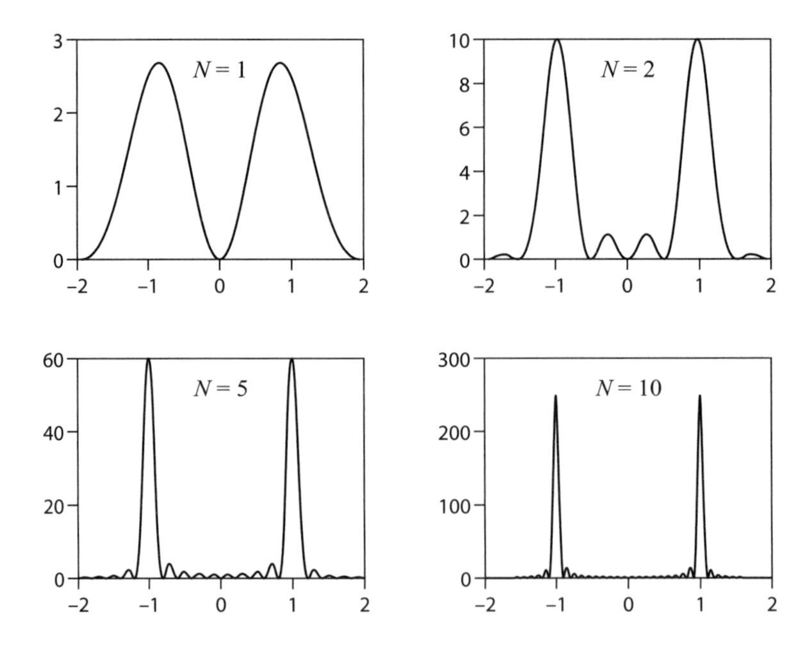

FIGURE 1.7.2. Relative energy spectral density of N-cycle sinusoidal bursts.

Now, to finish this section, let me show you a few more results involving impulses that we'll find invaluable in understanding radio. To start, what's the integral of the impulse? That is, if we write

$$u(t) = \int_{-\infty}^{t} \delta(x)\, dx,$$

what is $u(t)$? The impulse is located at $x = 0$, and so if $t < 0$, the impulse is not inside the interval of integration, which says $u(t) = 0$. If $t > 0$, however, then the impulse *is* inside the interval of integration, and so the integral is the area bounded by the unit-strength impulse. That is, $u(t) = 1$ if $t > 0$. Therefore, the answer to our question is

$$\int_{-\infty}^{t} \delta(x)\, dx = u(t) = \begin{cases} 0, & t < 0 \\ 1, & t > 0 \end{cases},$$

which is called the *unit step function*. Notice, carefully, that $u(0)$ is undefined. More generally, for t_0 any constant we have

$$\int_{-\infty}^{t} \delta(x-t_0)\,dx = u(t-t_0) = \begin{cases} 0, & t<t_0 \\ 1, & t>t_0 \end{cases}.$$

Since the step is the integral of the impulse, then the impulse is the derivative of the step:

$$\delta(t-t_0) = \frac{d}{dt}\{u(t-t_0)\}.$$

The step function in time is obviously important in radio electronics, as it allows us to mathematically model signals that turn on and off. For example,

$$s(t) = f(t)u(t-t_0) = \begin{cases} 0, & t<t_0 \\ f(t), & t>t_0 \end{cases}$$

is the description of a signal $s(t)$ that turns on at time $t=t_0$ to be $f(t)$. So, since $u(t)$ is important, let's calculate its Fourier transform, $U(j\omega)$. This is a particularly interesting calculation, because it will play an important role in our discussion, later in the book, on the theory of SSB (single-sideband) radio. Alas, to take the brute-force approach of just inserting $u(t)$ into the Fourier transform integral doesn't work, as we get

$$U(\omega) = \int_{-\infty}^{\infty} u(t)e^{-j\omega t}dt = \int_{0}^{\infty} e^{-j\omega t}dt = \left(\frac{e^{-j\omega t}}{-j\omega}\right)\Big|_0^\infty = ?$$

which has no meaning at the upper limit of infinity. That is, we can't assign a value to $e^{-j\infty}$, since both the real and imaginary parts of $e^{-j\omega t}$ oscillate forever between ±1 and never approach any particular limiting values as $t\to\infty$.

Fortunately, there is a clever, simple trick that gets around this difficulty, starting with the functions shown in Figure 1.7.3. There we see that $u(t)$ is the sum of $s_1(t)=\frac{1}{2}$ and $s_2(t)=\begin{cases} \frac{1}{2}, & t>0 \\ -\frac{1}{2}, & t<0 \end{cases}$. Since

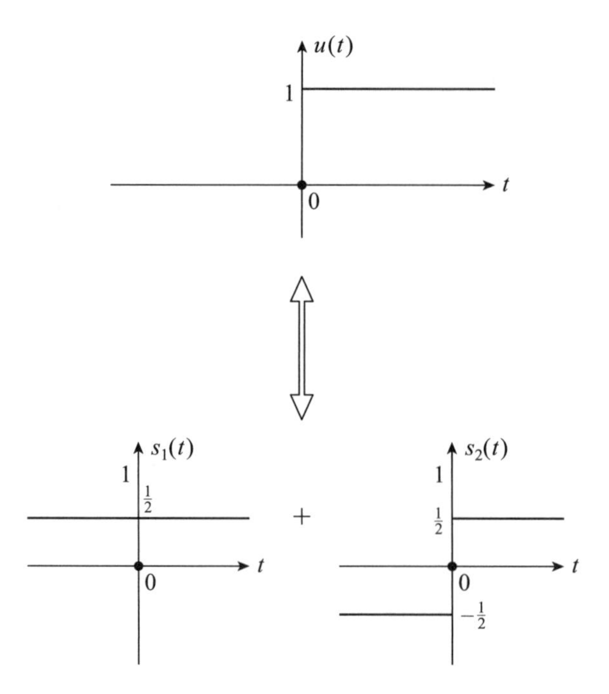

FIGURE 1.7.3. Constructing the unit step.

$u(t)=s_1(t)+s_2(t)$, then the transform of $u(t)$ is equal to the sum of the transforms of $s_1(t)$ and $s_2(t)$. As we've already established the pair $1 \leftrightarrow 2\pi\delta(\omega)$, we then have $\frac{1}{2} \leftrightarrow \pi\delta(\omega)$, and so $U(\omega)=\pi\delta(\omega)+F\{s_2(t)\}$.

To calculate $F\{s_2(t)\}$ we first write, for $\sigma>0$,

$$s_2(t,\sigma)=\begin{cases} \frac{1}{2}e^{-\sigma t}, & t>0 \\ -\frac{1}{2}e^{\sigma t}, & t<0 \end{cases}$$

and then argue that since $s_2(t)=\lim_{\sigma \to 0} s_2(t,\sigma)$, it seems plausible that

$$F\{s_2(t)\}=\lim_{\sigma \to 0} F\{s_2(t,\sigma)\}.$$

So,

$$F\{s_2(t,\sigma)\}=\int_{-\infty}^{\infty} s_2(t,\sigma)e^{-j\omega t}dt=\int_{-\infty}^{0} -\frac{1}{2}e^{\sigma t}e^{-j\omega t}dt+\int_{0}^{\infty}\frac{1}{2}e^{-\sigma t}e^{-j\omega t}dt$$
$$=-\frac{1}{2}\int_{-\infty}^{0} e^{(\sigma-j\omega)t}dt+\frac{1}{2}\int_{0}^{\infty}e^{-(\sigma+j\omega)t}dt,$$

which, with a little bit of algebra (for you to confirm), reduces to[47]

$$F\{s_2(t,\sigma)\}=-\frac{j\omega}{\sigma^2+\omega^2}.$$

Thus,

$$F\{s_2(t)\}=\lim_{\sigma\to 0}\left\{-\frac{j\omega}{\sigma^2+\omega^2}\right\}=-j\frac{1}{\omega},$$

and we arrive at the quite interesting Fourier transform pair

$$u(t)\leftrightarrow U(\omega)=\pi\delta(\omega)-j\frac{1}{\omega}.$$

I'll end this section (and the chapter) with one more enormously useful result from Fourier theory, a result that we'll use when we get to SSB radio. Suppose we have the transform pair $g(t)\leftrightarrow G(\omega)$. From the inverse transform we then have

$$g(t)=\frac{1}{2\pi}\int_{-\infty}^{\infty} G(\omega)e^{j\omega t}d\omega,$$

or if we replace t with $-t$ on both sides (which leaves the equality intact),

$$g(-t)=\frac{1}{2\pi}\int_{-\infty}^{\infty} G(\omega)e^{-j\omega t}d\omega.$$

47 Notice that $F\{s_2(t,\sigma)\}$ is purely imaginary, which is consistent with the statement in note 37 that purely imaginary transforms are associated with odd time functions (as is $s_2(t,\sigma)$).

Next, using the symbol interchange trick on ω and t that we used earlier (see note 43), we have

$$2\pi g(-\omega)=\int_{-\infty}^{\infty} G(t)e^{-j\omega t}dt.$$

The integral is the Fourier transform of the time function $G(t)$, and so we have the new, general Fourier transform pair $G(t)\leftrightarrow 2\pi g(-\omega)$, a result called the *duality theorem.*

As an example of the use of the duality theorem, since we showed earlier that

$$u(t)\leftrightarrow \pi\delta(\omega)-j\frac{1}{\omega},$$

then we have

$$\pi\delta(t)-j\frac{1}{t}\leftrightarrow 2\pi u(-\omega),$$

or

$$\frac{1}{2}\delta(t)-j\frac{1}{2\pi t}\leftrightarrow u(-\omega).$$

Notice, *carefully*, that $u(-\omega)$ is a *reversed* step in the frequency domain. That is,

$$u(-\omega)=\begin{cases} 1, & \omega<0 \\ 0, & \omega>0 \end{cases}.$$

So, the curious time function $\frac{1}{2}\delta(t)-j\frac{1}{2\pi t}$ has zero energy at positive frequencies. This equally curious spectrum goes with a complex time function, so you might wonder if such a strange thing actually occurs in "real life." The answer is *yes*, and I'll remind you of this transform pair when we get to SSB radio in chapter 4. (At this point, see Challenge Problem 1.8.)

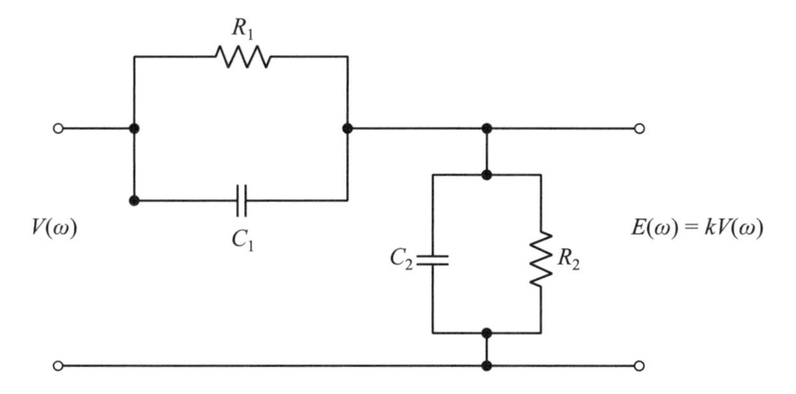

FIGURE CP1.1. A frequency-independent attenuator.

To end this first chapter on radio mathematics, here are some calculations for you to play with (solutions, partial answers, and/or more hints are at the end of the book).

Challenge Problem 1.1: If you haven't already solved the problem presented in the text for Figure 1.5.3, do it now. As another example of a "remarkable" circuit, consider Figure CP1.1. Show that if $R_1C_1 = R_2C_2$, then, *independent of frequency*, $E(\omega) = kV(\omega)$, where k is a real, positive constant less than 1 (specifically, what *is* k?).

Challenge Problem 1.2: In an 1874 study of mechanical vibrations, Lord Rayleigh (see note 36) anticipated (by *decades*) Van der Pol's equation for negative resistance oscillators when he encountered the differential equation

$$\frac{d^2y}{dt^2} - \varepsilon\left[\frac{dy}{dt} - \frac{1}{3}\left(\frac{dy}{dt}\right)^3\right] + y(t) = 0.$$

Show that this is equivalent to Van der Pol's equation by (1) making the change of variable $u = \dfrac{dy}{dt}$ in Van der Pol's equation and then (2) integrating indefinitely, term by term, the result from (1). What can you say about the constant of integration that appears in (2)?

Challenge Problem 1.3: An important concept in radio engineering is that of the *rms* value of a periodic (*not* necessarily sinusoidal) signal. Suppose $v(t)$ is the voltage drop across a resistor R. The instantaneous power is

$$p(t)=\frac{v^2(t)}{R}.$$

The total energy, W, dissipated (as heat) by the resistor, over one complete period of $v(t)$, is given by

$$W=\int_0^T p(t)\,dt=\frac{1}{R}\int_0^T v^2(t)\,dt.$$

Now, define V_{rms} as the constant (dc) voltage that would dissipate the same energy in the same time interval. Thus (notice that the value of R disappears)

$$\frac{1}{R}\int_0^T v^2(t)dt=\frac{V_{\text{rms}}^2}{R}T,$$

or

$$V_{\text{rms}}=\sqrt{\frac{1}{T}\int_0^T v^2(t)\,dt}.$$

You can now see where the subscript rms comes from: V_{rms} is the square ***r***oot of the ***m***ean of the ***s***quare of $v(t)$. Although this derivation of V_{rms} was done with the assumption that $v(t)$ is, physically, a voltage signal across a resistor (a very special situation), mathematicians simply extend this result and make it the definition of the rms value of *any* periodic $f(t)$:

$$F_{\text{rms}}=\sqrt{\frac{1}{T}\int_0^T f^2(t)\,dt}$$

whatever may be the physical nature of $f(t)$. If we know the details of $f(t)$, we can specifically calculate F_{rms}. For example, suppose $f(t)=F_M\sin(\omega_0 t+\varphi)$. Then, as $T=2\pi/\omega_0$, we have (after a bit of algebra)

$$F_{\text{rms}} = \sqrt{\frac{\omega_0}{2\pi}\int_0^{2\pi/\omega_0} F_M^2 \sin^2(\omega_0 t+\varphi)dt} = \frac{F_M}{\sqrt{2}}.$$

That is, the rms value of *any* sinusoidally time-varying signal $f(t)$ is simply the *peak value*[48] of $f(t)$ divided by $\sqrt{2}$ (that is, multiplied by 0.707), independent of frequency and phase. (I'll remind you of this result when we get to the discussion of FM radio in chapter 5.) Now, suppose that $f(t)$ is the *sawtooth waveform* shown in Figure CP1.3.

If this $f(t)$ is applied to a meter calibrated to display the correct rms value for an applied sinusoidal wave (see note 48), what is the error in the meter's displayed value?

Challenge Problem 1.4: Consider the circuit in Figure CP1.4, where $Z(\omega)$ denotes the frequency-dependent impedance of an arbitrary connection of an arbitrary number of resistors, inductors, and capacitors, all of arbitrary values. Calculate the impedance $Z_i(\omega)$ at the frequency $\omega = 1/\sqrt{LC}$. (Make sure your expression for $Z_i(\omega)$ is dimensionally correct, that is, has the units of impedance). Does your result explain why this circuit is called an *impedance inverter*? How does your result change if you replace the C with an L, and the two inductors with two capacitors of value C?

Challenge Problem 1.5: The circuit in Figure CP1.5 is called (for the obvious reason) a *lattice filter*. Define its transfer function as $H(\omega) = \frac{V_2}{V_1}$. Can you immediately see, *by inspection*, that $\lim_{\omega\to 0} H(\omega) = +1$, and $\lim_{\omega\to\infty} H(\omega) = -1$? Explain your reasoning.

Challenge Problem 1.6: Many different types of electric wave filters were developed over the first half of the twentieth century, with one of the most popular being the *Butterworth filter* (after its inventor, the

48 When a typical American home is said to be wired at 120 volts ac, that's an rms value. The peak value of the voltage of a wall outlet is actually about 170 volts. In the next chapter you'll see a simple circuit that measures the peak value of its input (a circuit that can be used to demodulate both AM and FM radio signals, in addition to making rms ac voltage meters).

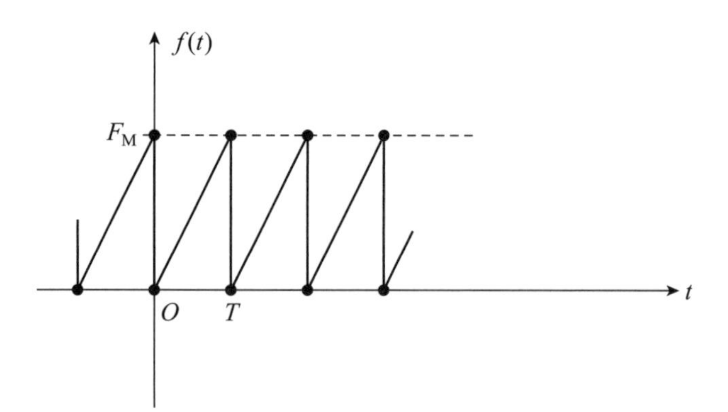

FIGURE CP1.3. A sawtooth signal.

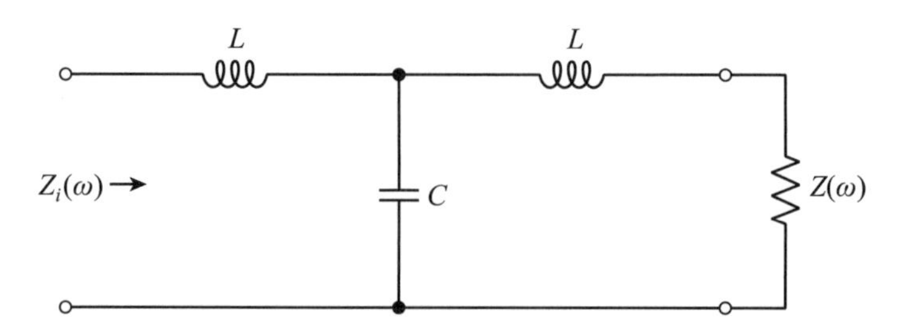

FIGURE CP1.4. An impedance inverter.

British mathematical engineer Stephen Butterworth (1885–1958)). As he wrote in a classic 1930 paper:[49] "[I]t is shown how to obtain the best results from a two-element filter and then how to combine any number of elementary pairs, separated from each other by valves [*valve* was the British name for *vacuum tube*; tubes were used to keep the operation of the individual stages independent], *so as to approach closer and closer to the ideal filter as the number of stages are increased* [my emphasis added]." An n-stage Butterworth low-pass filter has a transfer function of the form $|H(\omega)|^2 = \dfrac{1}{1+\left(\dfrac{\omega}{\omega_0}\right)^{2n}}$, where ω_0 is the frequency at which $|H(\omega)|^2$ is half its maximum value. This filter is said to be *maximally flat*, because the first $2n-1$ derivatives of $|H(\omega)|^2$

49 S. Butterworth, "On the Theory of Filter Amplifiers," *Experimental Wireless & The Wireless Engineer*, October 1930, pp. 536–541.

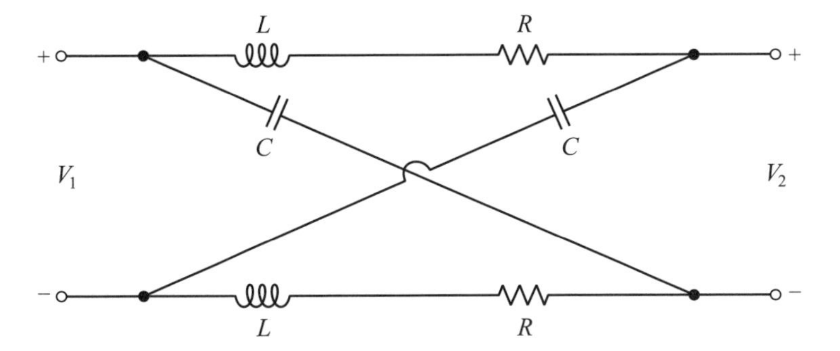

FIGURE CP1.5. A lattice filter.

all vanish at $\omega = 0$, and this results in a nearly constant (that is, *flat*) transfer function in the filter's passband (see Figure CP1.6). Show that this is so (this is a pretty straightforward calculation to do, but it is important to understand).

Challenge Problem 1.7: The circuit of Figure 1.5.3, redrawn as Figure CP1.7, has another remarkable property. When we get to SSB radio you'll see that a circuit of great interest is one that, over a wide interval of frequencies, generates from a single input an output of two equal constant-amplitude signals that *independent of frequency* maintain a constant phase difference of 90°. (Radio engineers call such a circuit a *phase splitter*.) Show that the circuit of Figure CP1.7 satisfies the phase-difference requirement for *all* frequencies from dc to infinity but fails the equal constant-amplitude output requirement. In the figure, the sinusoidal voltage source V (a complex value, in general, with a fixed amplitude and angle in the complex plane), operates at an arbitrary frequency ω, and the two output signals are A and B. Assume $R = \sqrt{\frac{L}{C}}$. *Hint*: Calculate $\frac{A}{B}$ (because the angle of a quotient of two complex quantities is the angle of the numerator minus the angle of the denominator) and show the result is, independent of frequency, pure imaginary (and so has angle 90° in the complex plane) for all ω. This calculation also shows that the outputs have amplitudes that depend on ω, but you should be able to see that by inspection, by simply recalling the impedance behavior, as a function of frequency, of inductors and capacitors.

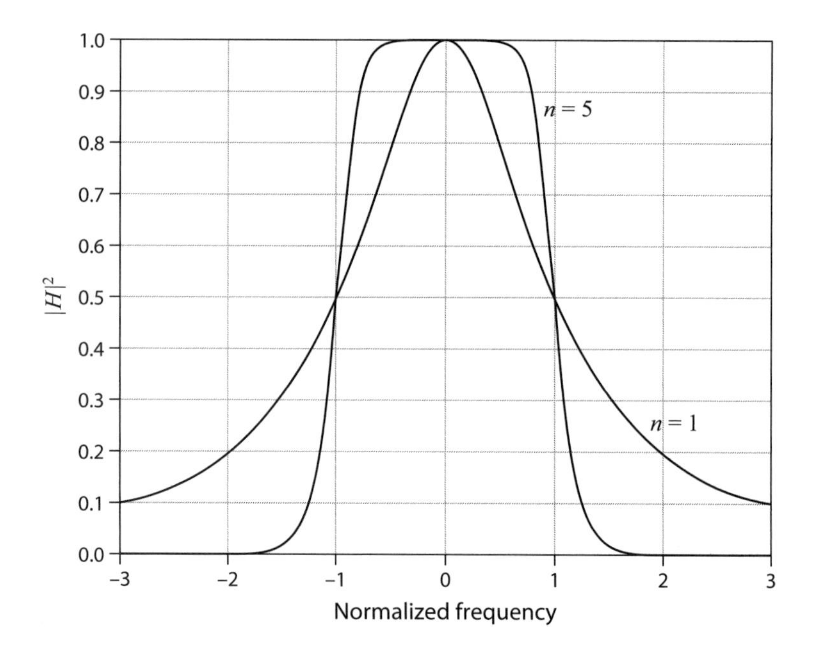

FIGURE CP1.6. The Butterworth LPF transfer function for two values of n.

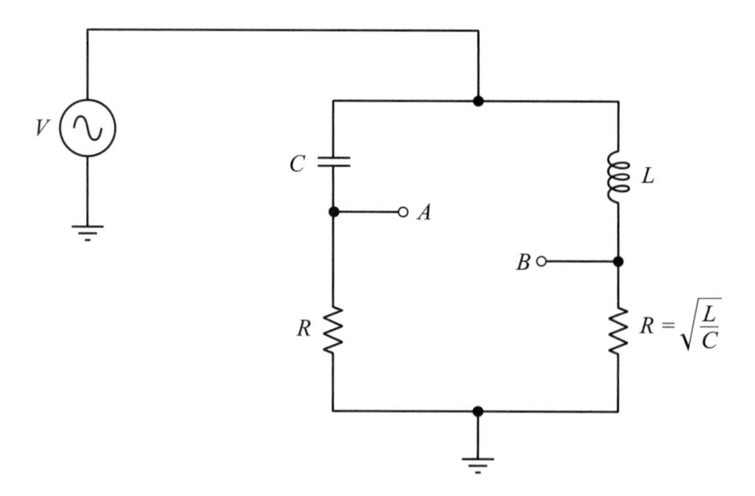

FIGURE CP1.7. A constant 90° phase-difference circuit.

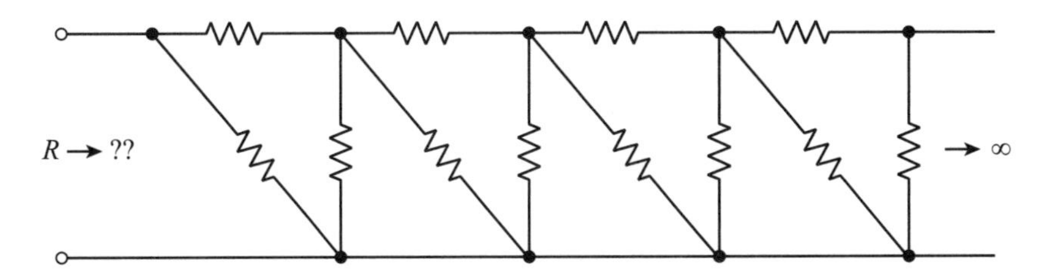

FIGURE CP1.10. Use symmetry to calculate R (each of the infinity of resistors is 1 ohm).

Challenge Problem 1.8: In the text we derived the pair $\frac{1}{2}\delta(t) - j\frac{1}{2\pi t} \leftrightarrow u(-\omega)$. What time function pairs with the transform $u(\omega)$? We'll return to this question in chapter 4. *Hint*: Start with $u(-\omega) + u(\omega) = 1$, and recall the pair $\delta(t) \leftrightarrow 1$.

Challenge Problem 1.9: Give an argument for why $\int_{-\infty}^{\infty} \frac{1}{t}\delta(t)\, dt = 0$, even though a naïve use of the sampling theorem, with $\varphi(t) = \frac{1}{t}$, says the answer is infinity.

Challenge Problem 1.10: Physicist Richard Feynman (1918–1988) (see the opening quote in the appendix) once explained how he chose physics for his life's work this way: Mathematics was too technical and electrical engineering was too practical. Physics was his Goldilocks compromise. There is, in fact, a large overlap of the three fields, and here's a problem that illustrates that overlap. It's a problem that, at first, looks pretty hard, but if you take advantage of the obvious symmetry (a favorite trick for mathematicians) of the infinite circuit of 1 ohm resistors in Figure CP1.10, it is actually pretty easy. What is the value of the resistance R? It is obviously less than 1, but, surprisingly (perhaps), the value is irrational. *Warning*: Writing Kirchhoff's equations is definitely *not* the approach to use!

Challenge Problem 1.11: Derive the trigonometric identity $A\cos(\alpha t) + B\sin(\alpha t) = \sqrt{A^2 + B^2}\cos\left\{\alpha t - tan^{-1}\left(\frac{B}{A}\right)\right\}$, where A, B, and $\propto$ are constants, which we will use in chapter 4 when we get to SSB radio. *Hint*: Use Euler's identity to write $A\cos(\alpha t) + B\sin(\alpha t)$ in complex exponential form.

Challenge Problem 1.12: Calculate the Fourier transform of the time function $x(t)=\frac{1}{1+t^2}$. *Hint*: You might find the integral at the end of the box in section 1.6 to be helpful.

Challenge Problem 1.13: Calculate the Fourier transforms of the time functions $e^{-t}u(t)$ and $e^{t}u(-t)$, and then use the duality theorem to derive the pairs $\frac{1}{1+jt} \leftrightarrow 2\pi e^{\omega}u(-\omega)$ and $\frac{1}{1-jt} \leftrightarrow 2\pi e^{-\omega}u(\omega)$. I'll remind you of these two pairs when we discuss the concept of *instantaneous frequency* when we get to FM radio in chapter 5.

Chapter 2

More Radio Mathematics
Circuits That Multiply

Multiplying two radio frequency time signals together isn't so hard. The only trick to it is to do it *without* multiplying.

—faded, barely legible pencil scrawl found on the inside cover of a library copy of an old radio engineering text[1]

2.1 Spectrum Shifting at the Transmitter

As everyone who owns a radio knows, a radio receiver can "tune in" many radio stations. Why don't all those multiple stations interfere with each other and reduce themselves to a massive babbling of incoherence? Well, of course, the answer lies in the words *tune in*—words that are synonymous with *spectrum shifting*.

To be specific, let's suppose Ann is speaking into her microphone at Station A and that Bill is speaking into his microphone at Station B, which is in the same town as Station A (perhaps both stations are even in the same building!). Since the voices of Ann and Bill are produced by the same physical mechanism (vibrating human vocal cords), the spectrums of their microphone signals will be located at generally the same frequencies (typically, a few tens of hertz up to a few thousand

1 As a simple example of what that long-ago anonymous writer perhaps meant, if you can add, subtract, and *square*, you can multiply. To see this, simply consider the identity

$$\frac{1}{4}\left[\{a(t)+b(t)\}^2-\{a(t)-b(t)\}^2\right]=a(t)b(t).$$

It is easy to build electrical circuits that can add and subtract (see *The Science of Radio*, note 8 in chapter 1), but how do you *square*? I'll return, *briefly*, to that point later, at the start of section 2.4. To divide a signal by 4 is easy: just apply the signal to a resistor of $3R$ that is in series with a resistor R (that connects, in turn, to ground). Then, Kirchhoff tells us the voltage across R to ground is one-fourth of the applied signal.

hertz). That is, the signal generated by Ann's microphone will occupy the same interval of frequencies that Bill's microphone signal does. You will recall from section 1.6 that this common interval of *audio frequencies* is called the *baseband spectrum.*

To apply the baseband signal of a microphone output directly to the transmitter antenna won't work, because, as discussed in the appendix, a quarter-wavelength antenna at audio frequencies is physically enormous (refer to note 2 in chapter 1). To have a reasonably sized antenna requires a transmitter signal at frequencies considerably higher than those of the baseband spectrum; that is, the baseband spectrum must be upshifted to the *radio frequencies*. After transmission of the upshifted baseband spectrum, and its reception by an antenna at a remote receiver, that receiver then downshifts the received spectrum back to the baseband frequencies we can hear. Each of these frequency shifts is performed by a multiplication process, as will be shown soon.

The frequency upshift at the transmitter, besides achieving transmission with a reasonably sized antenna, serves one other vital purpose. It allows Ann and Bill to avoid interfering with each other. That is, suppose at Station A Ann's baseband signal is upshifted by 640 kHz, while at Station B Bill's baseband signal is upshifted by 1,270 kHz. A radio receiver could then select which station to listen to by passing the antenna signal to a tunable bandpass filter, that is, to an adjustable—via a knob on the front panel of the radio—bandpass filter whose center frequency can be set to either 640 kHz or to 1,270 kHz.[2]

To accomplish the up and the down frequency shifts is as "easy" as doing a multiplication. But though "easy" in principle, doing brute-force multiplication by electronic circuitry at radio frequencies is *not* so easy. We will have to be a bit clever at doing such a multiplication, and Fourier theory will show us the way.

To see that implementing multiplication is our goal, let's suppose we *do* have a multiplier circuit, as shown in Figure 2.1.1. The two signals wc arc going to multiply are $m(t)$, which denotes the baseband signal out of a microphone, and $\cos(\omega_c t)$, where ω_c denotes the upshift

2 This explanation is conceptually okay, but building a tunable bandpass filter is not a trivial task. In chapter 3 I'll show you one way to do it (Challenge Problem 3.2).

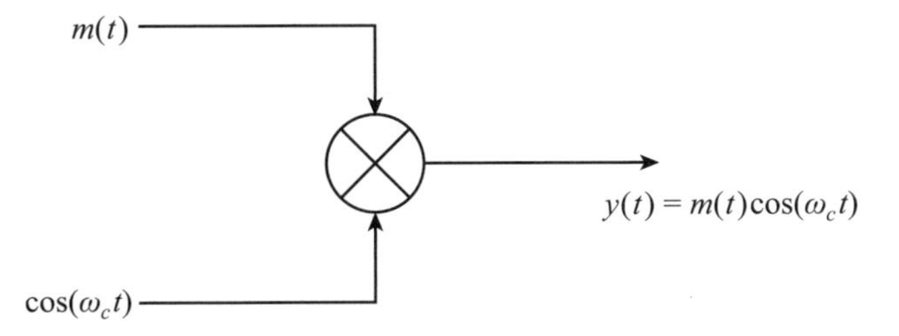

FIGURE 2.1.1. Block diagram of an imagined multiplier.

frequency (ω_c is the frequency of the radio station's carrier—refer to note 44 in chapter 1).

If we denote the baseband spectrum of $m(t)$ by $|M(\omega)|$, then the spectrum of $y(t)$ is $|Y(\omega)|$, where

$$\begin{aligned} Y(\omega) &= \int_{-\infty}^{\infty} y(t)e^{-j\omega t}dt = \int_{-\infty}^{\infty} m(t)\cos(\omega_c t)e^{-j\omega t}dt \\ &= \int_{-\infty}^{\infty} m(t)\frac{e^{j\omega_c t} + e^{-j\omega_c t}}{2}e^{-j\omega t}dt \\ &= \frac{1}{2}\left[\int_{-\infty}^{\infty} m(t)e^{-j(\omega-\omega_c)t}dt + \int_{-\infty}^{\infty} m(t)e^{-j(\omega+\omega_c)t}dt\right] \\ &= \frac{1}{2}\left[M(\omega-\omega_c) + M(\omega+\omega_c)\right]. \end{aligned}$$

The baseband spectrum of $m(t)$ has been shifted, *in both directions*, by ω_c along the frequency axis, as shown in Figure 2.1.2. This result is called the *amplitude modulation* (AM) or *heterodyne*[3] *theorem*. As shown in the figure, the spectrum of $m(t)$ is bandlimited, with all the energy of $m(t)$ confined to the frequency interval $-\omega_m < \omega < \omega_m$ (see Figure 1.6.1 again).

Once the station to be listened to has been selected, the frequency downshift at the receiver, back to baseband, can be accomplished with a second multiplication by $\cos(\omega_c t)$, which shifts the spectrum $|Y(\omega)|$ again, in *both* directions, by ω_c. That is, this shifted spectrum is that of a signal with its energy centered on $\pm 2\omega_c$ *and* on $\omega = 0$ (which

3 This interesting word, due to the Canadian-born American radio engineer Reginald Fessenden (1866–1932), is from the Greek *heteros* ("different"). Fessenden accomplished the first radio transmission of the human voice on Christmas Eve 1906 (but only by pushing spark-gap technology—technology eventually abandoned—far beyond its natural limits).

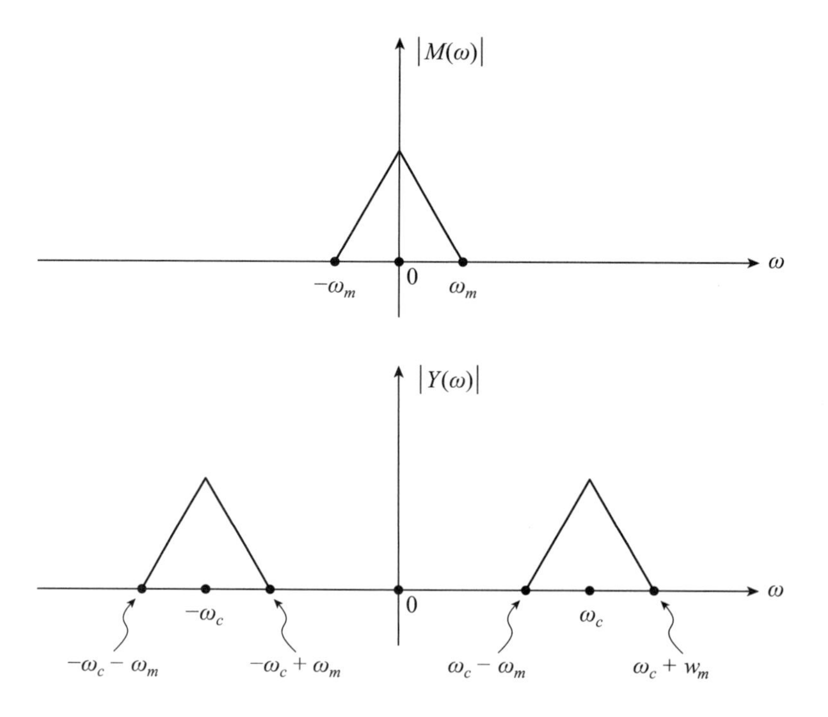

FIGURE 2.1.2. Spectrum of a bandlimited baseband signal (upper plot) and its heterodyned version (lower plot).

is, of course, the desired baseband signal). The baseband energy can then be selected by a low-pass filter that rejects the energy centered on $\pm 2\omega_c$. This filtering operation could be accomplished *mechanically* by the radio's loudspeaker, as its inherent inertia (the loudspeaker is made from a relatively massive magnet and vibrating cone) prevents it from responding to a frequency that, in commercial AM radio, is at least as high as 1 MHz.

You'll notice that I said all this *could* be done, in theory, but it is *not* how real AM radio receivers actually work. That's because while there is nothing wrong with the math, there is a subtle engineering stumbling block which becomes *glaringly obvious* once it's noticed. *The two multiplications using* $\cos(\omega_c t)$ *occur at different physical locations*. The first is at the transmitter (using what is called a *modulator*, which will be discussed later, in section 2.5), and the second is at the receiver (which may be dozens, hundreds,

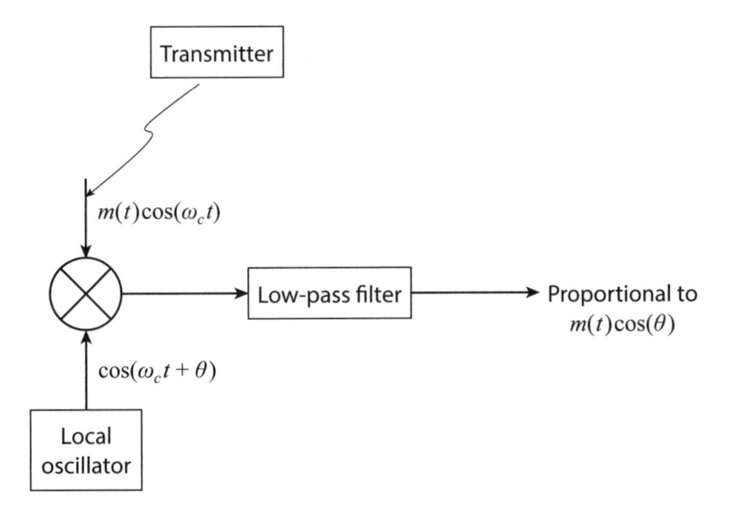

FIGURE 2.1.3. One way to demodulate (if $\theta = 0$, this is a *synchronous demodulation receiver*).

even thousands of miles distant). The receiver multiplication is accomplished by using a *local oscillator* to achieve what is called *demodulation*. When the station selection dial on such a radio receiver is turned, it changes the frequency of the local oscillator. For now, let's simply accept that such a local oscillator can be precisely tuned to ω_c, and after we've developed just a bit more theory, I'll show you (in section 3.3) one way to *automatically* determine ω_c *directly* from the received signal.

The math of all this assumes the receiver's $\cos(\omega_c t)$ is perfectly synchronized with the transmitter's $\cos(\omega_c t)$, when in fact a much more realistic view would be that the receiver's multiplication involves $\cos(\omega_c t + \theta)$, where θ is not necessarily zero. This is not to say a receiver couldn't achieve $\theta = 0$, just that it's not easy to do (that means it costs money). Receivers that (somehow) are able to ensure $\theta = 0$ are called *synchronous demodulation receivers*. The block diagram of such a receiver (with an arbitrary θ) is shown in Figure 2.1.3, and the multiplication output signal at the receiver is

$$r(t) = \left[m(t)\cos(\omega_c t)\right]\left[\cos(\omega_c t + \theta)\right] = \frac{1}{2}m(t)\left[\cos(\theta) + \cos(2\omega_c t + \theta)\right].$$

Again, the high-frequency term of $\cos(2\omega_c t+\theta)$ is rejected by low-pass filtering $r(t)$, but now the filter's output is not proportional to just $m(t)$ but rather to $m(t)\cos(\theta)$.

That is, the phase error θ appears as an amplitude attenuation factor, which might not be considered a fatal problem (unless $\theta=90°$, of course, because then there would be *zero* filter output!). For $\theta\neq 90°$, you might argue one could counter the attenuation by simply turning up the volume knob. The problem with that is θ is almost certainly not a constant but, rather, $\theta=\theta(t)$.[4] You could listen to such a receiver only by constantly adjusting the volume knob as the low-pass filter output faded in and out as $\theta(t)$ varied. Nobody would buy such a receiver! By now you've also almost certainly noticed that this more realistic analysis is itself not quite completely realistic. That's because we have continued to assume that the local oscillator is perfectly matched to the transmitter frequency. If there is the slightest mismatch in frequency, however, then we'll again get a time-varying amplitude (I'll let you work through the details). So, synchronous demodulation is beautiful mathematically but greatly flawed from the perspective of a radio engineer who wants to build (and sell) an inexpensive, easy-to-use receiver.

We end this section with two elementary examples of frequency shifting. Consider first the block diagram shown in Figure 2.1.4. This is an electronic speech scrambler, a personal (portable) gadget that clamps onto the mouthpiece *and* onto the earpiece of an ordinary telephone. That is, you carry two copies of this gadget in your purse or briefcase. (You'll see why *two* copies, soon.) This scrambler circuit is quite old, dating back to just after the First World War, and was first used commercially on the radio/telephone link connecting Los Angeles and the offshore casino and resort hotel on Santa Catalina Island, southwest from Los Angeles. (Numerous advances in radio technology made their first appearances in telephone circuitry.)

4 Think of a car radio, which can be constantly changing its distance from the transmitter. Even a stationary radio on the kitchen table will be constantly changing its "electrical distance" from the transmitter as the received signal reflects off the bottom of a fluctuating (in height) ionosphere (the famous *Heaviside layer*) as it rises and falls with temperature changes throughout the day and night.

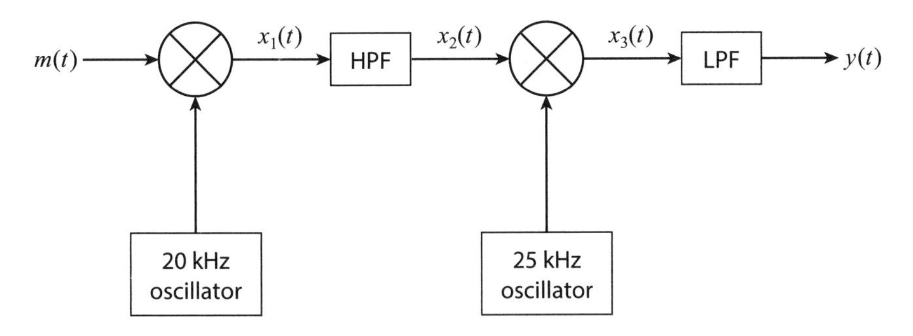

FIGURE 2.1.4. Electronic speech scrambler.

Both filters in Figure 2.1.4 are imagined to be ideal. That is, both have vertical skirts at 20 kHz, with the high-pass filter (HPF) passing all energy at frequencies above 20 kHz while rejecting all energy below 20 kHz, and the low-pass filter (LPF) passing all energy at frequencies below 20 kHz while rejecting all energy above 20 kHz. The input to the scrambler, $m(t)$, is the baseband signal generated by a telephone mouthpiece into which the user speaks (we'll assume the band limiting to be at 5 kHz, an assumption that can be assured by low-pass filtering the mouthpiece output signal). The signals that exist in the scrambler as we move from left to right have the spectrums shown in Figure 2.1.5. These spectrums are easily understood by recalling the heterodyne theorem: As the spectrum of the scrambler output, $y(t)$, shows, the scrambler has *inverted* the spectrum of $m(t)$. That is, energy at frequency f kHz at the input appears at frequency $(5-f)$ kHz in the output. This is sufficient garbling to render the speaker's voice unintelligible to a "casual" eavesdropper.[5]

We do, of course, now have an obvious question to answer because we don't want *everybody* to be baffled: that is, how does the person at the other end of the conversation understand what is being sent? *That* person clearly needs a *descrambler*. The fact that the scrambler works by inverting its input spectrum strongly hints at the amusing idea that the scrambler is its own descrambler (if you invert an inversion, you get back what you started with). Figure 2.1.6 shows how that works,

5 This scrambler would *not* defeat a serious hacker (a big-city police department, the FBI, and certainly not any of the world's numerous sophisticated spy agencies).

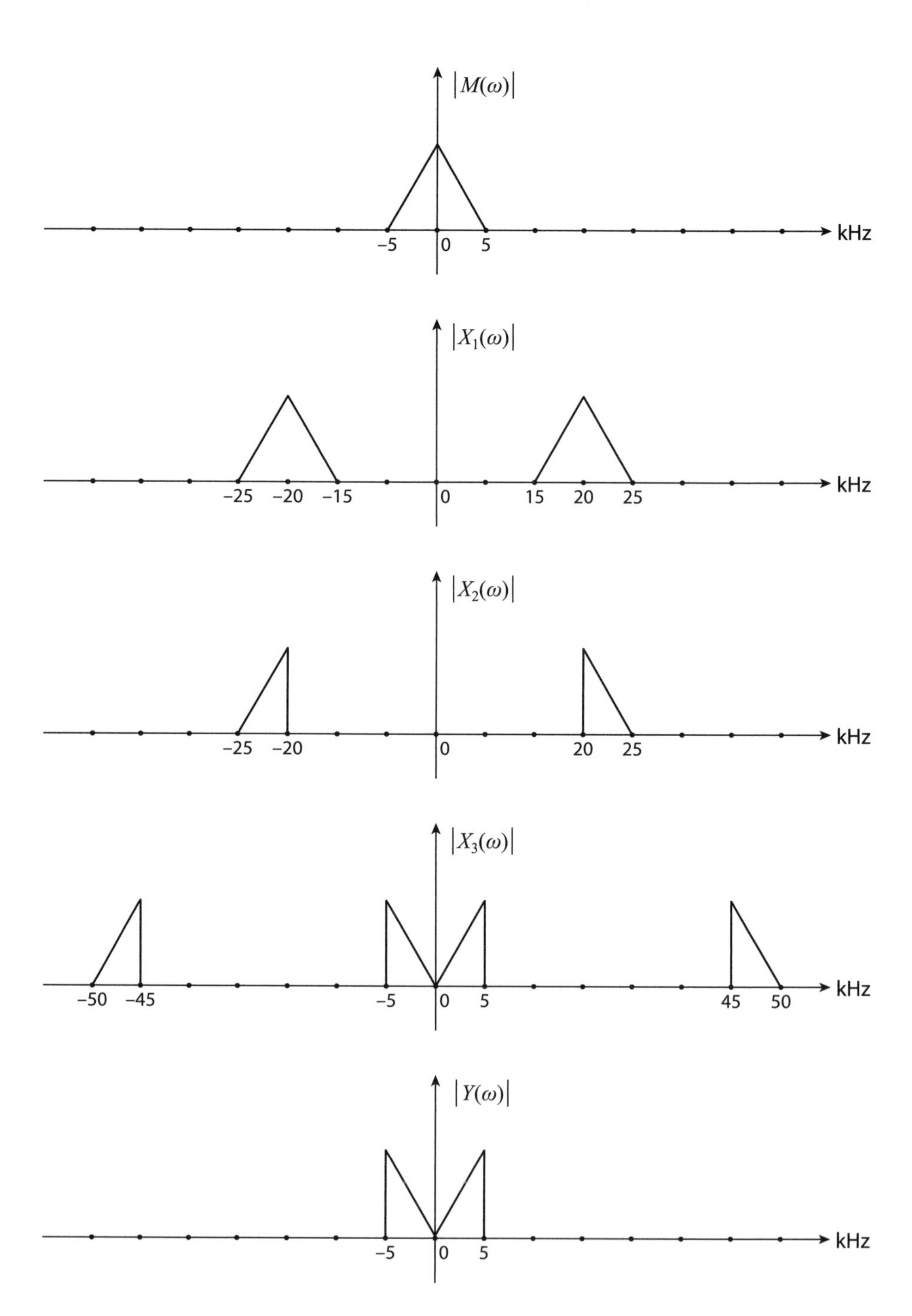

FIGURE 2.1.5. Signal spectrums in the speech scrambler.

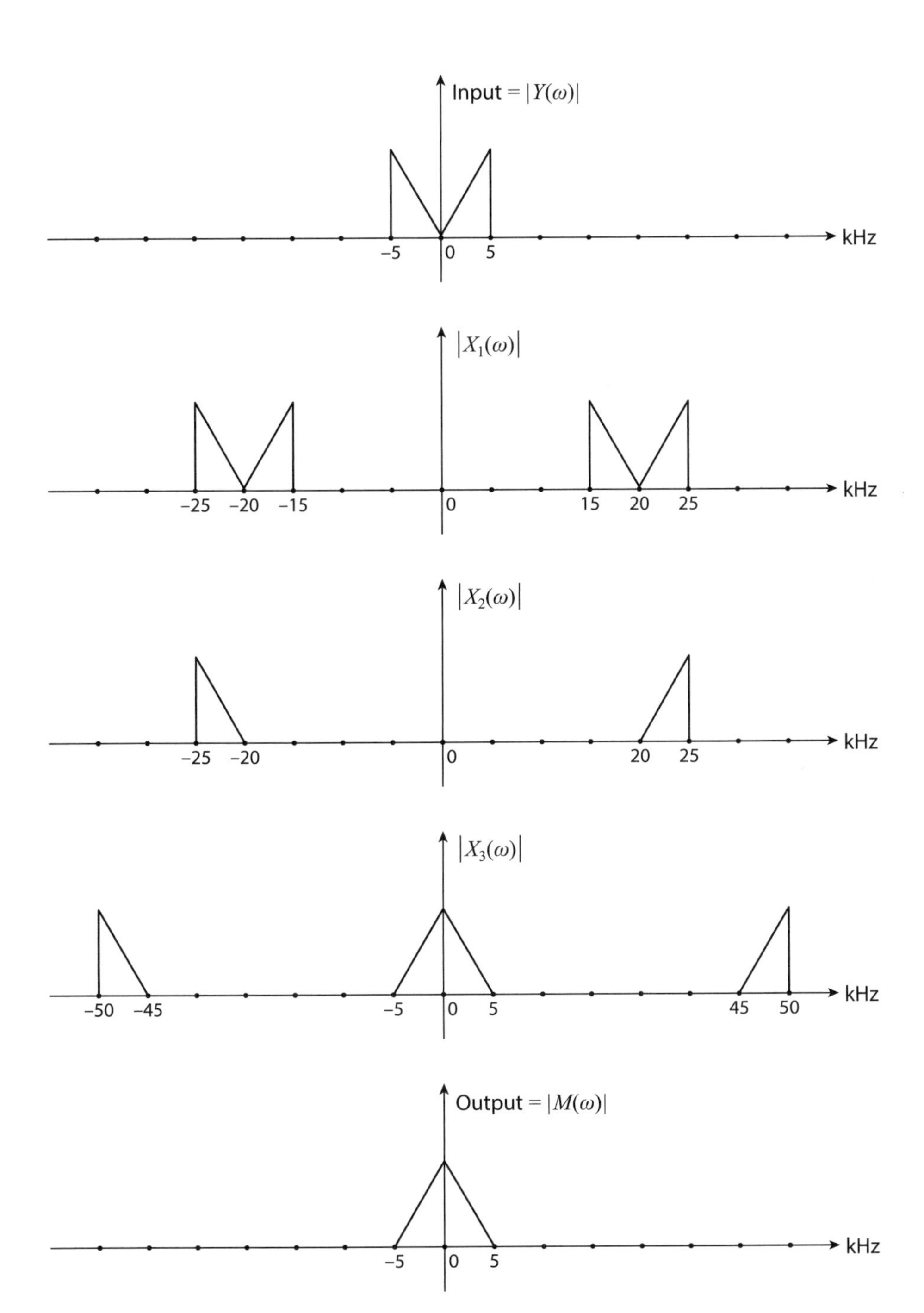

FIGURE 2.1.6. Signal spectrums in the speech descrambler.

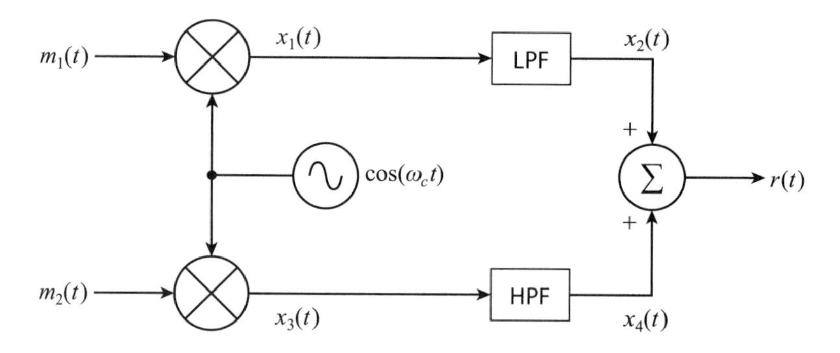

FIGURE 2.1.7. A dual-channel AM transmitter.

starting with the inverted output spectrum of Figure 2.1.5. So, each person simply attaches a second copy of the same scrambler circuit on their earpiece.

For my second example of the heterodyne theorem consider Figure 2.1.7, which, while as easily understandable as is the scrambler circuit, will at the end still seem (even to Hardy, I think) to be borderline miraculous. To see why I say this, recall the opening paragraphs of this chapter, where I said one of the reasons for using a different carrier frequency for each radio station is to avoid interference. That's still true but not because we otherwise would violate some fundamental law of physics. Rather, that's the way it's done in commercial radio because then it's easy to build *inexpensive* receivers. If you are willing to pay for a more expensive receiver, however, then it *is* possible to send more than one baseband signal on a single carrier frequency.

Figure 2.1.7 shows a *dual-channel* AM transmitter that does this, while Figure 2.1.8 shows the signal spectrums in the transmitter as we move from left to right. The two filters in the transmitter are both taken to be ideal, with each having a vertical skirt at $\omega = \omega_c$. Figure 2.1.9 shows a synchronous AM receiver circuit for dual-channel reception (again, all four filters of the receiver are ideal, with vertical skirts at $\omega = \omega_c$), while Figure 2.1.10 shows the signal spectrums as we move through the receiver. (Though not shown in Figure 2.1.10, it should be clear that low-pass filtering $x_2(t)$ recovers $m_2(t)$, while low-pass filtering $x_4(t)$ recovers $m_1(t)$.)

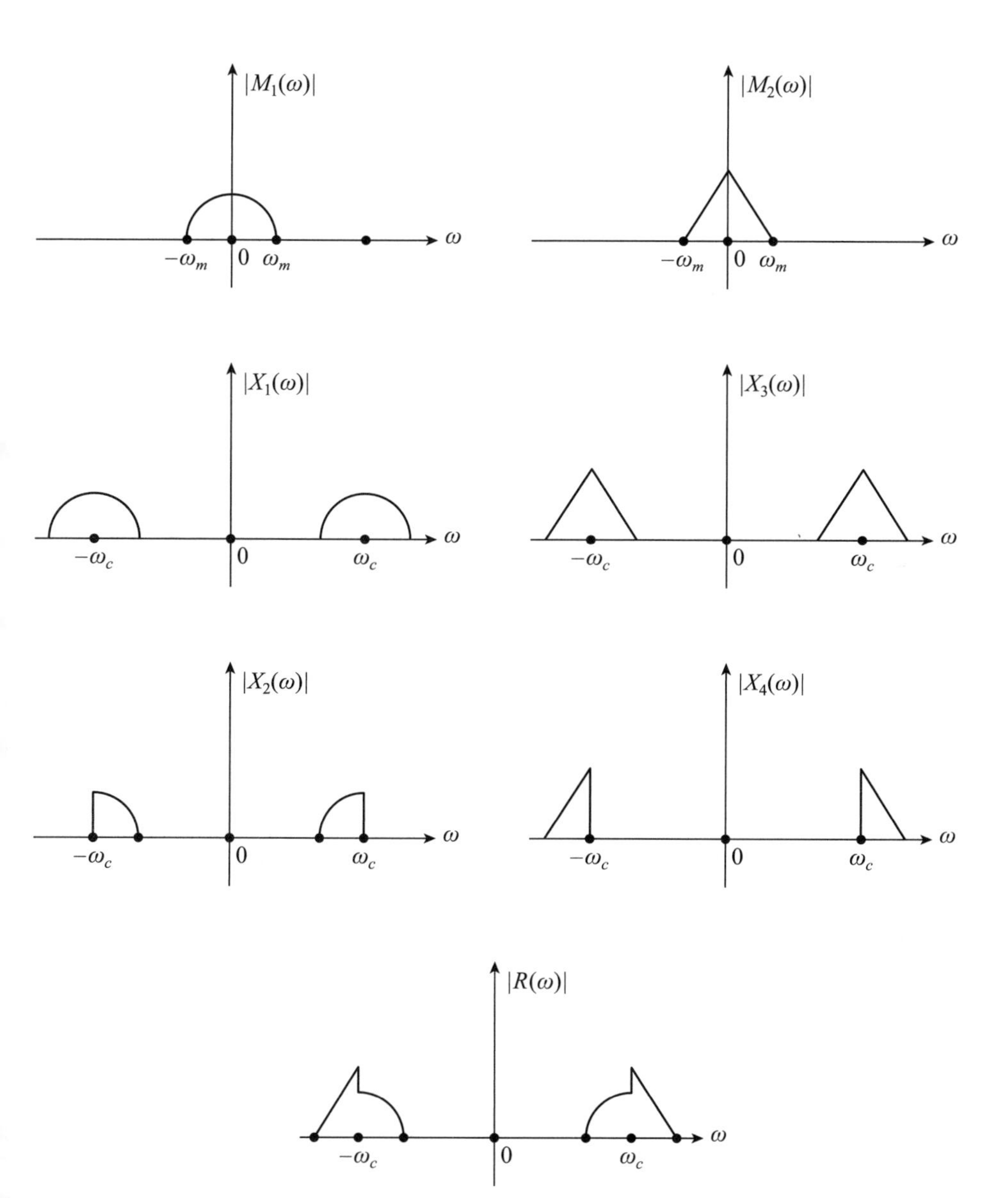

FIGURE 2.1.8. The signal spectrums in the dual-channel AM transmitter.

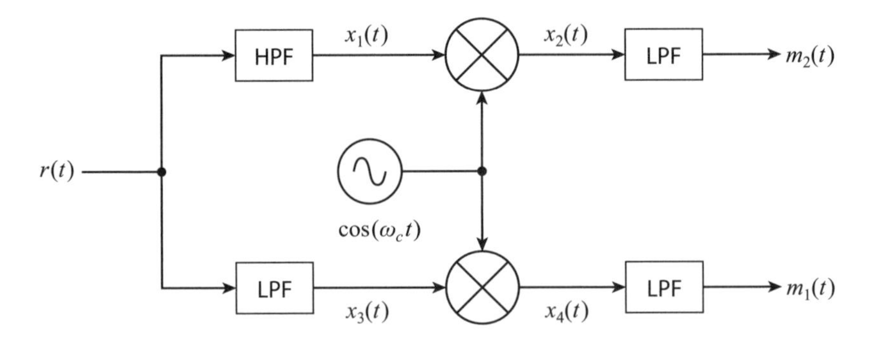

FIGURE 2.1.9. Dual-channel synchronous AM demodulation receiver.

A Puzzling Historical Mistake

As the spectrum $|R(\omega)|$ of the dual-channel AM transmitter in Figure 2.1.8 shows, to transmit baseband signals that have energy in the interval $-\omega_m < \omega < \omega_m$ on a carrier at frequency ω_c requires the use of frequencies in the intervals $-\omega_c - \omega_m < \omega < -\omega_c + \omega_m$ and $\omega_c - \omega_m < \omega < \omega_c + \omega_m$, that is, frequencies both smaller and larger in magnitude than the carrier frequency. The frequencies smaller (in magnitude) than $-\omega_c$ and $+\omega_c$ represent what is called the *lower sideband*, while the frequencies larger (in magnitude) than $-\omega_c$ and $+\omega_c$ represent what is called the *upper sideband*. As astonishing as it seems to the modern mind, some early radio pioneers denied that there was any physical reality to the sidebands. Rather, while admitting that the baseband signal uses the band of frequencies $-\omega_m < \omega < \omega_m$, they nevertheless asserted that after heterodyning, the *only* frequencies present in the transmitted signal are $\pm\omega_c$. The appearance of the sidebands in the heterodyne theorem was called a "mere mathematical artifice" and a "fiction." One such sideband denier was a former student of Maxwell's, Sir John Ambrose Fleming (1849–1945), a Fellow of the Royal Society and inventor of the diode vacuum tube, the device that started the electronic age (I'll tell you more about Fleming's diode later in this chapter). Clearly, Fleming was no fool, and yet he claimed (see his "The Wave-Band Theory of Wireless Transmission," *Nature*, January 19, 1930, pp. 92–93)

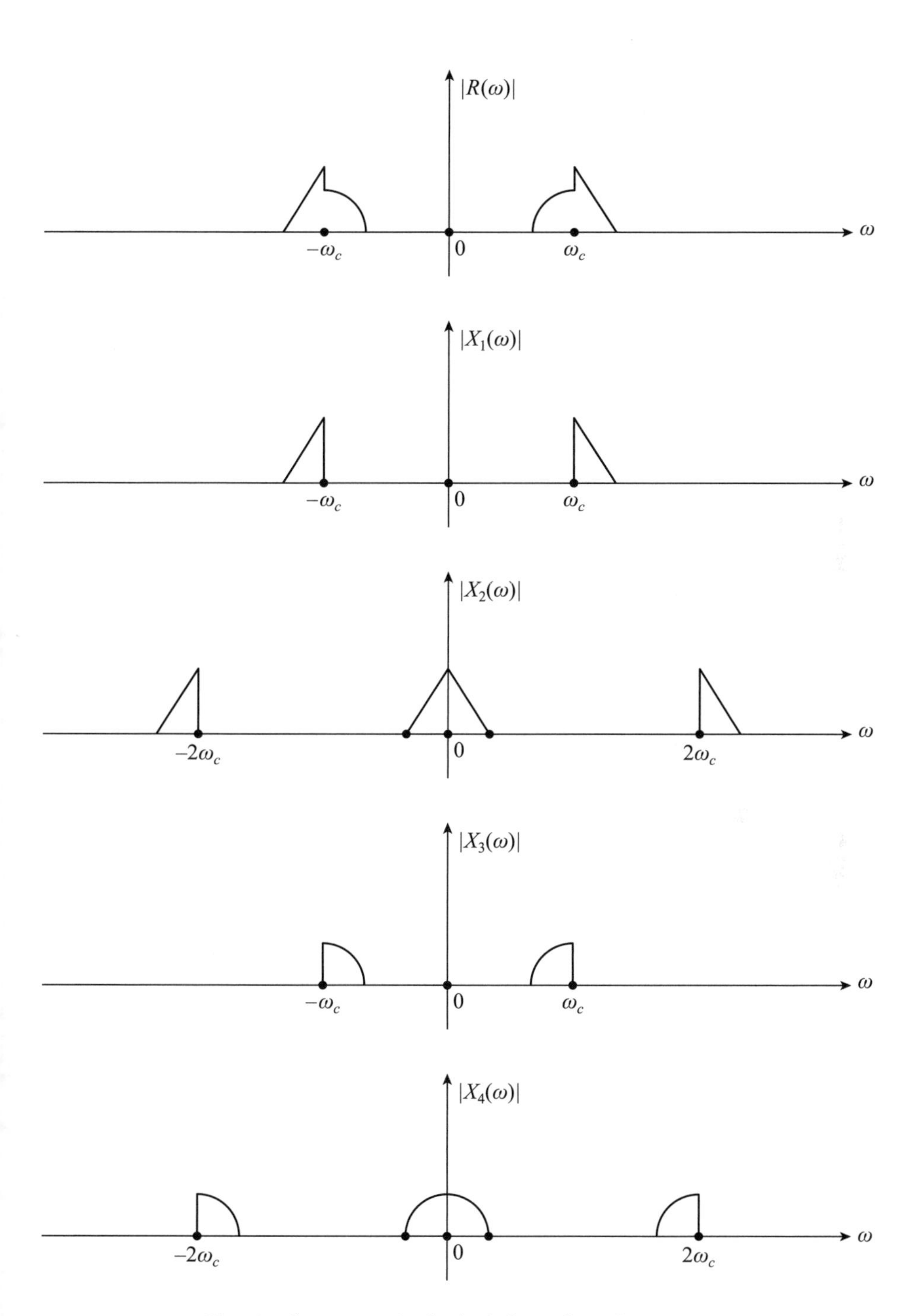

FIGURE 2.1.10. The signal spectrums in the dual-channel synchronous AM receiver.

that when the BBC's London-based station 2LO broadcast on a carrier of 842 kHz, that was the *only* frequency in its transmitted signal. Yes, the amplitude of the carrier did vary with time (he admitted), but so what? (he asked). This provoked a strong reaction from some readers, and you can find spirited replies to Fleming in the February 8, 1930, issue of *Nature* on pp. 198–199, along with Fleming's rebuttal. Fleming's refusal to admit to the existence of the modulation sidebands is particularly puzzling, because when he wrote, the first commercial single-sideband AM radio/telephone link between New York City and London had been operating for *three years*. According to Fleming, such a radio (discussed in chapter 4) simply could not have worked, and the fact that it did work apparently never caused him to reconsider.

The dual-channel AM transmitter/receiver circuitry *is* impressive, but the arrangements of Figure 2.1.7 and Figure 2.1.9 are *not* the only way to send two baseband signals on the same carrier frequency. In the final analysis of this section let's explore what is going on (in a purely analytic way) as opposed to simply drawing spectrum diagrams. We'll see amplitude attenuation when the phase error is not zero in a synchronous system just like we saw earlier, as well as a new problem of an entirely different nature. Consider, then, the circuits of Figure 2.1.11, which are the transmitter (upper half) and the receiver (lower half) of what is called *quadrature amplitude multiplex* radio (QAM).

The boxes labeled 90° in Figure 2.1.11 represents circuitry that produces an output that is the input shifted by 90° (this is the meaning behind the word "quadrature").[6] We can immediately write the transmitter signal as

$$r(t)=m_1(t)\cos(\omega_c t)+m_2(t)\sin(\omega_c t).$$

6 To build such circuits is easy if you remember that's what an integrator does: $\int\cos(x)dx=sin(x)$. For how to build an electronic integrator, with any specified gain, see *The Science of Radio* (note 8 in chapter 1), pp. 390–395.

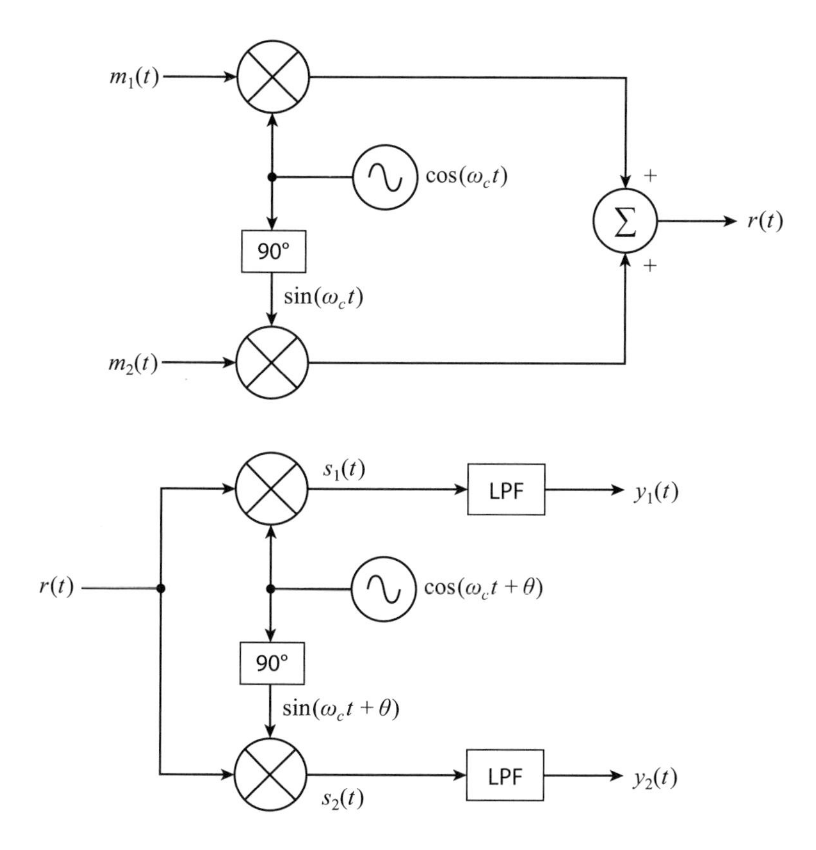

FIGURE 2.1.11. The transmitter (top) and the receiver (bottom) of a QAM radio, with a phase mismatch of θ at the receiver.

Thus, for the signals immediately after the multipliers in the upper and lower receiver channels, we have

$$s_1(t) = r(t)\cos(\omega_c t + \theta) = m_1(t)\cos(\omega_c t)\cos(\omega_c t + \theta) \\ + m_2(t)\sin(\omega_c t)\cos(\omega_c t + \theta)$$

and

$$s_2(t) = r(t)\sin(\omega_c t + \theta) = m_1(t)\cos(\omega_c t)\sin(\omega_c t + \theta) \\ + m_2(t)\sin(\omega_c t)\sin(\omega_c t + \theta).$$

Expanding these two expressions using the obvious trigonometric identities, we get

$$s_1(t)=\frac{1}{2}m_1(t)\left[\cos(\theta)+\cos(2\omega_c t+\theta)\right]$$
$$+\frac{1}{2}m_2(t)\left[-\sin(\theta)+\sin(2\omega_c t+\theta)\right],$$

and

$$s_2(t)=\frac{1}{2}m_1(t)\left[\sin(\theta)+\sin(2\omega_c t+\theta)\right]$$
$$+\frac{1}{2}m_2(t)\left[\cos(\theta)-\cos(2\omega_c t+\theta)\right].$$

Thus, after the indicated low-pass filtering shown in Figure 2.1.11, the signals in the receiver output channels are

$$y_1(t)=\frac{1}{2}\left[m_1(t)\cos(\theta)-m_2(t)\sin(\theta)\right],$$

and

$$y_2(t)=\frac{1}{2}\left[m_1(t)\sin(\theta)+m_2(t)\cos(\theta)\right].$$

If $\theta=0$ (if the receiver is perfectly phase-coherent with the transmitter[7]), then $y_1(t)=\frac{1}{2}m_1(t)$, and $y_2(t)=\frac{1}{2}m_2(t)$, and so we have achieved perfect recovery of the two baseband signals. If $\theta\neq 0$, however, then each baseband signal is attenuated by a factor of $\cos(\theta)$—an effect we saw earlier in this section—as well as suffers from "leakage" (called *crosstalk*) of the other baseband signal (an effect proportional to

7 To understand how to achieve this demanding level of phase coherence requires a bit more electronics and math than we'll pursue in this book. The classic approach, using what is called a *phase-locked* [*feedback*] *loop*, is discussed in a paper by J. P. Costas, "Synchronous Communications," *Proceedings of the IRE*, December 1956, pp. 1713–1718. The resulting receiver is called (not surprisingly) a *Costas demodulator* (after the General Electric electrical engineer John Peter Costas (1923–2008)).

sin(θ)). If $\theta = 90°$, each of the baseband signals appears alone and, once again, unattenuated—but now *in the wrong channel*!

An obvious question to ask is, How much can θ depart from zero and still have each baseband signal appear in its proper channel with an acceptable level of crosstalk(the less, the better, of course)? For example, suppose we wish the crosstalk of $m_2(t)$ on top of $m_1(t)$ in $m_1(t)$'s channel to be no more than −40 dB (a common level of unwanted signal rejection used in radio engineering).[8] That is, we demand that

$$20\ \log_{10}\left(\frac{m_2\sin(\theta)}{m_1\cos(\theta)}\right) = -40.$$

Then, if we further assume that m_1 and m_2 are of equal magnitude, we have

$$\frac{\sin(\theta)}{\cos(\theta)} = \tan(\theta) = 10^{-2} = 0.01 \text{ radian}.$$

Since $\tan(\theta) \approx \theta$ when θ is "small," which is clearly the case here, then $\theta = 0.57°$ is the maximum allowable phase mismatch. This is pretty small, and this calculation provides a good illustration of just how nearly perfect the phase coherence has to be for QAM radio to work.

8 The relative strength of two signals is measured by radio engineers in the dimensionless units of *decibels (dB)*. (A *deci*bel (dB, pronounced *dee-bee*) is one-tenth of a *bel* (B), a unit named after the inventor of the telephone, Alexander Graham Bell (1847–1922).) If the two signal powers are P_1 and P_2, their relative strength is $10\ log_{10}\left(\frac{P_1}{P_2}\right)$ dB. If $P_1 = P_2$ (the two signals are of equal strength), the two signals are said to be separated by 0 dB. If $P_1 = \frac{1}{2}P_2$ then, since $10\ log_{10}\left(\frac{\frac{1}{2}P_2}{P_2}\right) = -10\ log_{10}(2) = -3$ dB, P_1 is said to be 3 dB below P_2. Since power is voltage *squared*, if two signals have voltages V_1 and V_2, their relative strength is $20\ log_{10}\left(\frac{V_1}{V_2}\right)$ dB.

2.2 Large-Carrier AM

So far in this book I've used the term "AM radio" pretty casually, without really being precise about just what "AM" means. I've instead relied on your intuitive understanding of *amplitude modulation*. The first observation I'll now make is to point out to you something you may have not noticed—in the synchronous circuits of the previous section, *there is no energy at the carrier frequency in the transmitted/received signal*. That is, there is no $\delta(\omega - \omega_c)$ term in the transmitted/received signal spectrum. Rather, the only way for energy to appear at $\omega = \omega_c$ is for the dc energy of $m(t)$ to be heterodyned up to ω_c. But for music and speech, there *is* no energy at dc![9] Such synchronous radios are called *suppressed carrier* radio, and that is *not* how actual broadcast AM radio works. (We'll see this same situation again when we get to SSB radio in chapter 4.) Rather, real AM radio intentionally and constantly transmits a strong carrier signal, and what follows will explain why.

As previously argued, while synchronized radio is mathematically neat, it is quite challenging when it comes to the engineering behind it. To show how everyday AM radio actually works using (relatively) simple engineering, we make a crucial alteration to Figure 2.1.1, to arrive at Figure 2.2.1 (why the 1 is an input to the summer in Figure 2.2.1 will be explained soon). To keep the math simple, let's make the assumption that the carrier has unit peak amplitude, which simply means that whatever the actual peak amplitude may be, that amplitude will serve as our reference.

Now, suppose our message signal is the sound of a person whistling at the single frequency $\omega_m = 2\pi f_m$. Thus, $m(t) = A_m \cos(\omega_m t)$, where A_m is the amplitude of the whistle. Since the carrier and $m(t)$ are certainly not synchronized, we need to include an arbitrary phase angle φ to

9 Human speech and music consist of a multitude of time-varying sinusoidal frequency terms, and over the duration of each term, the average (dc) value of a term is zero. This is, in fact, an illustration of a perhaps misleading feature of Figure 1.6.1, our generic spectrum for a baseband signal, which seems to say there *is* energy at $\omega = 0$. There *might* be, but in any case it's not impulsive, which would be the mathematical signature of a continuously present carrier signal.

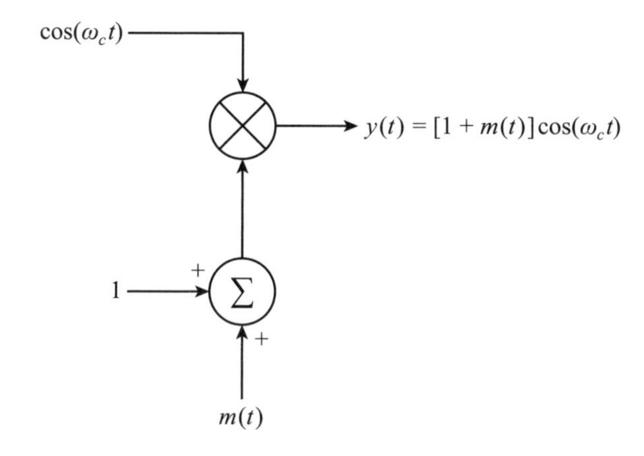

FIGURE 2.2.1. Generating an AM radio wave.

allow for the virtual certainty that $m(t)$ and the carrier are not simultaneously at their respective peak values at $t=0$. Thus, most generally,

$$y(t)=[1+A_m\cos(\omega_m t)]\cos(\omega_c t+\varphi)=\alpha(t)\cos(\omega_c t+\varphi),$$
$$\alpha(t)=1+A_m\cos(\omega_m t).$$

The 1 in $\alpha(t)$ is the dc term that is not present in $m(t)$ (see note 9).

We see that $y(t)$ is a high-frequency signal with the time-varying amplitude $\alpha(t)$ *if* $A_m\leq 1$. This proviso ensures that the amplitude of $y(t)$ is never negative. The upper plots of Figure 2.2.2 show $y(t)$ for the case of $A_m\leq 1$. As long as $\alpha(t)\geq 0$, it is the *envelope* of $y(t)$, and it varies (at the modulation frequency ω_m) between $1-A_m$ and $1+A_m$. What about the perfectly possible case of $A_m>1$? It would then appear that the transmitted signal would at times have a *negative* amplitude—and what could *that* mean?

Well, it *does* mean something, and the answer comes from Euler's identity, in particular from

$$e^{j\pi}=e^{-j\pi}=-1.$$

When $\alpha(t)$ goes negative (because $A_m>1$), then $y(t)$ is still the instantaneous value of a sinusoidal wave with *positive* amplitude, but the negative value of $\alpha(t)$ appears as a sudden phase shift of π radians. That's because $-\alpha(t)=|\alpha(t)|e^{\pm j\pi}$, and so

$$
\begin{aligned}
-\alpha(t)\cos(\omega_c t+\varphi) &= |\alpha(t)|(-1)\frac{\left[e^{j(\omega_c t+\varphi)}+e^{-j(\omega_c t+\varphi)}\right]}{2} \\
&= |\alpha(t)|\frac{(-1)e^{j(\omega_c t+\varphi)}+(-1)\,e^{-j(\omega_c t+\varphi)}}{2} \\
&= |\alpha(t)|\frac{e^{j\pi}e^{j(\omega_c t+\varphi)}+e^{-j\pi}\;e^{-j(\omega_c t+\varphi)}}{2} \\
&= |\alpha(t)|\frac{e^{j(\omega_c t+\varphi+\pi)}+e^{-j(\omega_c t+\varphi+\pi)}}{2} \\
&= |\alpha(t)|\cos(\omega_c t+\varphi+\pi).
\end{aligned}
$$

This effect is called *phase-reversal distortion* because, as the bottom plot of Figure 2.2.2 shows, the envelope of $y(t)$ is now *not* a mimic of the amplitude variation of $m(t)$. That is, the envelope of $y(t)$ is not $\alpha(t)$ but, rather, is $|\alpha(t)|$. Indeed, when $m(t)$ is *decreasing* (becoming more negative), then $|\alpha(t)|$ is *increasing*, and the envelope of $y(t)$ is reversed from what it should be. This problem is, fortunately, easy to avoid—don't overmodulate at the transmitter!

To end this section, it is interesting to calculate the *efficiency* of large-carrier AM. With single-tone modulation, as we wrote earlier, the transmitter signal is (where now I'll ignore the phase shift φ)

$$
\begin{aligned}
y(t) &= [1+A_m\cos(\omega_m t)]\cos(\omega_c t)=\cos(\omega_c t)+A_m\cos(\omega_m t)\cos(\omega_c t) \\
&= \cos(\omega_c t)+A_m\left\{\frac{1}{2}\cos[(\omega_c-\omega_m)t]+\frac{1}{2}\cos[(\omega_c+\omega_m)t]\right\} \\
&= \cos(\omega_c t)+\frac{1}{2}A_m\cos[(\omega_c-\omega_m)t]+\frac{1}{2}A_m\cos[(\omega_c+\omega_m)t].
\end{aligned}
$$

The first term is the carrier, and the second and third terms are the lower and upper sidebands, respectively. The average *power* of each sinusoidal term (the average of the *square* of the term) is half the square of the amplitude of the term, and so the total average power of the transmitted signal is

$$
\frac{1}{2}\left[1^2+\left(\frac{1}{2}A_m\right)^2+\left(\frac{1}{2}A_m\right)^2\right]=\frac{1}{2}+\frac{1}{4}A_m^2.
$$

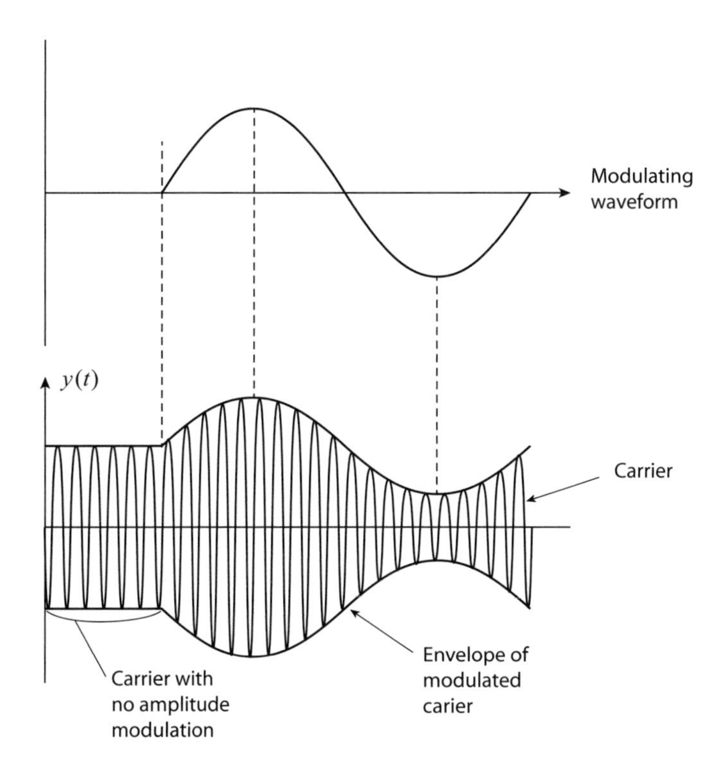

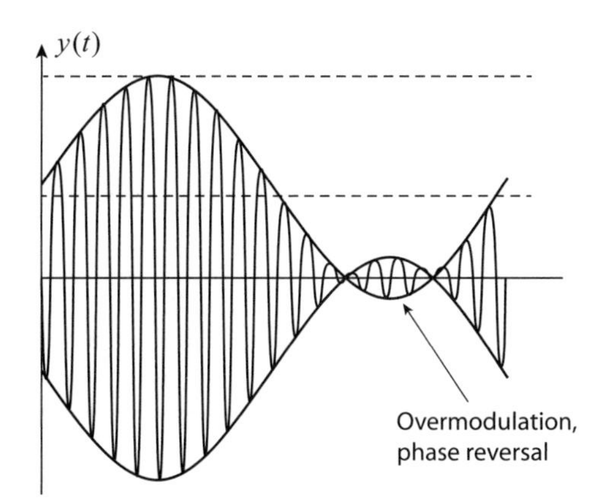

FIGURE 2.2.2. The envelope of a large-carrier AM signal, and phase-reversal distortion.

The fraction of this total average power that is in the carrier is

$$\frac{\frac{1}{2}}{\frac{1}{2}+\frac{1}{4}A_m^2}=\frac{2}{2+A_m^2}.$$

This fraction decreases with increasing A_m, but even with A_m as large as possible without causing overmodulation distortion ($A_m=1$), two-thirds(!) of the total average power is still in the non-information-bearing carrier. The theoretical maximum efficiency of large-carrier AM radio is, therefore, just 33%.[10] This is, you'd almost surely agree, grossly inefficient, and an obvious question to ask is, Why do radio engineers build such a seemingly deficient system? The answer, as will be further explored as we go along, is that the admitted inefficiency of the lone transmitter is compensated for, many times over, by the simplicity of the millions of receivers tuned to that transmitter. Keeping the receiver circuitry simple, which allows lots of people to buy and operate one, generates more than enough income to the station owner to pay for the transmitter inefficiency, as well as to realize a profit.[11] The presence of a high-energy carrier at the receiver is precisely what allows the simple, inexpensive AM radio receiver to work.

2.3 Convolution, and Multiplying by Squaring and Filtering

The circuits of the previous sections all assume we have available to us multipliers (or, remembering note 1, circuits that square their input). In this section you'll see the theory behind this assumption,

10 Actually, it's even worse than this. The one-third portion of the total average power that bears information is equally divided between the two sidebands, and *each* of the sidebands contains all the modulation information (this will be elaborated on in the chapter on SSB radio). So, the maximum efficiency of large-carrier AM radio can plausibly be argued to be just 17%!

11 The income to the AM radio station operator does not come directly from the millions of listeners. Rather, it comes from the selling of advertising time to merchants who are willing to buy that time precisely because there are millions of listeners (potential customers). I'll say a bit more on this point in chapter 6.

starting with a computation of the Fourier transform of the product of two arbitrary time functions, $m(t)$ and $g(t)$. This approach to multiplying is mathematically elegant, but it does have some practical engineering complications. Later in this chapter I'll show you an alternative approach that is elegant from both the mathematical *and* the engineering viewpoints.

By definition, the transform of $m(t)g(t)$ is

$$\int_{-\infty}^{\infty} m(t)g(t)e^{-j\omega t}dt = \int_{-\infty}^{\infty} m(t)\left\{\frac{1}{2\pi}\int_{-\infty}^{\infty} G(u)e^{jut}du\right\}e^{-j\omega t}dt,$$

where $g(t)$ has been written in the form of an inverse Fourier transform. (I've used u as the dummy variable of integration in the inner integral, rather than ω, to avoid confusion with the outer ω.) Continuing, if we reverse the order of integration, we have the transform of $m(t)g(t)$ as

$$\int_{-\infty}^{\infty} \frac{1}{2\pi}G(u)\left\{\int_{-\infty}^{\infty} m(t)e^{jut}e^{-j\omega t}dt\right\}du$$
$$= \frac{1}{2\pi}\int_{-\infty}^{\infty} G(u)\left\{\int_{-\infty}^{\infty} m(t)e^{-j(\omega-u)t}dt\right\}du,$$

or, as the inner integral is just $M(\omega-u)$, we have the Fourier transform pair

$$m(t)g(t) \leftrightarrow \frac{1}{2\pi}\int_{-\infty}^{\infty} G(u)M(\omega-u)\,du.$$

This last integral form occurs so often in mathematics, physics, and engineering that it has its own name: the *convolution integral*. Using the symbol *, *inline*,[12] the pair is commonly written

$$m(t)g(t) \leftrightarrow \frac{1}{2\pi}G(\omega) * M(\omega).$$

12 A *superscript* *, of course, denotes *conjugation*.

Since it doesn't matter which time function we call $m(t)$ and which we call $g(t)$, it is clear that convolution must be commutative. That is, we can also write

$$m(t)g(t) \leftrightarrow \frac{1}{2\pi} M(\omega) * G(\omega).$$

As a particularly interesting special case, if $m(t) = g(t)$, then

$$m^2(t) \leftrightarrow \frac{1}{2\pi} M(\omega) * M(\omega).$$

We can use this result to determine the spectrum of $m^2(t)$—the output of a squarer with input $m(t)$—which will tell us the frequency of the energy of $m^2(t)$. We'll assume (as before) that $m(t)$ is bandlimited, with all its energy confined to the interval $-\omega_m < \omega < \omega_m$. And once we have *that* result, then we can see how to build a multiplier using a *single* squarer (you'll recall that the approach sketched in note 1 used *two* squaring devices).

From our last transform pair, if we explicitly write out the convolution $M(\omega) * M(\omega)$, we have

$$m^2(t) \leftrightarrow \frac{1}{2\pi} \int_{-\infty}^{\infty} M(u)M(\omega - u)\, du.$$

The integral will certainly be zero if the integrand is zero, and that will be the case if ω is sufficiently positive (or negative) that $M(u)$ $M(\omega - u) = 0$. Now, by our bandlimited assumption for the baseband signal, $M(u) \neq 0$ only if $-\omega_m < u < \omega_m$ (see the shaded interval in the upper half of Figure 2.3.1). And similarly, $M(\omega - u) \neq 0$ only if $-\omega_m < \omega - u < \omega_m$, that is, only if $-\omega_m - \omega < -u < \omega_m - \omega$ which says $\omega_m + \omega > u > -\omega_m + \omega$ or, finally, only if $-\omega_m + \omega < u < \omega_m + \omega$ (see the shaded interval in the lower half of Figure 2.3.1).

Now, imagine that ω is increased, which shifts the lower shaded interval to the right. There will be an overlap of the two shaded intervals (and so a nonzero integrand) as long as $-\omega_m + \omega < \omega_m$ (that is, as long as $\omega < 2\omega_m$). And if *ω is decreased*, which shifts the lower shaded interval to the left, there will be an overlap of the two shaded

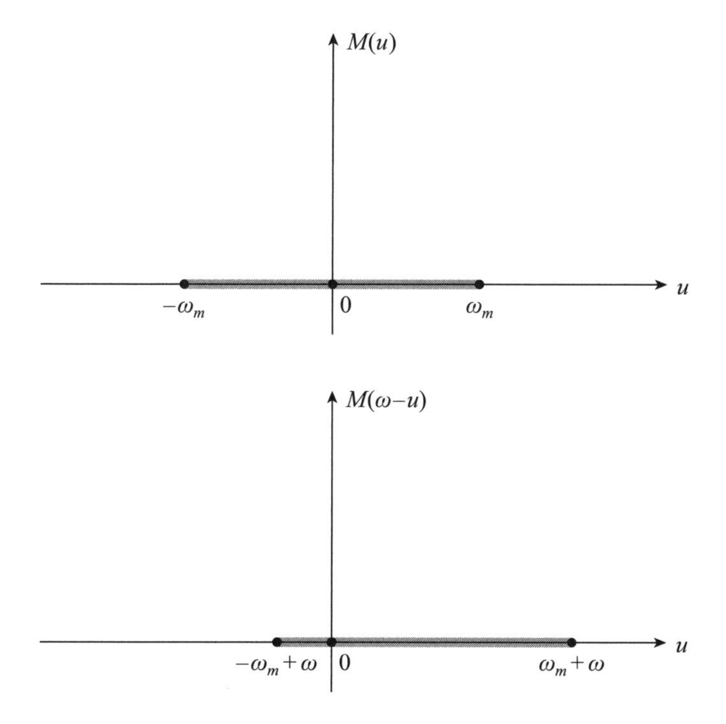

FIGURE 2.3.1. The integrand factors of the convolution integral for $m^2(t)$. The shaded segments are where $M(u)$ and $M(\omega-u)$ are nonzero.

intervals (and so a nonzero integrand) as long as $-\omega_m < \omega_m + \omega$ (that is, as long as $-2\omega_m < \omega$). Thus, there will be a nonzero integrand as long as $|\omega| < 2\omega_m$. In words, if all the energy of $m(t)$ is confined to the interval $|\omega| < \omega_m$, then the spectrum of $m^2(t)$ is twice as wide, with all the energy of $m^2(t)$ confined to the interval $|\omega| < 2\omega_m$.

With this theoretical result in our pocket, we can now understand how Figure 2.3.2 achieves multiplication. The inputs to the summing circle[13] are the bandlimited baseband $m(t)$ and $\cos(\omega_c t)$, and so the input to the squarer is $m(t) + \cos(\omega_c t)$; therefore, the output is

$$s(t) = [m(t) + \cos(\omega_c t)]^2 = m^2(t) + 2m(t)\cos(\omega_c t) + \cos^2(\omega_c t),$$

an expression that includes the desired product $m(t)\cos(\omega_c t)$. Alas, that product term is contaminated (or so it might seem) with the two

13 Recall note 1 and the claim there that such summing circuitry is easy to build.

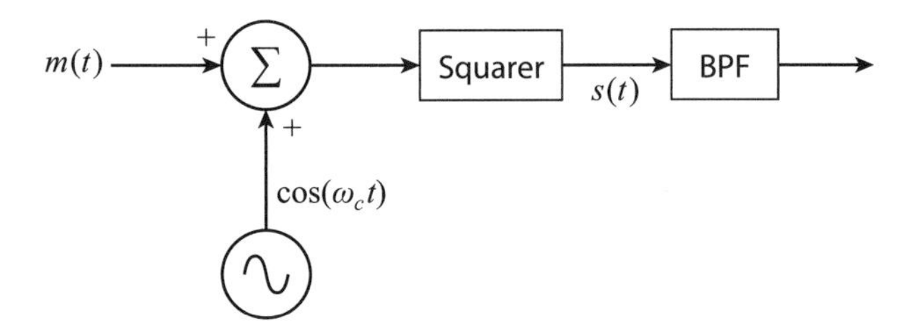

FIGURE 2.3.2. How to multiply with a squarer (and a filter).

additional terms $m^2(t)$ and $cos^2(\omega_c t)$. In general, this is true, but that is not *necessarily* fatal. Fourier theory will show us how to arrange things so it's *not* fatal. That is, we can arrange matters so that the energy of the product term is completely separate and distinct (in frequency) from the energies of the two other terms, and so the product term can be recovered, *uncontaminated*, through appropriate filtering.

By the heterodyne theorem, the energy of the $m(t)\cos(\omega_c t)$ term is simply the energy of $m(t)$ shifted up and down the frequency axis, to be centered on $\omega=\pm\omega_c$. Next, $cos^2(\omega_c t)$ can be expanded with a well-known trigonometric identity to give the equivalent expression (and associated Fourier transform)[14]

$$\frac{1}{2}+\frac{1}{2}\cos(2\omega_c t) \leftrightarrow \pi\delta(\omega)+\frac{1}{2}\pi\left[\delta(\omega-2\omega_c)+\delta(\omega+2\omega_c)\right].$$

That is, all the energy of $cos^2(\omega_c t)$ is at the three specific frequencies $\omega=0$ and $\omega=\pm 2\omega_c$. And finally, as we found earlier, all the energy of the $m^2(t)$ term is in the frequency interval $|\omega|<2\omega_m$. All this is displayed in Figure 2.3.3, which shows no overlapping of the individual energies of the three terms in the squarer output. It is clear from the figure that this will indeed be the case if $2\omega_m<\omega_c-\omega_m$, or $3\omega_m<\omega_c$, a condition easily satisfied in commercial AM radio ($3f_m$ is on the order of 15 kHz, while f_c is *at least* 540 kHz). The energy of the product term is therefore the *only* energy in the output of the bandpass filter of Figure 2.3.2 if the skirts of the filter are in the obvious location (which really don't have to be anywhere near vertical, given the "wide" frequency separations shown in Figure 2.3.3).

14 In section 1.7 we worked out the transform of $\sin(\omega_c t)$, not of $\cos(\omega_c t)$, but the two calculations are only trivially different.

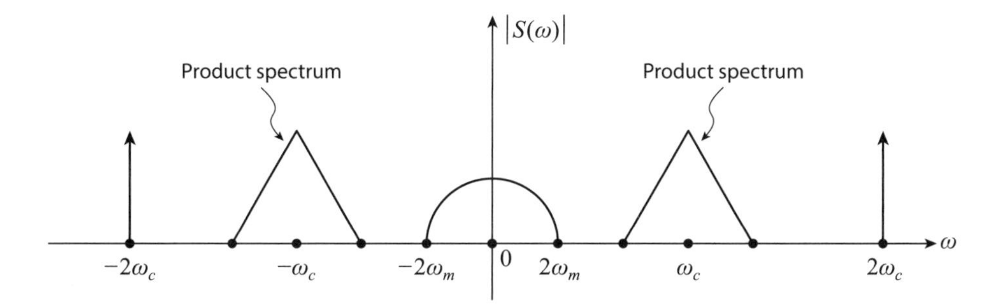

FIGURE 2.3.3. Location of the energy in the output $s(t)$ of the squarer in Figure 2.3.2.

2.4 Fleming's Vacuum Tube Diode and the Detection of Radio Frequency Waves

The idea of multiplying by squaring and filtering is, as I said earlier, *mathematically* elegant. But it leaves unanswered an obvious question: How do we build a squaring circuit? This is not an unsolvable problem,[15] but we'll not pursue it here. Instead, let's see how to directly make a *multiplier* (our ultimate goal, after all) using a different (but equally elegant) approach. Before jumping straight into the math, however, I need to introduce you to a new electrical component, the *diode*. The diode is mathematically even simpler than a resistor, and so I think mathematician Hardy would be willing, with little reluctance, to accept this new radio engineering element into our toolbox.

A perfect diode is a two-terminal (just like a resistor, capacitor, and inductor) circuit element that has the special property of conducting electrical current in only *one* direction (see Figure 2.4.1). When $v_a > v_b$ the diode acts like a short circuit, so v_a and v_b are actually equal in this case (called *forward biased*). The value of the resulting current i is completely determined by the surrounding circuitry in which the diode exists, and that current flows in the direction of the arrowhead. When $v_a < v_b$ (called the *reverse-biased* case) the diode acts like an open circuit, and the diode current is $i = 0$. Speaking poetically, what tries to be the diode current in reverse bias, flowing to the left in Figure 2.4.1 *against* the arrowhead, instead "runs straight into a brick wall," represented by the vertical line at the tip of the arrowhead.

15 See, for example, *The Science of Radio* (note 8 in chapter 1), pp. 241–243.

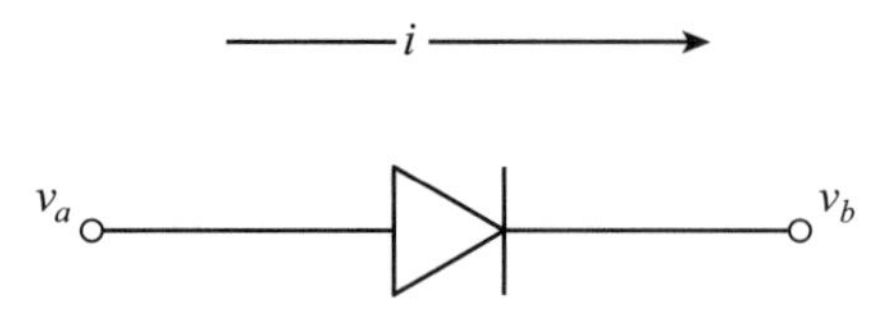

FIGURE 2.4.1. The diode current $i=0$ when $v_b > v_a$.

(Remember, the forwardbiased *i* is in the direction of what we call positive current, but the actual motion of the electron *negative* charge carriers is in the opposite direction.)

Later in this chapter I'll say a bit more about the physics of the diode, in the form of a vacuum tube,[16] but for now let's just accept that such a component exists. As a simple, quick example (which you'll soon see is not as frivolous as it might at first appear) of how important the diode was to the development of radio, consider that early radios were powered by dc voltages produced by bulky, expensive batteries. It was an important milestone in radio history when it became possible to use the ac electricity from a home wall outlet to provide the dc electricity for radios. As Figure 2.4.2 shows, with an ac voltage into what are called *half-wave* (top half) and *full-wave* (bottom half) *rectifier* circuits the resulting current in resistor *R* is *pulsating* dc (the pulsing can be smoothed with the addition of a capacitor across *R*). The pulse rate in a full-wave rectifier is twice that of a half-wave rectifier. The important thing in these rectifier circuits is that the current in *R* never reverses direction but is in the *same direction at every instant.*

Now, what did I mean when I said the circuits of Figure 2.4.2 are not as frivolous as you might perhaps think "simply" transforming ac into pulsating dc is? (Providing dc power to a radio is important, yes, but also pretty mundane!) I made that claim because those same circuits solved the Holy Grail problem that is central to radio: the *detection* of the presence of radio-frequency waves, which is the all-important preliminary step to demodulating the intelligence on

16 Modern electrical engineering students are quickly introduced to solid-state diodes based on the quantum-mechanical behavior of certain crystal structures. This is great stuff, but pure mathematician Hardy would, I fear, draw a big do-not-cross red line when confronted with such talk. Fleming's vacuum tube diode, however, avoids all such complications.

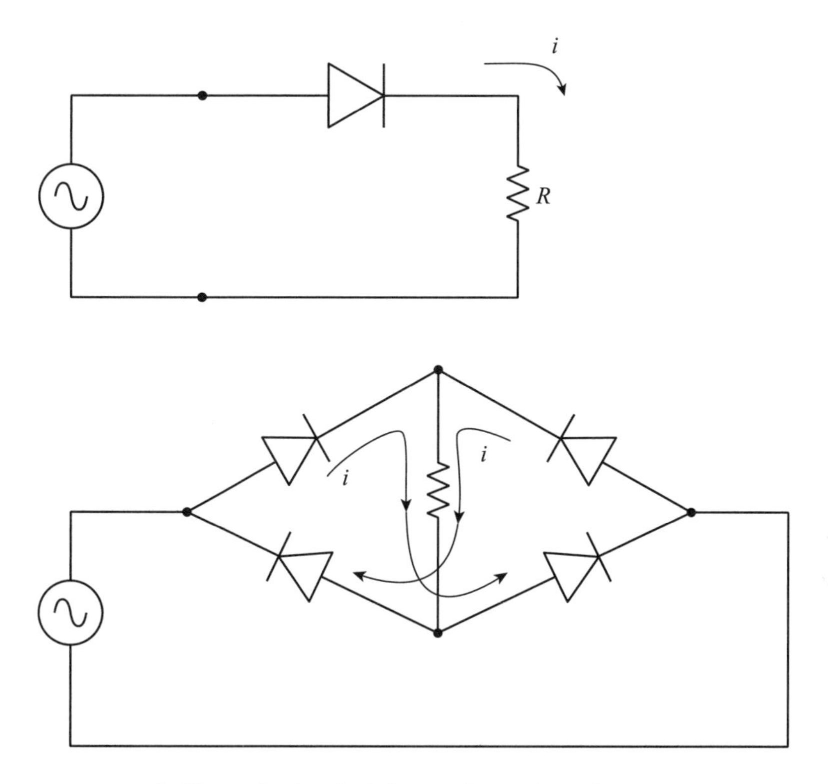

FIGURE 2.4.2. A half-wave (top) and a full-wave (bottom) rectifier.

those same waves. To understand this, imagine that the resistor in the rectifier circuits is the coil resistance of a current-indicating meter in which the deflection of the meter's needle is a measure of the instantaneous value of the ac current in the meter.

Without the diodes, at sufficiently low frequencies the needle will visibly swing back and forth around zero as the current goes back and forth from positive to negative to positive, and so on. As the frequency increases, however, the back-and-forth swinging of the needle will be reduced to a mere quivering and then eventually (as the frequency goes even higher) will fail to display any deflection at all! That's because of the mechanical inertia of the needle and its associated magnetic coil.

With the diode rectifier circuits of Figure 2.4.2, however, the meter needle *will* deflect in just one direction, no matter how high the frequency, because the current pulses in the meter are *always in the same direction*. The diode circuits of Figure 2.4.2 allow the meter to indicate (or *detect*) the presence of electric waves at any frequency. At high frequency, the magnitude of the steady meter needle deflection will be a measure of the average value of the current pulses.

If we introduce another new gadget, the magnetically coupled, center-tapped transformer (we'll see it again when we discuss single-sideband radio in chapter 4, and the demodulation process in FM radio in chapter 5), we can achieve full-wave rectification with just *two* diodes. Back in Figure 1.1.3 I slyly slipped in a transformer (to couple a received radio-frequency signal into a circuit) using Faraday's discovery of electromagnetic induction. An ac signal into one coil (the *primary*) of a transformer induces an ac signal in the second coil (the *secondary*) of the transformer (the two coils are both wrapped around a common core of magnetic material, typically iron).[17] Now, suppose we add a *center-tap* wire to the secondary, which is located at the precise physical center of the secondary, as shown in Figure 2.4.3. On the half-cycle where $v_a > v_b$, then v_a will also be greater than the center-tap voltage, and when on the alternate half-cycle where $v_b > v_a$, then v_b will also be greater than the center-tap voltage. With this in mind, you should be able to see that the *two* diodes are alternately forward biased (radio engineers call this a "push-pull" circuit), and their currents flow *in the same direction* in R (to the left).

Before the invention of the electronic vacuum tube diode, by the British radio engineering pioneer John Ambrose Fleming, the early radio-frequency detectors were of a suspect nature whose workings

17 If the primary has N_p turns of wire around the magnetic core, and the secondary has N_s turns of wire around the same core, then the induced secondary voltage is $v_s = \frac{N_s}{N_p}$ times the primary voltage v_p. ($\frac{N_s}{N_p}$ is called the *turns ratio* of the transformer). In a *lossless* transformer (which means the instantaneous power in the primary equals the instantaneous power in the secondary, with no internal energy loss) the currents in the primary and the secondary scale in the opposite way from the voltages. That's because *lossless* means $v_p i_p = v_s i_s$, and so $i_s = \frac{v_p}{v_s} i_p = \frac{i_p}{v_s / v_p} = \frac{i_p}{N_s / N_p} = \frac{N_p}{N_s} i_p$. Note, carefully, that any dc present in the primary does *not* induce a voltage in the secondary (only *time-varying* signals make it through a transformer); radio engineers routinely use this feature to remove the dc component from a time-varying signal.

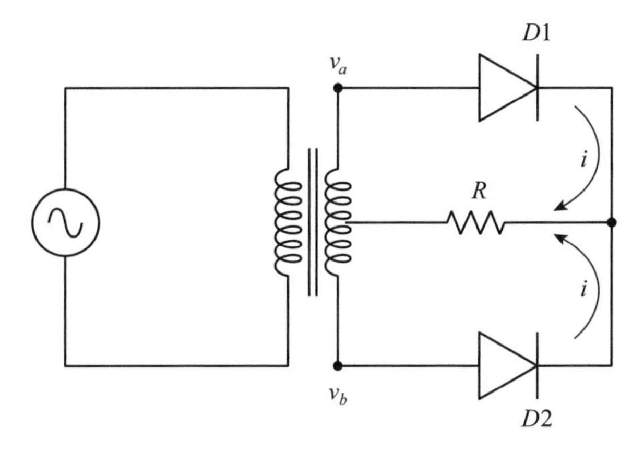

FIGURE 2.4.3. Full-wave rectifier using two diodes and a center-tapped transformer. Diode D1 conducts when the input ac signal is executing a positive half cycle ($v_a > v_b$), and diode D2 conducts when the input ac signal is executing a negative half cycle ($v_b > v_a$).

were 90% mystery.[18] Then, in 1904, Fleming discovered his detector and fully explained how it worked. His invention found its origin, interestingly, in a puzzling problem encountered decades earlier by the American experimentalist Thomas Alva Edison (1847–1931) during his famous development of the incandescent electric lightbulb. In February 1880 Edison performed a simple experiment that, had he realized what he had stumbled upon, would have been worthy of a Nobel Prize in Physics.

In earlier experiments with his lightbulbs Edison had observed that the glass envelopes of the bulbs quickly became covered with a black deposit. Assuming that the blazing-hot carbon filaments (operating at temperatures from 1,800°*F* to over 4,500°*F*) were ejecting electrically charged particles of carbon that collected on the glass, Edison looked for a way to prevent these assumed particles from making their way

18 You can read more on the nature of the early pre–vacuum tube radio-wave detectors in *The Science of Radio* (note 8 of chapter 1), pp. 54–58, 62–63. Just *what* was physically going on in these first-generation detectors is still a bit of a puzzle, even today. It was, in fact, the mystery of those detectors that inspired Kipling's tale "Wireless," mentioned in the frontmatter of this book. They worked (sort of) but just *how* was the subject of long and vigorous debates. Fleming's vacuum tube diode detector, however, is easily understood, and that transparency (even Hardy would, I think, have found Fleming's diode quite straightforward) is the reason I've included as much discussion of it here as I have.

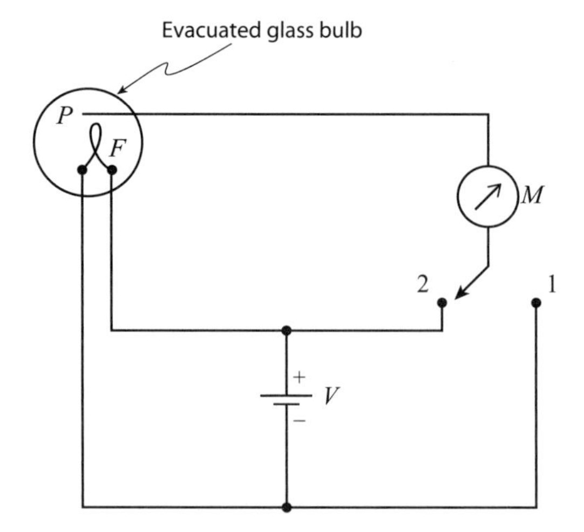

FIGURE 2.4.4. The Edison-effect circuit. The battery V heats filament F to a very high temperature in an evacuated glass bulb. When the switch is in position 1, no current is indicated by meter M. When the switch is in position 2, however, M indicates a current is flowing across the vacuum of the bulb from the probe P to F.

from the filament to the glass. The circuit of Figure 2.4.4 was the result, with a wire probe inserted through the glass envelope.

Edison found that if he connected a current-indicating meter between the probe and the *positive* terminal of the filament battery, there was a current. If, however, he connected the meter to the probe and the *negative* terminal of the battery, there was no current. This observation eventually became known as the *Edison effect,* but as it failed to help Edison solve his immediate problem of bulb darkening, he failed to pursue the matter.

Unknown to Edison was that the thermal agitation in the intensely heated filament is such that electrons are literally "'boiled out" of the filament and into the vacuum of the bulb.[19] These electrons form a negatively charged (see note 9 in the appendix) "cloud" around the filament. The size of this space-charge cloud depends on the filament temperature (the higher the temperature the larger the cloud), but for a given temperature the cloud quickly self-limits. That is, the total

19 One can't really blame Edison for not realizing this, as the electron wasn't "discovered" until *17* years later (1897)!

negative charge of the cloud soon reaches a level where the repulsion force of the cloud prevents any further electrons in the filament from joining the cloud (recall, from high school physics, like charges repel). But when the switch in Figure 2.4.4 is in position 2 (the *positive* terminal of the filament battery), then probe P *attracts* the *negative* electrons in the space-charge cloud, thus creating a current (a current flowing in the direction opposite the direction of electron motion) in meter M. Edison had stumbled upon the first *electronic* device—the vacuum tube diode—but didn't realize his momentous discovery and saw no application for it (it *was* 1880, after all, and radio lay decades in the future).

By 1904, however, radio engineer Fleming *did* realize what was going on in Edison's enhanced lightbulb, as well as how to use it as a radio-wave detector. In November of that year he wrote to the Italian experimenter Guglielmo Marconi (1874–1937) that "I have been receiving signals on an aerial with nothing but a [current meter] and my device."[20] I've drawn what Fleming sent to Marconi in Figure 2.4.5 but will defer explaining it until the end of this chapter. Right now I want you (and our newly resurrected Hardy) to simply ponder the obvious simplicity of Fleming's circuit. It is central to the operation of modern AM *and* FM radio.[21]

20 Fleming was well acquainted with Marconi, as Fleming was the designer of the spark-gap transmitter (called "The Big Thing": see *The Science of Radio*, pp. 51–52) that was used by Marconi in a 1901 attempt to send Morse code radio signals across the Atlantic. A modern evaluation of that attempt, and of Marconi's claims concerning what he said occurred, is J. A. Ratcliffe, "Scientific Reaction to Marconi's Transatlantic Experiment," *Proceedings of the IEE* (British), September 1974, pp. 1033–1038.

21 In the circuit of Figure 2.4.5, point a corresponds to the probe P shown in Figure 2.4.4, but point b does *not* correspond to the filament F in Figure 2.4.4. Rather, filament F is inside a surrounding structure (called the *cathode*) that is *indirectly* heated by F, and it is that surrounding structure that creates the space-charge cloud and so corresponds with point b. In this way, point b is not connected to the battery in Figure 2.4.4 and so is electrically independent of that battery. This is important, because in early-1920s radios that used household ac to directly heat a filament, the space-charge cloud around the filament undulated in rhythm with the power-line ac frequency (60 Hz in the US, 50 Hz in continental Europe and Great Britain). This undulation, in turn, showed up in the audio output as an annoying hum. Using batteries avoided this, but batteries were expensive. The development in mid-1927 of indirectly heated vacuum tubes solved the ac-hum problem because the thermal inertia of the comparatively massive cathode structure was too sluggish to follow the ac line frequency. (The ac-hum problem in directly heated ac filament radios was sufficiently vexing that Nobel Prize–winning physicist Richard Feynman mentioned it in his very funny essay "He Fixes Radios by *Thinking*!" in his famous 1985 collection, *Surely You're Joking, Mr. Feynman!*) For another way to eliminate ac hum, see Challenge Problem 2.4.

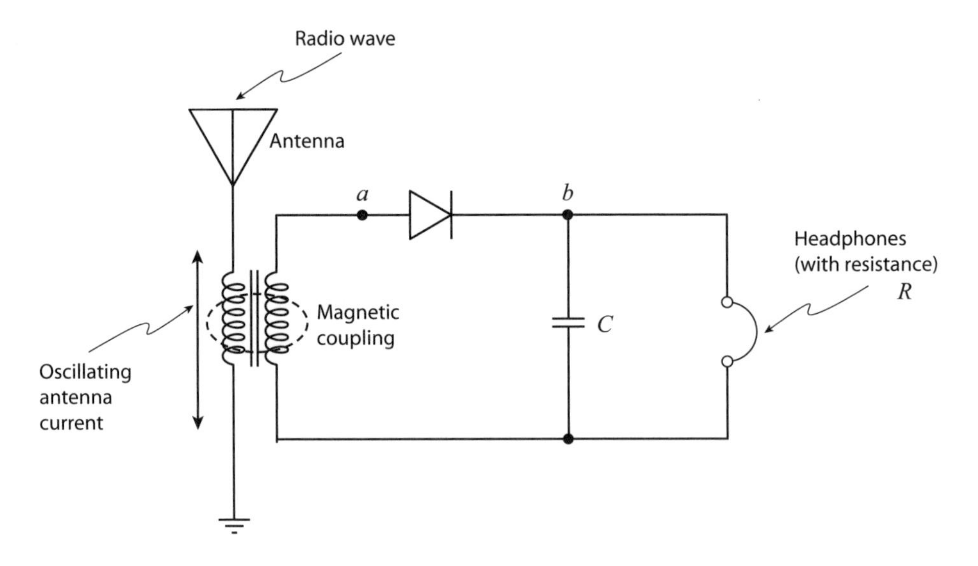

FIGURE 2.4.5. Fleming's circuit for a radio-wave detector.

2.5 Multiplying by Sampling and Filtering

Fourier theory is the mathematics behind a surprising way to frequency shift a baseband signal $m(t)$ (with maximum frequency ω_m) up to radio frequencies. It all starts with the idea of *sampling*. A sampler is a circuit that at regular intervals of time (called the *sampling period*, T seconds) passes $m(t)$. A mechanical sampler is easily visualized as a rotating switch, as shown in the upper half Figure 2.5.1. We can mathematically model this sampler with the equation $m_s(t)=m(t)$ $s(t)$, where $m_s(t)$ is the output of the sampler, and $s(t)$ is the function shown in the lower half of the figure.

Since $s(t)$ is a periodic function, we can write it as a Fourier series, with $\omega_s=2\pi f_s$ as the sampling frequency (where $f_s=\frac{1}{T}$), and so we have

$$m_s(t)=m(t)\sum_{n=-\infty}^{\infty} c_n e^{jn\omega_s t},\ \omega_s=\frac{2\pi}{T}.$$

The Fourier transform of $m_s(t)$ is, therefore,

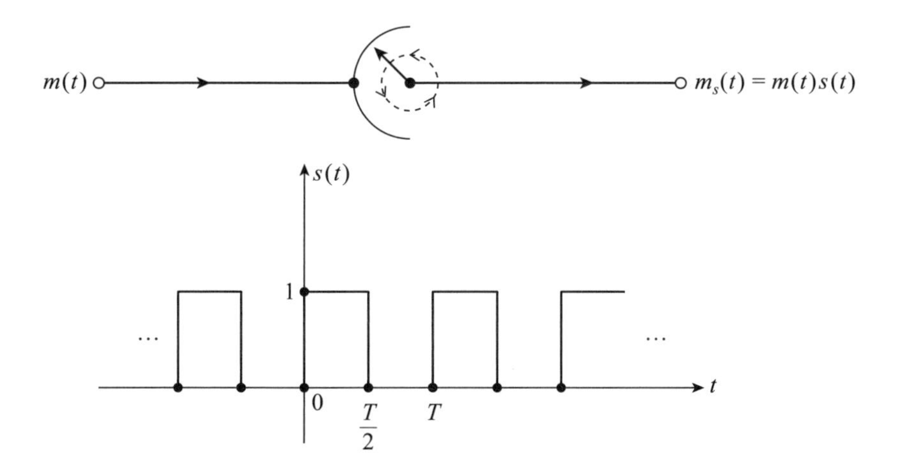

FIGURE 2.5.1. A mechanical sampler: the switch rotates once each T seconds, with the electrical path through the sampler completed as long as the switch touches the curved contact.

$$M_s(\omega) = \int_{-\infty}^{\infty} \left\{ m(t) \sum_{n=-\infty}^{\infty} c_n e^{jn\omega_s t} \right\} e^{-j\omega t} dt$$
$$= \sum_{n=-\infty}^{\infty} c_n \int_{-\infty}^{\infty} m(t) e^{-j(\omega - n\omega_s)t} dt.$$

Recognizing the last integral as $M\{(\omega - n\omega_s)\}$, we have

$$M_s(\omega) = \sum_{n=-\infty}^{\infty} c_n M\{(\omega - n\omega_s)\}.$$

This simple-looking expression says that the transform of the sampler output signal is just the transform of the input signal repeated endlessly, up and down the frequency axis, at intervals of ω_s. Figure 2.5.2 shows what such a sampler output transform looks like, under two assumptions: (1) $m(t)$ is a baseband-limited signal, and (2) $\omega_m < \omega_s - \omega_m$ (that is, $\omega_s > 2\omega_m$). These two assumptions guarantee that there is no overlap of the infinite copies of the transform of $m(t)$ in the sampler output. Each copy of the replicated baseband transform comes with its own amplitude factor, c_n, but the specific

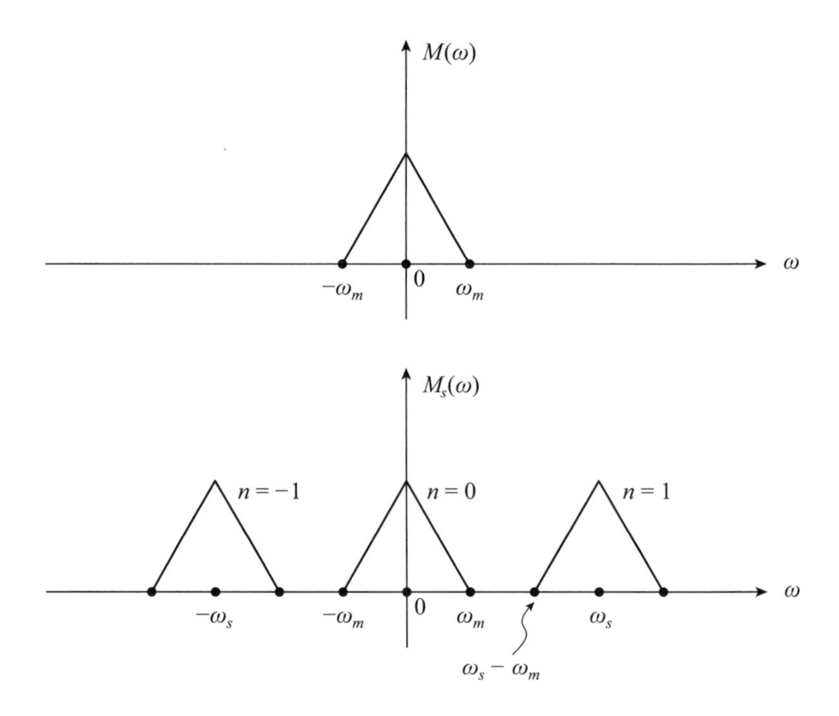

FIGURE 2.5.2. The transforms of a baseband signal $m(t)$ and of its sampled version $m_s(t)$.

values of these scaling factors are unimportant (more precisely, the values are unimportant in understanding the *theory*).

The process of sampling has, in particular, shifted the baseband transform up ($n = 1$) and down ($n = -1$) by ω_s. Bandpass filtering the sampler output $m_s(t)$ (with the passband centered on $\omega = \omega_s$) makes the filter output, to within a scaling factor, equal to the product $m(t)$ $\cos(\omega_s t)$. If we identify the sampling frequency ω_s with the AM carrier frequency, then we have achieved the desired spectrum shift of $m(t)$ up to radio frequencies. Since the sampler can be said to physically "chop up" the input signal $m(t)$ in the time domain, the sampler/filter combination is often called a *chopper modulator*. In Figure 2.5.1 the sampling time is shown as half the sampling period T: even at the lowest AM carrier frequency of 540 kHz, T is less than 2 μs (microseconds), and so the sampling time (the time interval during which the sampler passes $m(t)$) is less than 1 μs in duration. The highest frequency in $m(t)$ is ω_m, typically 5 kHz in AM radio, corresponding to a time interval of 200 μs. Lower frequencies will have even longer time intervals. Thus, during each sampling time, $m(t)$ remains essentially a constant.

If the sampler output is low-pass filtered to select the energy of the $n=0$ term of $m_s(t)$, then the output spectrum of the filter is, to within a scaling factor, equal to the original input's spectrum. *This means that the sampling process has not lost any information.* That is, if the original input signal is bandlimited at $\omega=\omega_m$, then sampling at a rate greater than $2\omega_m$ allows *perfect* recovery of the original continuous signal from just *samples* taken at discrete intervals of time. This perhaps nonintuitive conclusion is called the *sampling theorem*, and it is generally associated with the names of four people: the British mathematician Edmund Whittaker (1873–1956), who stated it in 1915; the Swedish American electronics engineer Harry Nyquist (1889–1976), who rediscovered it in 1928; the Russian electronics engineer Vladimir Kotel'nikov (1905–2005), who discovered it yet again in 1933; and the US mathematician and electrical engineer Claude Shannon (1916–2001), who stated it in his classic 1948–1949 papers on information theory (in which he closed the "discovery loop" by citing Whittaker's 1915 work).

A mechanically rotating switch is okay for *thinking* about sampling, but it clearly isn't a practical way to actually implement sampling at AM radio frequencies. What we need is an *electronic* circuit that is very fast but has *no moving parts*,[22] and Figure 2.5.3 shows an ingenious way to achieve that.[23] Here is how that circuit works.

If transformer T2 (with the $\cos(\omega_s t)$ signal source) and the ring of diodes were not present, then $m(t)$ would be transmitted (via transformers T1 and T3) straight through to appear as $m_s(t)$. That is, $m_s(t)=m(t)$. Now, we restore T2 and the $\cos(\omega_s t)$ signal source, and the ring of diodes, and then, to keep things really simple, we suppose the four diodes are perfect (are short circuits when forward biased, and open circuits when reversed biased). On the positive half cycles of $\cos(\omega_s t)$—the polarity shown in Figure 2.5.3—the series diodes D1 and

22 Would Hardy have felt safe—would *you* feel safe—being next to radio circuitry containing a hunk of metal rotating at more than half a million times per second?

23 Invented in 1934 by Frank A. Cowan (1898–1957), who was an electronics engineer at AT&T (American Telephone and Telegraph). Cowan's invention was inspired by an earlier (1929), similar circuit due to Clyde R. Keith (1900–1994), an electronics engineer at Bell Telephone Laboratories.

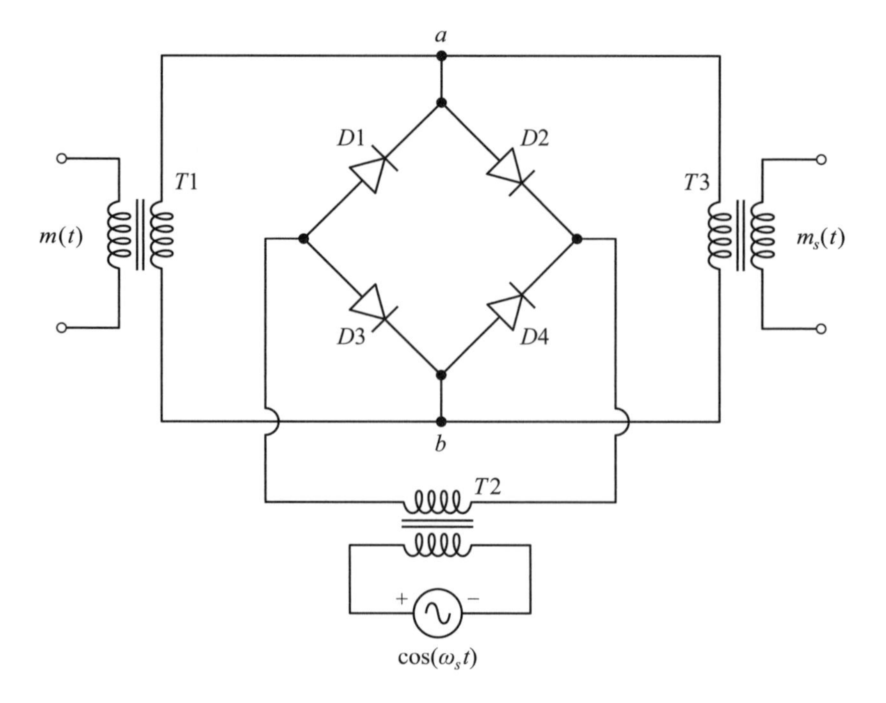

FIGURE 2.5.3. Diode-ring sampler (chopper modulator) for multiplying at radio frequencies.

D2, and the series diodes D3 and D4, are forward biased. This brings points a and b together (electrically), and so $m_s(t)=0$. On the negative half cycles of $\cos(\omega_s t)$ the series diode pairs are reverse biased, and so the diodes don't conduct. Points a and b are now electrically isolated, and so $m_s(t)=m(t)$. Thus, $m_s(t)$ is a sampled version of $m(t)$, with the sampling occurring at the rate of f_s (which we identify with the AM carrier frequency, f_c) samples per second. Since the carrier frequency for AM radio is at least 540 kHz, and as $f_m=5$ kHz, the sampling theorem requirement that $f_c>2f_m$ is easily satisfied. Finally, as mentioned earlier, appropriate bandpass filtering of $m_s(t)$ gives us $m(t)\cos(\omega_s t)$.

2.6 AM Envelope Detection, Tuning, and Impedance Matching

To end this chapter let's return to Figure 2.4.5, Fleming's detector circuit, and see how, with a few modifications, it demodulates the AM radio signal of Figure 2.2.2. Recall that this signal consists of a carrier (at frequency ω_c) that is amplitude modulated by the single tone at frequency ω_m. Now, how does Fleming's circuit extract a signal at frequency ω_m from a received signal that does *not* contain a term at that frequency? To see this, refer to the equation immediately following Figure 2.2.2, which indicates that the only terms in the AM signal are at frequencies ω_c (the carrier frequency) and $\omega_c \pm \omega_m$ (the sidebands). There is no term at frequency ω_m. The *envelope* of the received signal, however, does vary at frequency ω_m, and so Fleming's circuit of Figure 2.4.5 is called an *envelope detector*.

Everything at this point hinges on the *time constant* of the product of the values of the *C* and the *R* (of the headset phones).[24] What this means is we want the capacitor voltage (the input to the headphones) to "follow the envelope" but *not* to follow the carrier or sideband frequencies. That is, the *RC* time product should be *long* compared with the carrier period but *short* compared with the modulation frequency period. The lowest AM carrier frequency (540 kHz) has a longest period of less than 2 μs, while the highest modulation frequency (5 kHz in AM broadcast radio) has a period of 200 μs. So, suppose we make the *RC* time product 20 μs (which is both more than 10 times longer than the longest carrier frequency period *and* 10 times shorter than the maximum modulation frequency period). Since a typical value for the resistance *R* of early radio headphones was around 2,000 ohms, we have $RC = 2 \times 10^3\, C = 20 \times 10^{-6} = 2 \times 10^{-5}$, and so $C = 10^{-8}$ F, or $C = 0.01\mu$F, a commonly available capacitor value.

This calculation needs to be modified if we replace the high-resistance headphones (which was okay for early radio buffs—see Figure 2.6.1—but not for a group of people listening together to a news broadcast or to a radio drama) using a receiver with a loudspeaker. Loudspeakers have a far smaller resistance (typically 8 or 16 ohms), and to directly replace the headphones with a loudspeaker would, to keep the *RC* time constant unchanged, require a correspondingly large increase in the value of *C*. One way to sidestep this is to use a transformer as an *impedance changer* (that is, to make the 8 ohms of a speaker coil "look like" the 2,000 ohms of the headphones).

24 When the capacitor *C* receives a burst of charge (thus raising its voltage), it then begins to discharge through the resistor *R* (it can't discharge back through the diode), with the capacitor voltage decreasing exponentially as $e^{-t/RC}$. Since the exponent must be dimensionless (do you see why?), we instantly know the product *RC* has the unit of time.

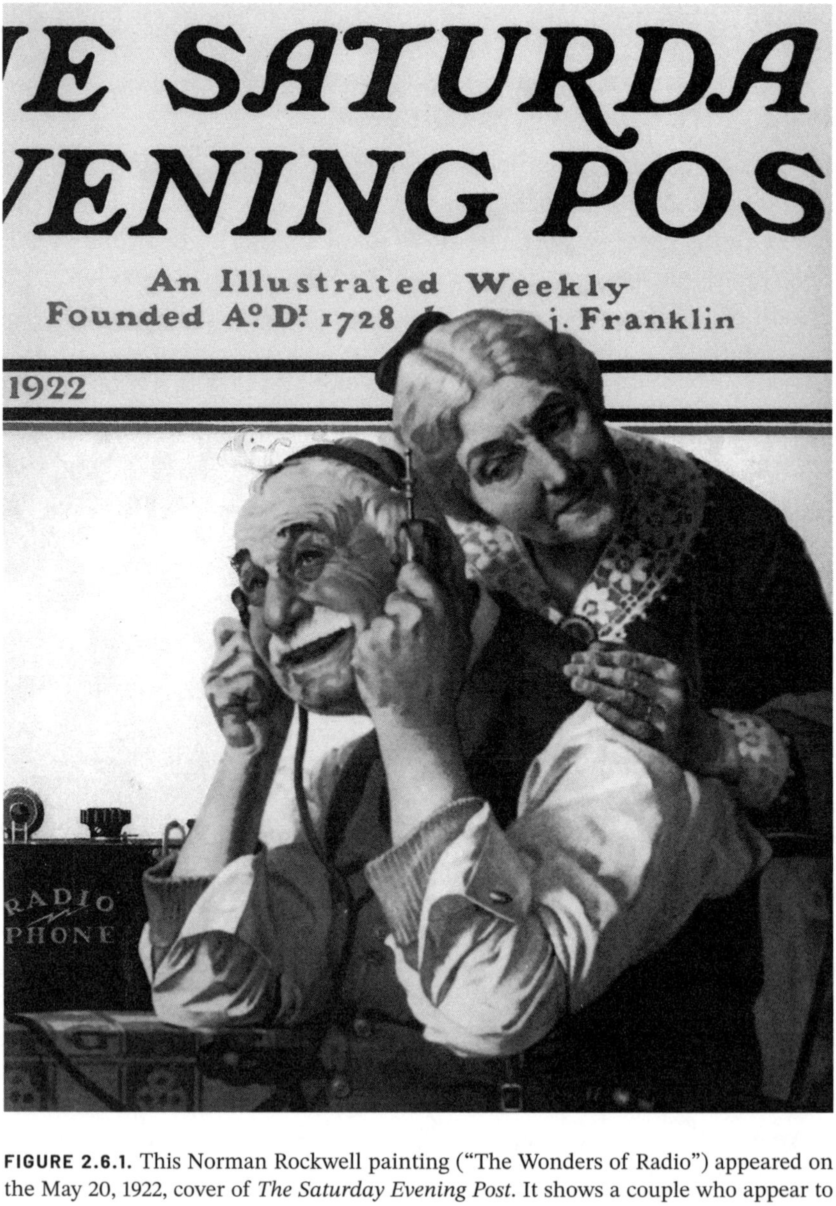

FIGURE 2.6.1. This Norman Rockwell painting ("The Wonders of Radio") appeared on the May 20, 1922, cover of *The Saturday Evening Post*. It shows a couple who appear to be in their late 70s or early 80s (and thus who were born long before the formulation of Maxwell's electromagnetic field equations) trying to listen to an opera program on a single set of headphones.

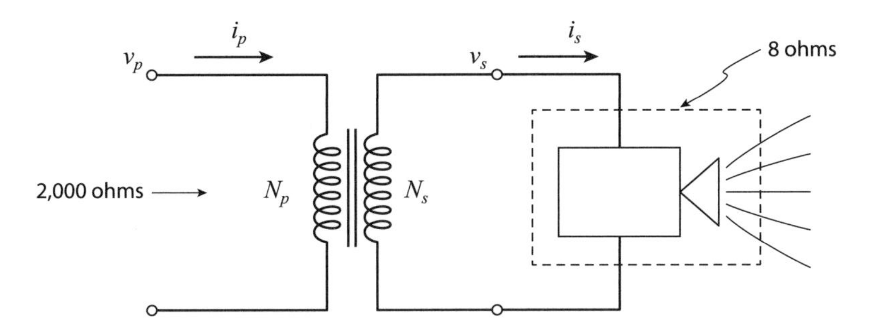

FIGURE 2.6.2. Changing impedance with a transformer.

In Figure 2.6.2 you see a transformer with a turns ratio of $n = N_s/N_p$ (refer to note 17). Putting a loudspeaker with an 8 ohm coil in the secondary, we claim there is a value of n such that the impedance "looking into" the primary appears to be 2,000 ohms. Here is how that works.

We have

$$v_s = \frac{N_s}{N_p} v_p = n v_p,$$

and

$$i_s = \frac{1}{n} i_p,$$

and so

$$\frac{i_s}{i_p} = \frac{1}{n}.$$

Now, we claim

$$v_p = 2{,}000 i_p,$$

and since

$$v_s = 8 i_s,$$

then

$$\frac{v_s}{v_p} = n = \frac{8 i_s}{2{,}000 i_p} = \frac{8}{2{,}000}\left(\frac{1}{n}\right).$$

Thus,

$$n^2 = \frac{8}{2{,}000},$$

or

$$n = \sqrt{\frac{8}{2{,}000}} = \frac{N_s}{N_p}.$$

So,

$$\frac{N_p}{N_s} = \sqrt{\frac{2{,}000}{8}} = \sqrt{250} \approx 16.$$

If we use a transformer with approximately 16 times more turns in the primary winding than in the secondary, then 8 ohms in the secondary appears as 2,000 ohms in the primary.

It should be noted that speaker use had to await the development of electronic amplifiers (that is, the invention of the triode vacuum tube[25]), because the raw output of the vacuum tube diode detector in Fleming's receiver was not sufficiently powerful. The invention of electronic amplification was of enormous importance for the commercial development of radio. Indeed, it was crucial. Before the triode, the only energy available to the listener was the energy intercepted by the antenna. This is not much energy, and so the need for headphones, as in the Rockwell painting (the Radio-Phone was a crystal[26] radio receiver, with no amplification). *With* amplification, however, the puny signal out of an antenna could be magnified by a truly heroic

25 The triode was the result of experiments done (less than two years after Fleming's diode) by the American Lee De Forest (1873–1961). De Forest was a trial-by-error tinkerer (some historians are even harsher in their judgment) who seems to have had a hazy (and that's being generous) understanding of what he had stumbled upon. I'll say a bit more at the start of chapter 3 about how triodes can be used to make amplifiers.

26 The crystal was a lump of galena (lead sulfide) or carborundum (silicon carbide), both of which have a crystalline structure, that played the role of Fleming's vacuum tube diode detector. The early radio experimenters didn't have the quantum-mechanical explanation we have today for a crystal's diode action, but by trial-and-error probing of a crystal's surface with a sharp-pointed wire (the so-called cat's whisker), one could eventually find a spot where diode action occurred. There is some irony in this, as it means the early crystal radios were *solid-state* gadgets, a foreshadowing of the modern solid-state transistor radio.

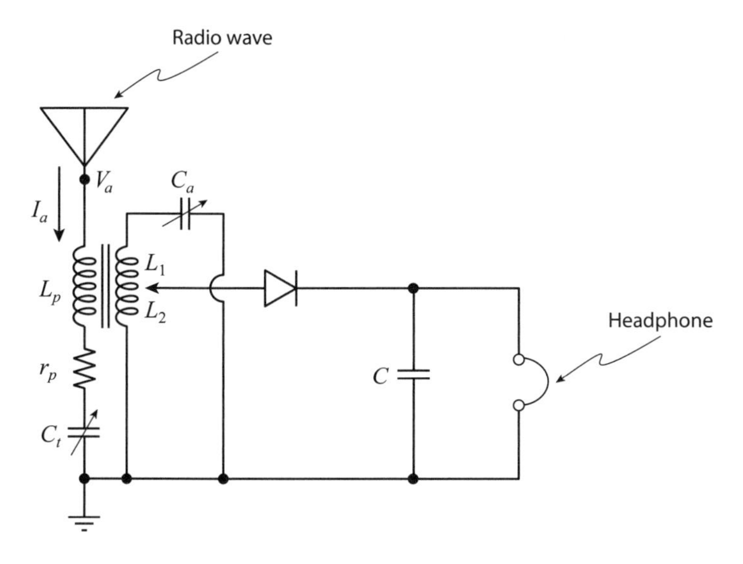

FIGURE 2.6.3. Modified version of Fleming's circuit of Figure 2.4.5.

factor, providing energy sufficient to power a loudspeaker that could entertain a roomful of listeners rather than just one with headphones clamped to the head. With a one-tube triode amplifier, a voltage gain of 30 was possible, and so three such stages of amplification could produce a gain factor of $30^3 = 27{,}000$.

To finish this chapter, consider Figure 2.6.3, which shows an elaboration of Fleming's circuit of Figure 2.4.5. With the addition of the variable capacitors C_t in the primary and C_a in the secondary, the claim is that we can *tune* the receiver to a specific frequency ω_0 and also that with the adjustable location of the tap on the secondary coil we can attempt to apply maximum power into the headphones.

Let's start our analysis with the primary side of the antenna coupling transformer, where L_p denotes the inductance of the primary coil. The resistance of that coil is represented by r_p, which is the actual ohmic resistance of the wire that forms the coil *plus* the effect of what is called the *skin effect*. This refers to the fact that, at radio frequencies, electrical current in a metal wire is *not* uniformly distributed over the wire's cross section but, rather, is almost totally confined to a thin region near the surface of the wire. In general, the higher the frequency the thinner is this "skin," and so r_p increases with increasing frequency.[27]

Our goal is to get the largest antenna current (the current I_a) in the primary as we can, for a given antenna voltage V_a, as that will result in the largest voltage drop across L_p. The idea is that this maximum primary coil voltage drop will then induce the maximum voltage in the secondary. By varying C_t, we can have the maximum voltage drop across L_p occur at the frequency of the radio station we wish to "tune in." Here's how we do that. We start by writing the ac impedance in the primary, at any frequency ω, as

$$Z_p = r_p + j\omega L_p + \frac{1}{j\omega C_t},$$

and so

$$Z_p = r_p + \frac{1-\omega^2 L_p C_t}{j\omega C_t}.$$

The magnitude of $Z_p(\omega)$ is minimized at the value r_p if the frequency is $\omega = \omega_0 = \frac{1}{\sqrt{L_p C_t}}$, called the *resonant frequency* (the frequency of the signal we wish to tune in). If the voltage powering this series circuit is $V_a(\omega)$—the antenna voltage due to an incoming radio wave—then *at resonance* the identical current in all three series elements is $I_a(\omega_0) = \frac{V_a(\omega_0)}{r_p}$. Thus, the voltage drops across each of the three individual elements, *at resonance*, are

$$\text{voltage drop across } r_p \text{ is } I_a(\omega_0) r_p = V_a(\omega_0),$$

$$\text{voltage drop across } L_p \text{ is } I_a(\omega_0) j\omega_0 L_p = \left(\frac{V_a(\omega_0)}{r_p}\right) j\omega_0 L_p$$

$$= j\left(\frac{\omega_0 L_p}{r_p}\right) V_a(\omega_0),$$

27 This phenomenon was first commented on by Maxwell in his famous 1873 *Treatise*, and you can find more discussion on the physics and history of the skin effect in my book *Oliver Heaviside*, The Johns Hopkins University Press, 2002.

voltage drop across C_t is $I_a(\omega_0)\dfrac{1}{j\omega_0 C_t}$

$$=\left(\frac{V_a(\omega_0)}{r_p}\right)\frac{1}{j\omega_0\dfrac{1}{\omega_0^2 L_p}}=\left(\frac{V_a(\omega_0)}{r_p}\right)(-j\omega_0 L_p)=-j\left(\frac{\omega_0 L_p}{r_p}\right)V_a(\omega_0).$$

Thus, the three individual voltage drops add to $V_a(\omega_0)$, in accordance with Kirchhoff's voltage law. Notice, however, that the drops across L_p and C_t, *at resonance*, are equal in magnitude but opposite in sign, and so together those two drops add to zero. Individually, however, they can be quite different from the voltage drop across r_p. If the factor $\dfrac{\omega_0 L_p}{r_p}$ is called Q, then, in particular, the drop across L_p is, at resonance, Q times the voltage drop across r_p.

To fully understand what this means, it's a great help to have a physical interpretation of Q. The current in the primary at *any* frequency is

$$I_a(\omega)=\frac{V_a(\omega)}{Z_p(\omega)}=\frac{V_a(\omega)}{r_p+j\omega L_p+\dfrac{1}{j\omega C_t}}=\frac{V_a(\omega)}{r_p+j\left(\omega L_p-\dfrac{1}{\omega C_t}\right)},$$

and so the magnitude of the primary current is (at *any* frequency)

$$I_{\text{amag}}=\frac{V_{\text{amag}}}{\sqrt{r_p^2+\left(\omega L_p-\dfrac{1}{\omega C_t}\right)^2}},$$

where V_{amag} is the *magnitude* of the antenna voltage $V_a(\omega)$. Now, *at resonance* ($\omega=\omega_0$) this magnitude is maximum (which I'll write as I_0), given by

$$I_0=\frac{V_{\text{amag}}}{r_p},$$

and so, at *any* frequency, we can write

$$I_{\text{amag}} = \frac{I_0 r_p}{\sqrt{r_p^2 + \left(\omega L_p - \dfrac{1}{\omega C_t}\right)^2}}.$$

Now, since $\omega_0^2 = \dfrac{1}{L_p C_t}$ we have

$$\frac{I_{\text{amag}}}{I_0} = \frac{r_p}{\sqrt{r_p^2 + \left(\dfrac{\omega^2 L_p C_t - 1}{\omega C_t}\right)^2}} = \frac{1}{\sqrt{1 + \dfrac{1}{r_p^2}\left(\dfrac{\omega^2 \dfrac{1}{\omega_0^2} - 1}{\omega \dfrac{1}{\omega_0^2 L_p}}\right)^2}}$$

$$= \frac{1}{\sqrt{1 + \dfrac{1}{r_p^2}\left(\dfrac{\omega^2 - \omega_0^2}{\dfrac{\omega}{L_p}}\right)^2}} = \frac{1}{\sqrt{1 + \dfrac{1}{r_p^2}\left(\dfrac{L_p}{\omega}(\omega^2 - \omega_0^2)\right)^2}}$$

$$= \frac{1}{\sqrt{1 + \dfrac{1}{r_p^2}\left(\dfrac{L_p}{\omega}\omega^2\left(1 - \dfrac{\omega_0^2}{\omega^2}\right)\right)^2}} = \frac{1}{\sqrt{1 + \dfrac{1}{r_p^2}\left(\omega L_p\left(1 - \dfrac{\omega_0^2}{\omega^2}\right)\right)^2}}$$

$$= \frac{1}{\sqrt{1 + \left(\dfrac{\omega^2 L_p^2}{r_p^2}\right)\left[1 - \left(\dfrac{\omega_0}{\omega}\right)^2\right]^2}} = \frac{1}{\sqrt{1 + \dfrac{\omega^2}{\omega_0^2}\left(\dfrac{\omega^2 L_p^2}{r_p^2}\right)\left[1 - \left(\dfrac{\omega_0}{\omega}\right)^2\right]^2}},$$

or, as $Q = \dfrac{\omega_0 L_p}{r_p}$, we have

$$\frac{I_{\text{amag}}}{I_0} = \frac{1}{\sqrt{1 + \left(\dfrac{\omega}{\omega_0}\right)^2 Q^2 \left[1 - \dfrac{1}{\left(\dfrac{\omega}{\omega_0}\right)^2}\right]^2}}.$$

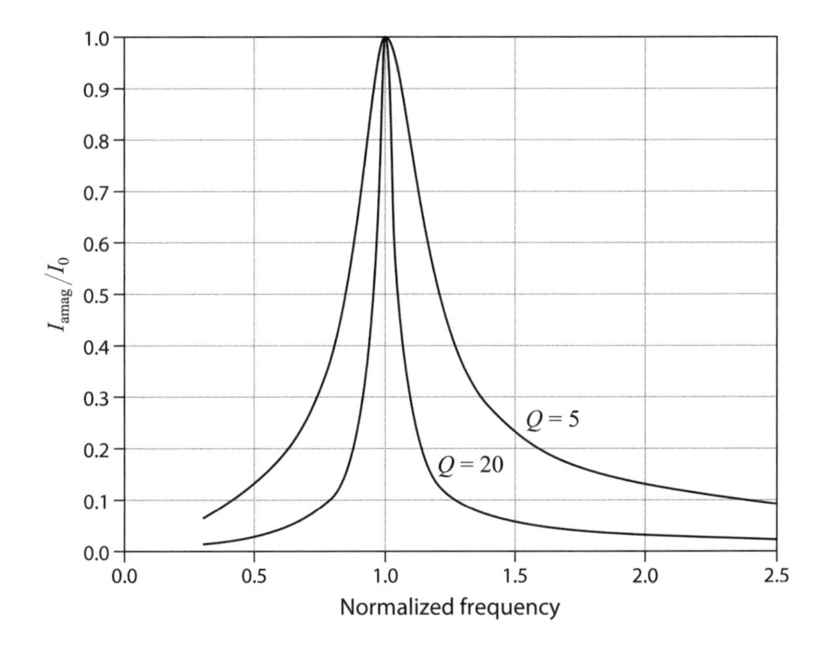

FIGURE 2.6.4. The influence of Q on the sharpness of resonance (the normalized frequency of 1 is $\omega=\omega_0$).

Figure 2.6.4 shows $\frac{I_{amag}}{I_0}$ for two values of Q, and it is clear that the *sharpness* of resonance in the primary (the selectivity of the tuning as we vary C_t with a control knob) increases with increasing Q.

Let's now move our attention to the secondary of the antenna coupling transformer. If we call the total inductance of the secondary coil L, and the coil's total resistance r, then the adjustable tap divides the secondary into two sections, an upper section consisting of L_1 in series with r_1, and a lower section consisting of L_2 in series with r_2, where $L_1+L_2=L$, and $r_1+r_2=r$. We start by writing the induced voltage in the secondary due to a received radio wave as E, distributed uniformly along the secondary coil (see Figure 2.6.5). Since the induced voltage is uniform along the coil, we put a voltage source of $E\frac{r_1}{r}$ in the upper half of the secondary coil and a voltage source of $E\frac{r_2}{r}$ in the lower half of the coil.

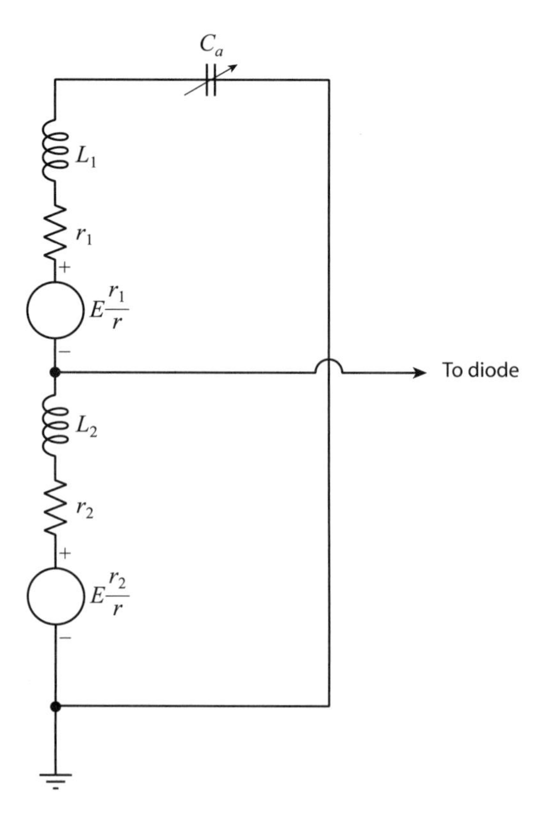

FIGURE 2.6.5. The adjustable tap divides the secondary coil of Figure 2.6.3.

To understand what happens as we adjust the knobs that control the values of C_a, L_1, and L_2 in Figure 2.6.5, it is necessary to consider what electrical engineers call the *maximum average power transfer theorem*. This theorem says[28] that if a sinusoidal voltage source is series connected to two impedances Z_1 and Z_2, one of which is fixed (let's say it's Z_2), then to maximize the average power delivered to Z_2 we must have $Z_1 = Z_2^*$. So, suppose Z_2 is the impedance of the headphones, and Z_1 is the impedance of everything in the receiver to the left of the headphones in Figure 2.6.3. Roughly speaking (you'll see why I say *roughly* in just a moment), we adjust the knobs until we *experimentally* get the loudest sound in the headphones at the selected frequency ω_0 that we adjusted C_a to achieve.

28 You can find proofs of this theorem in any good electrical engineering circuits text. See, for example, W. R. LePage and S. Seely, *General Network Analysis*, McGraw-Hill 1952, pp. 80–82 for derivations of several interesting generalizations of the theorem.

It should be clear that as we adjust C_a we vary the frequency at which the current in the secondary coil is maximized (notice that some of the secondary coil current is the diode current that (partially) ends up in the headphones. And as we adjust the tap location on the secondary coil, we vary the impedance "seen" looking back through the diode to the secondary coil.[29] In this way, we hope to achieve, as best we can, the condition for maximum power transfer to the headphones. The adjustments of C_t, C_a, and the secondary tap require the listener to twiddle multiple knobs back and forth to get the optimal balance between station selectivity and sound volume (see Figure 2.6.6). This was a common (and most undesirable) feature of early 1920s radios, which the development of the AM superheterodyne receiver circuit brilliantly resolved. And that's the subject of the next chapter.

During the Second World War radio technology was put to numerous clever uses, in applications that would have been evidence in Hardy's eyes, of the evils of technology. In the 1930s great advances had been made in developing what were called *radio beam riders* that allowed pilots to execute safe landings, even when weather reduced visibility to zero. (You can read about how such beams worked in Alfred Price's 1967 book *Instruments of Darkness*.) The Battle of Britain in 1940 showed that German scientists had adapted such radio direction guidance to allow bombers, flying from their bases in Germany (and later France), to pass right over London even when the ground was totally obscured by clouds. The question, however, was, *When* did a bomber arrive over London? That is, where *along* the guide beam was the bomber? To determine that location, a high-frequency carrier (around 45 MHz), modulated by a low-frequency (f) tone was transmitted from the bomber's base along the guide beam. The bomber, distance d from the base, carried a receiver that demodulated the high-frequency carrier to extract the tone. With the extracted tone, the bomber modulated its

29 As we adjust the position of the secondary coil tap, the effective induced voltage source seen by the diode also changes, which complicates the statement of the maximum average power transfer conditions, and that is why I used the word *roughly* in the text.

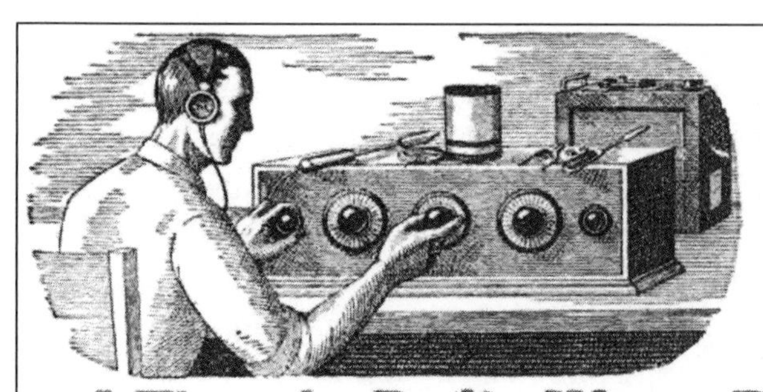

FIGURE 2.6.6. This illustration of a multidial, battery-powered (notice the *big* battery behind the receiver) radio with headphones appeared in numerous 1920s ads from the National Radio Institute correspondence school.

own transmitter (which operated at a slightly different carrier frequency) and sent the resulting signal back to its base. There the received signal was demodulated to extract the tone, which would, because of the round-trip time delay, show a phase shift with respect to the tone originally sent. This phase shift was a direct measure of distance d When the phase shift produced a value equal to the distance from the base to London, a command would be sent to the pilot to release his bombs. If c is the speed of light, the round-trip time delay is $2d/c$. The period of the modulating tone is $1/f$. Thus, the time delay is this fraction of a period: $\frac{\left(\frac{2d}{c}\right)}{\left(\frac{1}{f}\right)} = \frac{2df}{c}$. Since there are 2π radians of phase in a period, the resulting phase shift is $2\pi\left(\frac{2df}{c}\right) = \frac{4\pi df}{c}$. The maximum *unambiguous* distance D this system can handle is reached when the phase shift has just reached 2π. So, $\frac{4\pi Df}{c} = 2\pi$, or $f = \frac{c}{2D}$. If D is selected to be 500 km (about 310 miles), then $f = \frac{300{,}000\ km\,/\,second}{1{,}000\ km} = 300$ Hz. At a distance of 250 miles the range accuracy of the system was ±100 yards. The defeat of this system was one of the "battle of the beams" victories mentioned in note 13 of the preface.

Challenge Problem 2.1: Imagine that you have recently acquired a gadget that has two input terminals (one marked *plus* and the other *minus*) and one output terminal. The *plus* input voltage is v_1, the *minus* input voltage is v_2, and the output voltage is v_0. With A some constant, the gadget is defined by the equation $v_0 = A(v_1 - v_2)$. That is, the output voltage is A times the *differential input* $v_1 - v_2$ (and so the gadget is called a *differential amplifier with gain A*). Such gadgets were originally constructed using vacuum tubes,[30] and then later with transistors, and now with integrated circuits. Modern electronics simply couldn't exist without differential amplifiers. With all this in mind, consider Figure CP2.1a, which shows an antenna transformer-coupled into a differential amplifier, with the antenna voltage v_a from the transformer secondary as the *plus* input of the differential amplifier and the output voltage v_0 as the *minus* input. This is an example of what radio engineers call *negative feedback*.

From the defining equation of the differential amplifier we have, since $v_1 = v_a$ and $v_2 = v_0$, $v_0 = A(v_a - v_0)$, or $v_0 = \frac{A}{A+1} v_a$. So, if the differential amplifier is high gain (that is, if $A \to \infty$), we see that $\lim_{A \to \infty} v_0 = v_a$. This is the reason the circuit of Figure CP2.1a is called a *voltage follower* (the output v_0 *follows* the antenna signal v_a). What makes this so interesting is that as $A \to \infty$ we see that the differential input voltage goes to zero (because of the negative feedback). This means that even with a finite input impedance, the differential amplifier's input currents also go to zero. In other words, in the limit $A \to \infty$ the antenna needs to provide vanishing current (energy) to the differential amplifier,

30 De Forest's triode, the first vacuum tube that could amplify, was Fleming's diode with a third electrode added to the diode's two. Experimenters after De Forest added even more electrodes to make *tetrode* and *pentode* vacuum tubes, all with the common goal of making tubes that could amplify at ever-higher frequencies.

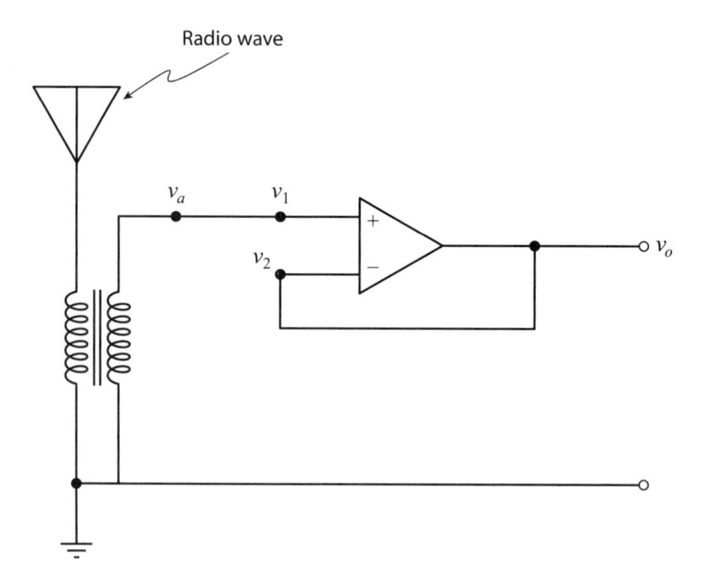

FIGURE CP2.1A. Using a high-gain differential amplifier *with negative feedback* to isolate a radio antenna from subsequent circuitry.

and this can be very important when one considers the *tiny* energy in the antenna due to a weak incoming radio signal. This circuit isolates the antenna because (in the limit of unbounded gain) the differential amplifier requires *zero* energy input. The energy required by any circuitry after the differential amplifier is supplied by the differential amplifier's power supply, under the control of the antenna voltage. In general, when we have a high-gain differential amplifier *with negative feedback present* we can assume (1) the differential input voltage is zero (the voltages at the *plus* and *minus* input terminals are the same, and (2) there is *zero input current* to the differential amplifier. The *output* current of the differential amplifier is a different matter: *that* current is, in general, not zero. Now, with that in mind, consider the circuit of Figure CP2.1b. Under the assumption of very high gain, that is $A \to \infty$, what is the transfer function of this circuit?

This is called a *Sallen-Key filter* (after the electrical engineers Roy Pines Sallen (1927–2018) and Edwin Lee Key (1926–2019) at MIT's Lincoln Laboratory, who published it in 1955). By properly choosing the four impedances, one can make low-pass filters (Z_1 and Z_2 resistors, and Z_3 and Z_4 capacitors), high-pass filters (Z_1 and Z_2 capacitors, and Z_3 and Z_4 resistors), and bandpass filters (Z_1 and Z_3 resistors, Z_2

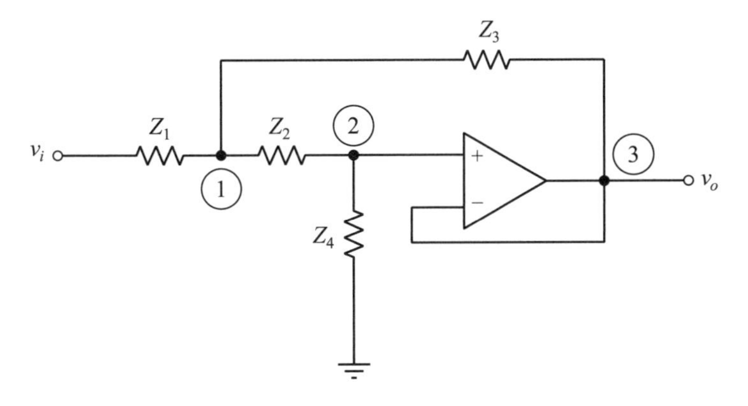

FIGURE CP2.1B. What does this circuit do?

a capacitor, and Z_4 the combination of a resistor in parallel with a capacitor). *Hint*: Apply Kirchhoff's current law at nodes ① and ②. Since the differential amplifier output current is unknown, do *not* attempt to sum currents at the output node ③.

Challenge Problem 2.2: If you were a new, young employee in a 1930s radio repair shop, you might be the target of the old joke of being asked to find a left-handed screwdriver to use in disassembling a radio cabinet. (If the setting was instead an auto repair shop, the request might be for a left-handed wrench.) Today, probably very few would fall for such a prank, but suppose you were asked, instead, for a negative capacitor? Would *that* deserve just a laugh and a sarcastic reply, too? *No*, because there actually is such a thing! Or at least there is if you have access to a high-gain differential amplifier. Specifically, consider the circuit of Figure CP2.2.

Show that $i = C_{eq}\dfrac{dv}{dt}$ the defining equation for a capacitor of value C_{eq}. Specifically, find an expression for C_{eq} in terms of R_1, R_2, and C (a "normal" *positive* capacitor), and observe that $C_{eq} < 0$. That is, the circuitry to the right of the vertical dashed line behaves like a negative capacitor.[31] *Hint*: Apply Kirchhoff's current law to nodes ①

31 One can only imagine how Professor Twombly (in the preface) would react if he did this challenge problem. Would he next try to make an *imaginary* capacitor? If he succeeded, would he warp himself into the fourth (or even the fifth) dimension? If so,

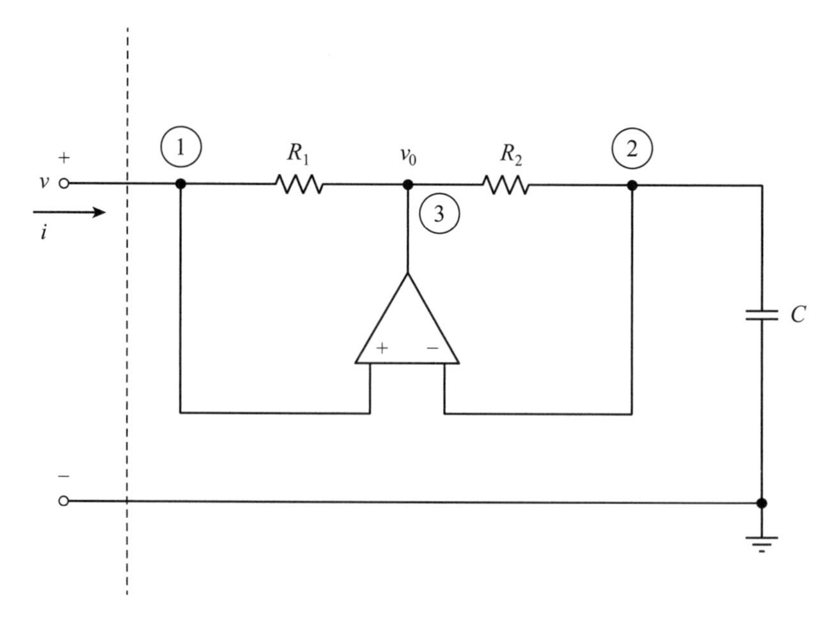

FIGURE CP2.2. A negative capacitor isn't a fantasy.

and ②—but *not*, of course, to node ③—and make use of the assumptions (because there is negative feedback present) that the differential input voltage to the amplifier is zero, as are the input currents to the amplifier.

Challenge Problem 2.3: Consider the circuit of Figure CP2.3. Define $\omega_0 = \frac{1}{\sqrt{LC}}$ and $Q = \frac{\omega_0 L}{R}$.

(a) Show that the impedance Z is purely real at frequency

$$\omega = \omega_0 \sqrt{1 - \frac{1}{Q^2}}.$$

(b) Show that the magnitude of Z is maximum at frequency

$$\omega = \omega_0 \sqrt{\sqrt{1 + \frac{2}{Q^2}} - \frac{1}{Q^2}}.$$

Notice that both these frequencies are less than ω_0.

---

could Professor Tweedle save him? For the electrifying answers, be sure to listen in to next week's exciting episode! (Read chapter 6—particularly note 6—for what *this* is all about.)

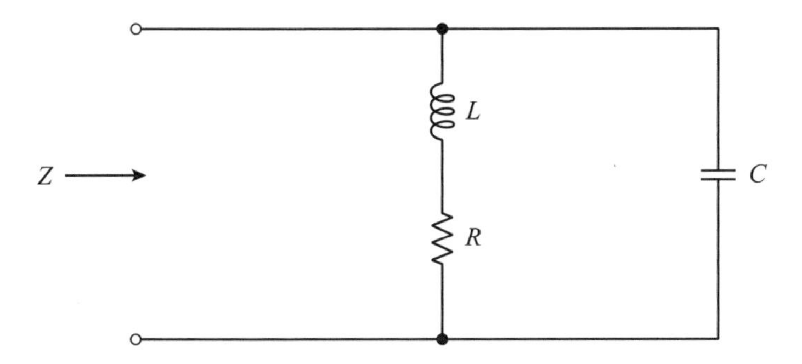

FIGURE CP2.3. Two questions involving Q.

Warning: These calculations have the potential to rapidly degenerate into algebraic nightmares, so *carefully* plan your analytic approach.

Challenge Problem 2.4: In note 21 I mentioned the annoying problem of 60 Hz ac hum making its way to the audio output of a pre-1927 radio powered by electrical outlets. One interesting way to eliminate that hum (even before the advent of indirectly heated vacuum tubes) would have been to use the filter shown in Figure CP2.4. Alas, this clever circuit wasn't invented until 1934 (by Herbert Augustadt (1906–1993), an electrical engineer at Bell Telephone Laboratories). Famous today as the *twin tee* or, alternatively, the *parallel tee* (for what I think an obvious reason), it is now discussed in all undergraduate textbooks on electric filters. Find the magnitude of the filter's transfer function, $\left|\frac{V_0(\omega)}{V_i(\omega)}\right|$, and then, using the definition $\omega_0 = 2\pi f_0 = \frac{1}{RC}$, confirm that a semilog plot of the transfer function magnitude for $0.01 \leq \left(\frac{\omega}{\omega_0}\right) \leq 100$ looks like Figure CP2.4b. If you set $f_0 = 60$ Hz, do you see why this circuit (sometimes also called a *notch* filter) is so famous? *Hint*: Sum currents at nodes ①, ②, and ③.

Challenge Problem 2.5: You'll recall that at the end of section 2.1, in the box "A Puzzling Historical Mistake," I told you of the apparent lack of understanding of electrical sidebands by one well-known early radio engineer. I called this "puzzling" because, among physicists, the existence of sidebands in modulated *sound* waves had been known for decades before 1900. The US physicist Alfred Marshall Mayer

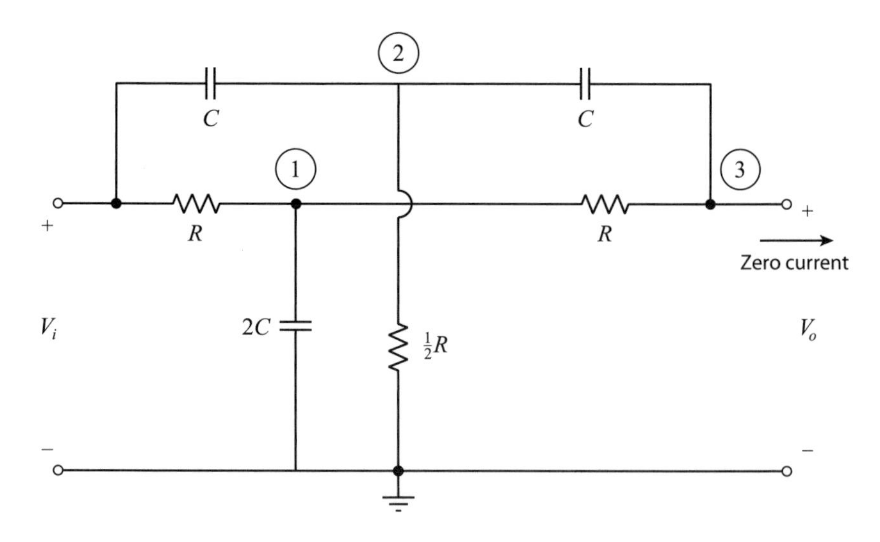

FIGURE CP2.4A. The twin-tee filter.

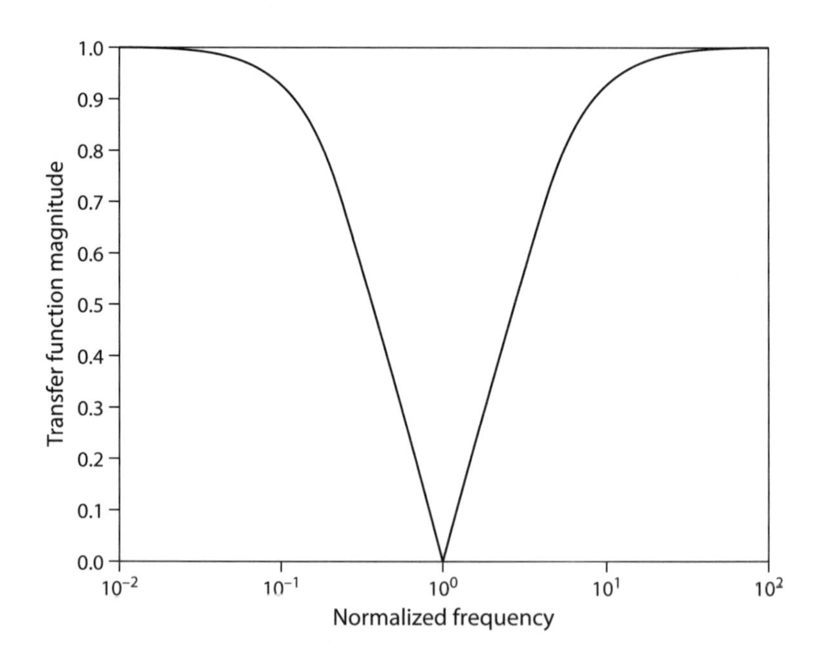

FIGURE CP2.4B. Can you explain this curve?

(1836–1897) had experimentally observed sound wave sidebands in 1875, and Lord Rayleigh (see note 36 in chapter 1) had provided the theoretical basis for acoustic sidebands in his famous 1894 book *The Theory of Sound*.[32] One of Rayleigh's analyses, for example, studied the modulation of $\cos(\omega_c t)$ by $\cos^4(\omega_m t)$. Follow in Rayleigh's footsteps and find the amplitudes of all the frequencies present in the signal $\cos^4(\omega_m t)$ $\cos(\omega_c t)$. *Hint*: Use Euler's identity and the binomial theorem.

Challenge Problem 2.6: If $m(t) \leftrightarrow M(\omega)$ and $g(t) \leftrightarrow G(\omega)$, derive the Fourier transform pair $\int_{-\infty}^{\infty} m(u)g(t-u)du \leftrightarrow M(\omega)G(\omega)$. That is, show the *time convolution theorem* that says $m(t) * g(t)$ and the spectrum product $M(\omega)G(\omega)$ form a transform pair. Use this result to show that a special case is $\delta(t) * g(t) = g(t)$. *Hint*: Review the derivation (at the start of section 2.3) of the pair $m(t)g(t) \leftrightarrow \frac{1}{2\pi} M(\omega) * G(\omega)$, and use the same line of reasoning. It may also help to recall (from section 1.7) the pair $\delta(t) \leftrightarrow 1$.

Challenge Problem 2.7: Figure CP2.7 shows what is called a *Colpitts oscillator* (after the electrical engineer Edwin Colpitts (1872–1949), who invented it in 1918), constructed around a high-gain differential amplifier. Unlike the *RC* phase-shift oscillator of Figure 1.5.2, the Colpitts oscillator uses what radio engineers call a *tank* (the inductor L and the two capacitors C_1 and C_2) in which electrical energy sloshes (like water in a tank, hence the name) back and forth between the magnetic field of the L and the electric fields of the C's.[33] Such an energy-exchange mechanism does *not* exist in the phase-shift oscillator. If the two points marked v_f in the figure are connected, thus completing the feedback

32 An understanding of sidebands, in the context of *telephone circuits and radio*, appears to have originated with the AT&T mathematician and electrical engineer John Renshaw Carson (1886–1940). Carson filed a patent application in December 1915 that gives a complete mathematical discussion of electrical sidebands, and he will appear again, prominently, when we get to SSB and FM radio. Carson wasn't alone in his understanding of electrical sidebands, however, and some interesting historical discussion can be found in Arthur A. Oswald, "Early History of Single-Sideband Transmission," *Proceedings of the IRE*, December 1956, pp. 1676–1679.

33 A similar tank oscillator circuit with the L replaced with a C, and the two C's replaced by two L's, is called a *Hartley oscillator*, after Ralph Vinton Lyon Hartley (1888–1970), who invented it in 1915. We will encounter Hartley again, in chapter 4, when we get to SSB radio.

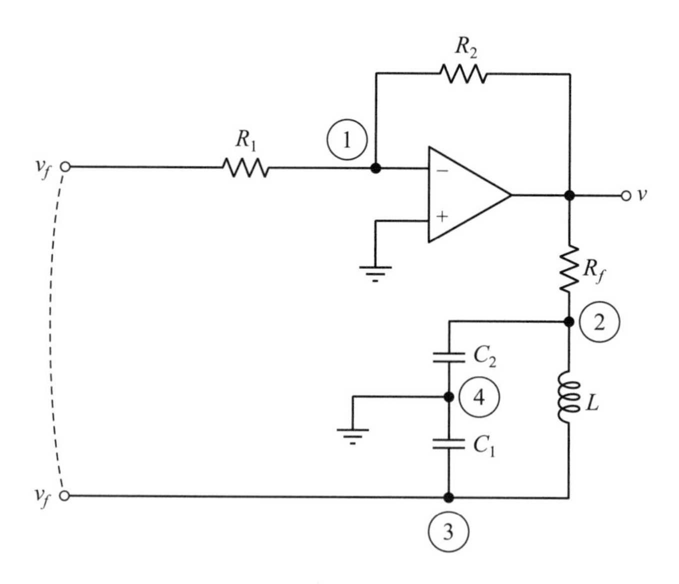

FIGURE CP2.7. Colpitts oscillator using a high-gain differential amplifier.

loop, calculate the frequency at which the circuit's output v will oscillate. If you look on the Web or in older textbooks for discussions of the Colpitts oscillator, you will usually find presentations saying the answer is $\frac{1}{\sqrt{LC}}$, where $C = \frac{C_1 C_2}{C_1 + C_2}$. *This is not completely correct*. The correct answer is $\sqrt{\frac{1}{LC} + a\ second\ term}$. What is that "second term"? *Hint*: Sum currents at ①, ②, and ③, and notice that you *cannot* sum currents at ④ because you don't know the value of the current flowing to ground (that is, the current in C_1 is *not* the current in C_2, even though the two capacitors "look like" they are in series—that ground connection makes a difference!) Summing at ① says $v_f = -\frac{R_1}{R_2} v$ (that is, v_f is v multiplied by a negative real number). Summing at ② and ③ must give a result that agrees with that condition.

Challenge Problem 2.8: Consider the circuit in Figure CP2.8a (called a *bridged-T* filter). Filters like this are used in a variety of radio circuits. With $\omega_0 = \frac{1}{RC}$, what is the transfer function $H(\omega)$ of this filter?

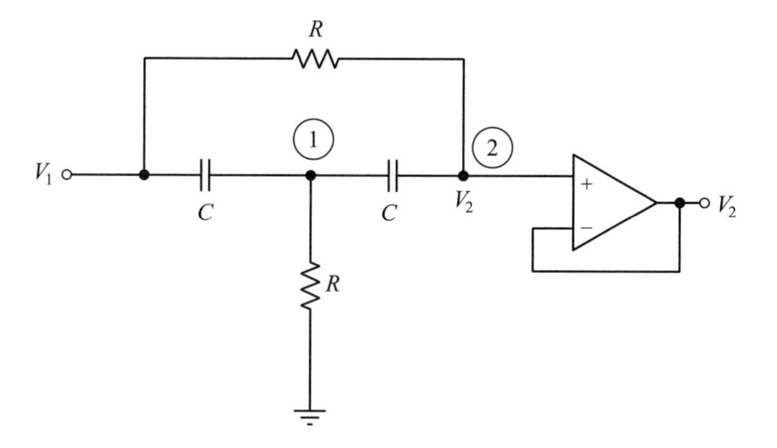

FIGURE CP2.8A. A bridged-T filter.

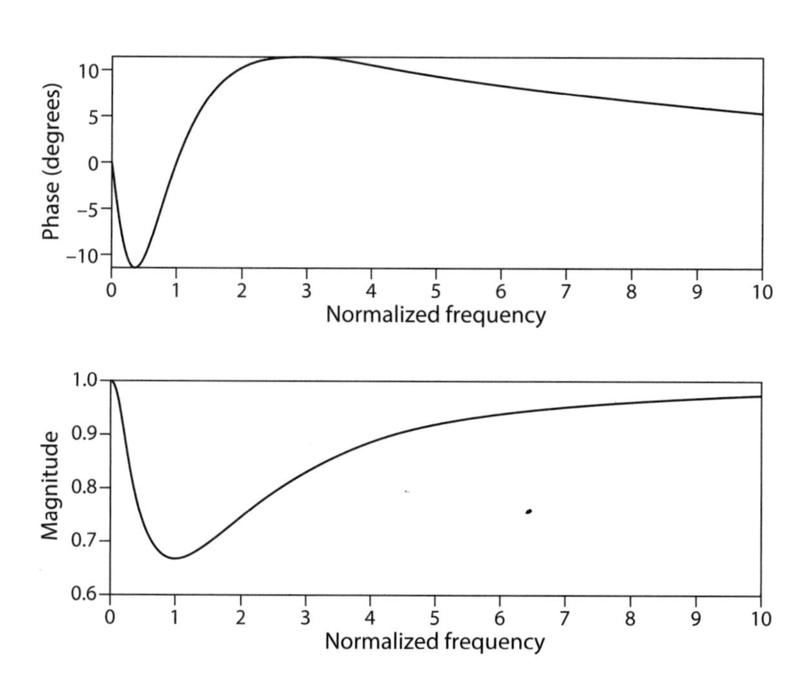

FIGURE CP2.8B. The phase function of the bridged-T filter in Figure CP2.8a.

Hint: Sum currents at ① and ② and use the fact that the current into the high-gain differential amplifier is zero (refer to Challenge Problem 2.1). The phase and magnitude functions associated with your $H(\omega)$ should look like the curves in Figure CP2.8b, where both plots are functions of the *normalized frequency* $\frac{\omega}{\omega_0}$.

Chapter 3

The AM Radio Receiver

> The superheterodyne[1] radio receiver was invented in the early 1920s [and its] design was so superior that within a decade it took over all but a few radios, and it is still today the basic design of all AM and FM radio receivers.
>
> —from J. J. Carr's erudite 1991 book *Old Time Radios! Restoration and Repair*

3.1 Prelude to the Superheterodyne

The circuit of Figure 2.6.3, the enhanced version of Fleming's radio-wave detector circuit of Figure 2.4.5 (featuring tuning and impedance matching) was the ultimate form of crystal radio. Such receivers were, in the early 1920s, gadgets of enormous astonishment. As one 80-year-old enthusiast put it in 1990, remembering back to when he was 13 years old in 1924,

> Living in the steel city of Youngstown, Ohio, I delivered *The Youngstown Vindicator* to fifty homes after school. A special weekly section in the paper reported the exciting developments in the new field of radio, including wiring diagrams for making one's own receiver. My paper delivery dollars made it possible for me to buy a crystal, an earphone, and the necessary wire. The primitive receiver I duly assembled picked up the messages from KDKA [in Pittsburgh, 70 miles from Youngstown] . . . what joy![2]

1 The *super* adjective was possibly the work of an imaginative advertising copy writer; it has no technical significance. What I like to imagine is that when a long-ago, unknown copywriter first heard a superheterodyne receiver, he reacted by leaping to his feet and exclaiming, "*Super*, by damn, that's a *super* heterodyne radio!" (to refresh your memory of what *heterodyning* means, refer to the opening section of chapter 2). For alternative speculations on the origin of *super*, see notes 14 and 27 in this chapter.

2 John Archibald Wheeler, *A Journey into Gravity and Spacetime*, Scientific American Library, 1990. Wheeler (1911–2008) soon left his paper delivery route far in the past and

Such reactions were common in the early 1920s, but, still, a crystal radio with headphones struck many as being more a plaything than anything else (refer to Figure 2.6.6). The central problem was that received antenna voltages are measured in the range of microvolts (or, to be generous, *maybe* millivolts) per meter of antenna length, and to make radio a business and not just a plaything, such weak signals needed to be amplified. Only then could radio be elevated from one radio/one listener with headphones to one radio/multiple listeners *without* headphones. Before 1906 there simply was no way to do that. Then, almost overnight, everything changed with De Forest's invention of the triode vacuum tube (he called it the *audion*), with its amazing ability to amplify electrical signals (see notes 25 and 30 in chapter 2). It is not hyperbole to declare the triode to be one of the great inventions in history—*it started the age of electronics!*—and so let's take a few moments to get an idea of how it works.

In Fleming's diode (the enhanced Edison-effect tube of Figure 2.4.4) the heated filament (the *cathode*) and the probe P (the *anode*, usually called the *plate* by radio engineers) were joined by a third electrode placed between the cathode and the plate, called the *grid* (because it is often in the form of a fine-meshed screen, through which the space-charge electrons that *are* the cathode/plate current flow).[3] Unlike the plate, which is operated at a *positive* voltage with respect to the cathode (often several hundreds of volts), the grid is operated at a *negative* voltage with respect to the cathode.[4] Since the

---

eventually became a world-famous physics professor at Princeton University, but he never forgot the thrill of his first radio.

3 I'm writing in the present tense because vacuum tubes are still manufactured today (not in America, perhaps, but certainly in China, Russia, and the Czech and Slovak Republics). Vacuum tubes are still able to outperform solid-state devices in applications involving high power (megawatts) and/or high frequencies (gigahertz). Vacuum tubes are better able, too, to survive in physically hostile environments involving high nuclear radiation levels that would destroy the crystal structure integrity of a solid-state device (atomic power plants), or the electromagnetic pulses (EMPs) generated by nuclear explosions that could disable critical infrastructures such as electric power grids. In a nuclear war, it is almost certain your car radio, with all its solid-state chips cremated by EMPs, will no longer work—but old-time radios with vacuum tubes *will*!

4 This is the usual case for an amplifier, but in more sophisticated vacuum tube circuits the grid was sometimes operated at positive voltages with respect to the cathode, causing electrons to be attracted to the grid. The accumulating negative charge on the grid could eventually become large enough to cut off the tube current. To prevent that,

grid is physically closer to the cathode (and its space charge) than is the plate, the tube current is more responsive to changes in the grid-to-cathode voltage drop than it is to equal changes in the plate-to-cathode voltage drop. Indeed, if the grid-to-cathode voltage drop is made sufficiently negative, the tube current can be driven to zero, even with a quite large, positive plate voltage (the tube is then said to be *cut off*). This occurs because a negative grid repels the space-charge electrons back toward the space charge, rather than letting them through to continue onward to the plate.

Now, consider Figure 3.1.1, which shows a triode connected as an amplifier, along with the notation we will be using (notice, in particular, the use of k to denote the cathode rather than the perhaps more obvious c, which is done here simply to be consistent with tradition). The input signal (v_i) is applied to the grid, and the output signal (v_p) is taken from the plate. To start, imagine that the input voltage $v_i(t)$, the voltage we wish to amplify, is zero. The tube current $i_p(t)$ is then a steady (that is, dc) current, formed by electrons flowing from cathode to plate, and so i_p is shown as flowing from plate to cathode.[5] The current i_p in resistor R_L results in a voltage drop of i_pR_L, and so the plate voltage is, with respect to ground, $v_p = V - i_pR_L$, where V is the dc power supply voltage. In addition, the current i_p flows in the cathode resistor R_k, resulting in a voltage drop of i_pR_k; the cathode is therefore i_pR_k volts *above* ground voltage. Since $v_i = 0$, we see that the grid, at ground voltage (zero), is i_pR_k volts *less* than the cathode voltage, that is, the grid is *negative* with respect to the cathode. In radio jargon, the grid is *negatively biased* by the voltage drop across the cathode resistor.

If we write v_{gk} to denote the grid-to-cathode voltage drop, then, according to Kirchhoff's voltage law, the sum of the voltage drops around a closed loop is zero. Thus, if we start at the cathode and sum around a clockwise loop, we have

$$i_pR_k - v_i + v_{gk} = 0,$$

---

a large resistor would be connected from grid to ground, providing what was called the *grid-leak* discharge path.

5 Recall that electrons are negative-charge carriers and so, since the "conventional current" is imagined as the motion of positive-charge carriers, i_p flows in the direction opposite the electron flow.

or

$$v_{gk} = v_i - i_p R_k.$$

This is a general relationship that holds for *any* triode. To continue our analysis, we need to next describe the particular behavior of whatever specific tube we use, and the following describes an actual triode[6] once in popular use: with a plate-to-cathode voltage drop (v_{pk}) of 250 volts, the tube current (i_p) for the 12AX7 is, over a wide interval of negative values for the grid-to-cathode voltage drop (v_{gk}), given by $i_p = (6 + 2.5v_{gk}) \times 10^{-3}$ A (ampere). (Recall that $v_{gk} < 0$, *always*, for our amplifier.) For example, if $v_{gk} = -0.8$ volts, then $i_p = 4$ mA (milliamperes), and if $v_{gk} = -2$ volts, then $i_p = 1$ mA.

So, suppose we decide to operate the tube so that when no input signal v_i is present, we have $v_{gk} = -1.5$ volts, which means $i_p = 2.25$ mA. The $v_i = 0$ case is called the *quiescent point*. Since v_{gk} is, for $v_i = 0$, just the voltage drop across R_k, we have

$$R_k = \frac{1.5}{2.25 \times 10^{-3}} = 667 \text{ ohms}.$$

Since the cathode is 1.5 volts above ground voltage, and since we want a 250 volt drop across the tube, the plate voltage must be 251.5 volts. That means if we (arbitrarily) pick the dc power supply voltage to be $V = 300$ volts, there must be a 48.5 volt drop across R_L when the current in R_L is 2.25 mA. Thus,

$$R_k = \frac{48.5}{2.25 \times 10^{-3}} = 21{,}556 \text{ ohms}.$$

Now, we want to calculate how much the plate voltage (which is where the signal v_p is generated) *changes* in response to a *change* in v_i. That is, in mathematical terms that Hardy would appreciate, what is the *voltage gain*

6 The 12AX7 was first manufactured by RCA in 1946 and is still available today. The 12AX7 is an indirectly heated *dual* triode, with two identical but independent triode structures contained within a common glass envelope (see Challenge Problem 3.1).

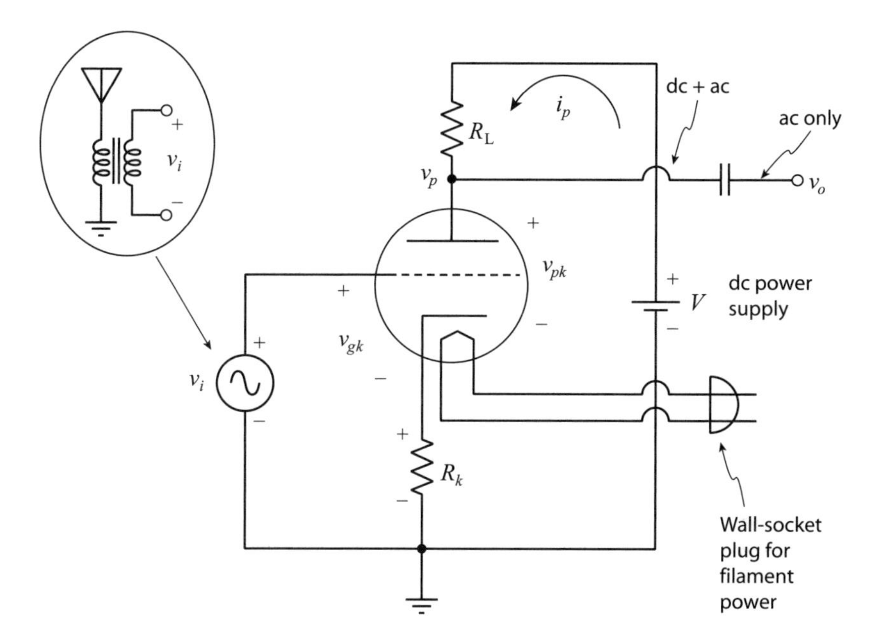

FIGURE 3.1.1. The triode operated as an amplifier (the input signal v_i is the signal in the secondary of a transformer with its primary connected to the antenna).

$$\frac{dv_p}{dv_i} = ?$$

To summarize, the equations that describe the circuit of Figure 3.3.1 are

(a) $v_{gk} = v_i - i_p R_k$,
(b) $i_p = (6 + 2.5 v_{gk}) \times 10^{-3}$,
(c) $v_p = 300 - i_p R_L$.

Substituting (a) into (b) gives

$$i_p = [6 + 2.5\,(v_i - i_p R_k)] \times 10^{-3},$$

or

$$1{,}000 i_p = 6 + 2.5 v_i - i_p\, 2.5 R_k,$$

and so

$$i_p = \frac{6+2.5v_i}{1{,}000+2.5R_k}.$$

Substituting this last result into (c), we have

$$v_p = 300 - \frac{6+2.5v_i}{1{,}000+2.5R_k}R_L,$$

which says

$$\frac{dv_p}{dv_i} = -2.5\frac{R_L}{1{,}000+2.5R_k} = -2.5\frac{21{,}556}{1{,}000+2.5(667)}$$
$$= -2.5\frac{21{,}556}{2{,}667} = -20.2.$$

The minus sign indicates that there is a 180° phase shift from input at the grid to the plate (the input voltage change and the plate voltage change are always in opposite directions). Finally, v_p is a dc value with an ac signal superimposed on it. To extract just the ac part (the amplified version of v_i), we use a capacitor connected to the plate to give the output v_0 (recall that the impedance of a capacitor to dc is infinite).

With the availability of simple amplifiers such as the one we just analyzed, the performance of crystal radio receivers was greatly enhanced, allowing headphone reception of very weak signals from distant stations. Some of the commercially sold crystal radios actually incorporated one or even two vacuum tubes (a Fleming diode instead of a solid-state crystal for detection, and perhaps a triode for some amplification) but, *at heart*, they were still crystal radios. The next real step beyond crystal radio was taken in 1912 by Edwin Howard Armstrong (1890–1954) while still an undergraduate student in electrical engineering at Columbia University.

Experimenting with a De Forest audion, Armstrong "stumbled" (his characterization) across the *regenerative receiver*. It used *positive feedback* to vastly increase the amplification of the receiver, by returning some of the energy in the plate circuit back to the input at the

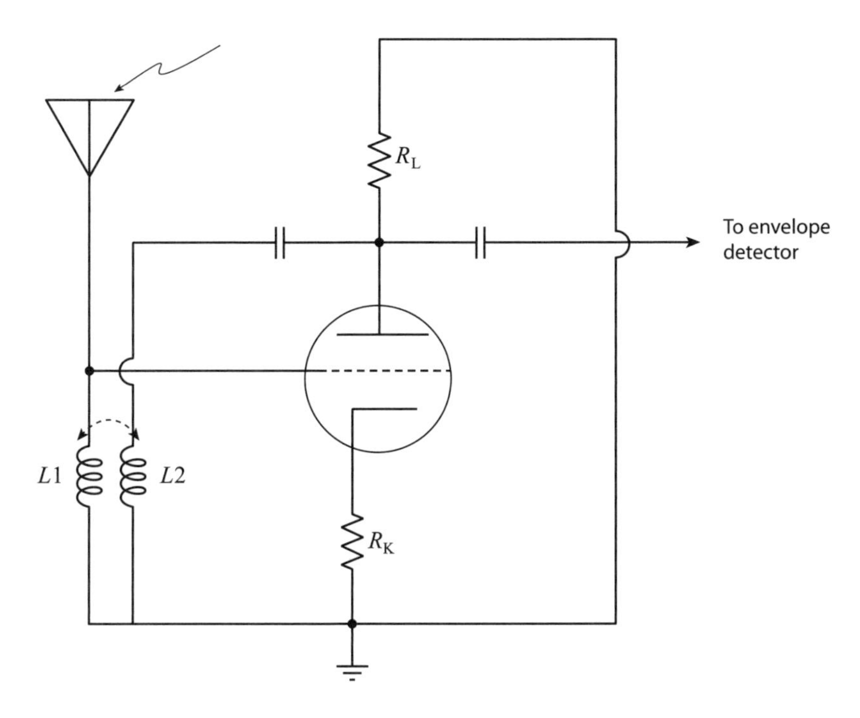

FIGURE 3.1.2. Radio frequency amplifier with positive feedback.

grid. By a *very* delicate adjustment of the feedback path, Armstrong was able in this way to achieve enormous amplification at radio frequencies (even the crude audion triode would work at frequencies as high as several hundred kilohertz). Figure 3.1.2 shows one way to do this. The voltage developed across the antenna coil L1 is the input signal to the grid of the triode amplifier. The amplified ac voltage developed at the plate is fed to coil L2 (called the "tickler coil"), which is magnetically coupled to the antenna coil L1. L2 is located inside L1 and can be rotated (via a control panel knob) to vary the coupling (the amount of feedback).[7]

The positive-feedback adjustment was a delicate one, because it was quite easy to go too far with the feedback and suddenly transform

7 In radios sold in the 1920s using regeneration, the tickler coil was rotated with the ingenious *variometer*, which was a beautiful piece of mechanical engineering. You can find discussion (and photos) of variometers in David Rutland, *Behind the Control Panel: The design & Development of 1920's Radios*, Wren, 1994.

the entire circuit from a receiver into an *oscillator*.[8] This was actually, however, a second huge advance by Armstrong, as he had discovered how to easily generate *radio-frequency* oscillations with just a single small triode vacuum tube, as opposed to existing (and very cumbersome) approaches: the technologies of spark, arc, and ac alternators, which we have not really said much about in this book, because they belong to the "obsolete past" of radio. Recall (from the preface) that Einstein was particularly impressed with this ability of the triode to oscillate, enough so to mention it in his 1930 speech at that year's Berlin Radio Exhibition.

De Forest later falsely claimed that it was he, rather than Armstrong, who was the discoverer of both the regenerative receiver and the triode oscillator. The legal dispute between the two men went all the way to the Supreme Court, where, in 1934, despite De Forest's well-known fundamental ignorance of how a triode even works, the decision went in his favor—to the complete astonishment of all radio engineers who knew who did what, when.[9] Professor Tucker accurately described the reaction of the radio engineering community to the Court's action when he wrote:

> It is interesting to speculate on the soundness of the principles of law which led to the award of the priority (and hence profits) of invention to [D]e Forest who did not "invent" feedback (although his assistants may have done so in a limited sense[10]) and who did not understand, exploit or patent the principle

8 The possibility of oscillatory behavior in an electrical system, in which a portion of the output is returned to the input, had been known long before such behavior was observed in radio. For example, it had routinely been discovered, over and over, that when a telephone receiver and its microphone were placed near each other, a hard-to-miss howl could result.

9 If this seems to be a rather harsh judgment of De Forest, you might find interesting reading in the paper by D. G. Tucker, "The History of Positive Feedback: the oscillating audion, the regenerative receiver, and other applications up to 1923," *Radio and Electronic Engineer*, February 1972, pp. 69–80. Two lines in Professor Tucker's paper speak particularly to my point: "In contrast to Armstrong's very professional and scientific approach to radio, De Forest appears almost as a fumbling amateur. In his patent specifications as in his published papers, he shows little understanding of what he is doing."

10 It isn't at all clear just how much "hands-on, at the workbench" time De Forest actually put in. The real work was done by his assistants Charles Veyne Logwood (1881–1927) and Herbert Briant Van Etten (1882–1958).

> until long after several others did understand, did exploit and did patent it.

It is difficult to avoid concluding that De Forest was, *at best*, an ethically challenged and shameless self-promoter far more than he was a radio engineer.[11]

With the regenerative receiver the trick was to take the circuit up to *almost* the point of oscillation but not *to* oscillation. In 1922 Armstrong developed what he called the *super*-regenerative receiver, which purposely *did* take the circuit even closer to full-spirited oscillation than did the regenerative receiver. To quote Professor Tucker (note 9):

> This . . . is achieved by using an auxiliary circuit . . . that as soon as the oscillation starts to build up it is quenched by a brief alteration of a parameter of the circuit, [oscillations then start] to build up again, and so on. If the quenching is caused to occur at intervals which are too close to cause any *audible* [my emphasis] modulation of the output audio signal, but which are sufficiently spaced in relation to the cycles of the radio signal[12] then the envelope of the oscillation waveform reproduces the modulation (audio) signal with extremely high amplification and usually acceptable distortion.

To put all that in mechanical terms (perhaps just a bit melodramatically), imagine driving your car right up to the edge of a cliff but

11 De Forest was tireless in proclaiming himself to be the Father of Radio, and he used those very words as the title for his 1950 autobiography. A second book (supposedly written by his fourth wife) had the title *I Married a Genius*, but it failed to find a publisher. The early days of radio involved numerous talented people, and it isn't surprising that many of the important discoveries were almost simultaneously made by multiple people. And so, of course, patent fights were a routine occurrence. Generally, however the courts might rule, both sides often presented compelling cases, but a glaring exception appears (to me, in any case) with anything involving De Forest, who was quick to claim the work of others as his own, even when it was soon clear he had little understanding of the matter at hand. De Forest, himself, was challenged on his audion claims by Fleming, who thought his vacuum diode gave him priority in anything involving a vacuum tube, and it seems in turn that Edison had an unhappy opinion of Fleming, thinking that the vacuum diode was simply a trivial extension of the Edison effect.

12 A quenching frequency of 20 kHz was typical, well above audible frequencies and yet well below radio frequencies.

coming to a complete stop just short of the front wheels going over the edge. That's the regenerative receiver. Then imagine driving up to the edge and letting the front wheels actually go halfway over the edge, followed (instantly!) by slamming the car into reverse, backing up a foot, then shifting back to drive to again go halfway over the edge, then back into reverse, . . . and so on. That's the *super*-regenerative receiver. Armstrong is generally thought to be the inventor of super-regeneration, but his 1922 patent actually came well *after* a 1919 patent granted to the British military radio engineer John Bruce Bolitho (1889–1965) for essentially the same invention. Unlike in his nasty dispute with De Forest over regeneration, however, Armstrong praised Bolitho's work, and the two men avoided getting into a legal dispute.[13]

Armstrong revealed his super-regenerative receiver in a 1922 paper,[14] and it illustrates, perfectly, Armstrong's thinking. *There is not a single equation in the paper.* It is, *in its entirety*, just text and circuit diagrams. Armstrong had a highly developed, intuitive "feeling" about how circuits work and could, somehow, simply "see" how stuff should be wired together. Oddly, this intuitive, math-free approach to radio might, contrary to a first reaction, have been appreciatively understood by pure mathematician Hardy. Such intuitive genius had, in fact, been one of the defining romantic characteristics of Hardy's famous friend the Indian math genius Ramanujan.[15]

This isn't to say that Armstrong didn't appreciate the power of mathematical analysis. At the end of his 1922 paper, for example, he wrote: "I wish to express my very great indebtedness to Professor L. A. Hazeltine for much valuable aid in connection with the theoretical side."[16] *What* theoretical side? a casual reader might have wondered, seeing only engineering circuit diagrams and words in Armstrong's paper. Actually, the opening of the paper explains this,

13 You can find a comprehensive description of what Bolitho did in "The Bolitho Circuit," *Wireless World and Radio Review*, June 2, 1923, pp. 266–267.

14 "Some Recent Developments of Regenerative Receivers," *Proceedings of the IRE*, August 1922, pp 244–60. It is in this paper that Armstrong explicitly called the circuit the *super*-regenerative receiver. So, in this case, *super* is not just ad hyperbole (refer to note 1) but, rather, was due to Armstrong. Perhaps the *super* in superheterodyne is his, too.

15 See the 1991 book *The Man Who Knew Infinity*, by Robert Kanigel, for the story of Hardy and Ramanujan.

16 Louis Alan Hazeltine (1886–1964) was professor of electrical engineering at Stevens Institute of Technology in New Jersey.

as well as illustrates what Professor Tucker meant when he wrote of the super-regenerative receiver's oscillations being "quenched by a brief alteration of a parameter of the circuit." As Armstrong wrote:

> It is well-known that the effect of regeneration (that is, the supplying of energy to a circuit to reinforce the oscillations existing therein) is equivalent to introducing a negative resistance reaction in the circuit, which neutralizes positive resistance reaction, and thereby reduces the effective resistance of the circuit.

You'll recall we saw the role of negative resistance in oscillators in chapter 1, in the discussion of the electric arc (section 1.3).

What Armstrong was referring to with the words "well-known" was an earlier paper[17] by Hazeltine, which, in contrast with Armstrong's writing, is packed with equations that established Armstrong's assertion. Armstrong's "auxiliary circuit" (to use Professor Tucker's words) that quenched the oscillations in the super-regenerative circuit was a second oscillator that periodically (see note 12 again) introduced enough positive resistance to overcome the negative resistance of the rf oscillator (and so suppress its oscillations).

Regenerative receivers were sold throughout the 1920s, but their operation was plagued by several serious problems. Most irritating was their tendency to cross the line from high-gain amplifier to oscillator. Then the receiver became a transmitter, and it would emit loud howls and shrieks that bothered not only the owner of the radio but also all the owners of nearby receivers. It worked both ways, of course. Even if a receiver provided good listening, that could quickly end when a neighbor turned on *their* radio, and it started to misbehave.

Somewhat less irritating (but not by much) was the wandering nature of the tuning. A favorite station would not consistently appear at the same spot on the tuning dial, because the regenerative tickler feedback loop was quite sensitive to multiple variables (power supply voltage, the extent of the magnetic coupling in the tickler feedback path, and the temperature). In addition, the art of achieving a good (so-called hard) vacuum in a tube was still in its early stages, and so tubes often contained enough residual gas to significantly influence

17 "Oscillating Audion Circuits," *Proceedings of the IRE*, April 1918, pp. 63–97.

their characteristics. (This was one of the glaring faults in De Forest's misunderstanding of his own audion—he thought the presence of some gas was *necessary* for triode operation.) So, each tube of any particular model could be quite different from others of the same model. That meant if a tube in a regenerative receiver burned out, its replacement virtually guaranteed that all the station dial settings would change.

An alternative to using positive feedback to achieve high gain at radio frequencies is to simply cascade several amplifier stages, each of relatively low gain, with each amplifier stage independently tuned to the frequency of interest. This eliminated the problem of feedback path variations, thereby mostly (but not completely) avoiding the oscillation problem. The major problem with doing this, however, was that these *TRF* (tuned radio frequency) *receivers* presented their users with a proliferation of control panel knobs, one for each adjustable capacitor and inductor coil in each stage of amplification. It required a lot of patience to operate such radios.

And there was, alas, *still* an oscillation problem. This problem for TRF radios was due to a far less obvious reason than it was in regenerative receivers (the delicate nature of the tickler coil feedback path). While it isn't difficult to build a stable (that is, nonoscillating) amplifier that has good gain at any particular, *fixed* frequency, it is much more difficult to do so for a *tunable* amplifier. There are almost certain to be some frequencies in the tuning interval at which the amplifier will oscillate. The conditional stability of amplifiers wasn't theoretically understood until 1932, with the work of Harry Nyquist (1890–1976) at Bell Telephone Laboratories. Nyquist's work was motivated by the discovery of Harold Black (1898–1983), also at Bell Labs (where he invented the negative-feedback amplifier discussed in Challenge Problems 2.1 and 2.2), who found he could construct circuits that satisfied the Barkhausen criterion but, nevertheless, did *not* oscillate (see Figure 1.5.1 and note 29 in chapter 1).

The culprits behind the oscillation problem in TRF receivers were found to be the interelectrode capacitances of the amplifier tubes. Although measured in the micromicro (or *pico*farad (pF) range, these tiny capacitances could, at certain frequencies, form unwanted positive-feedback paths in one or more of the amplification stages and cause oscillation. One way around this came from Armstrong's friend

Louis Hazeltine (note 16), who made a fortune with his *neutrodyne* TRF radio, which used *negative* feedback to neutralize the instability of tunable rf amplifiers. This was accomplished by intentionally *adding* some capacitance to cancel the plate-to-grid electrode capacitance,[18] which probably strikes you as counterintuitive. But you've already seen a similar example of this ingenious trick—or you have if you did the second part of Challenge Problem 1.1.

If you refer to Figure CP1.1, imagine that C_1 and C_2 are not components actually connected across R_1 and R_2, respectively. Rather, they are what are called *parasitic* capacitances resulting from interaction of the resistors with the surrounding circuitry. At dc and "low" frequencies it is clear the circuit is simply a voltage divider, with $k = \frac{R_2}{R_1 + R_2}$. As the frequency increases, however, C_1 and C_2 tend to bypass their respective resistors, and k becomes a frequency-dependent parameter. An analysis of the circuit shows, however, that if $R_1C_1 = R_2C_2$, then k remains equal to $\frac{R_2}{R_1 + R_2}$ at *any* frequency. So, to neutralize the effects of C_1 and C_2, you simply determine which resistor has the smaller *RC* product and intentionally connect (in parallel with the resistor) enough additional capacitance to make the two *RC* products equal. That is, *adding more capacitance* can remove (neutralize) the effect of *all* the circuit capacitance!

Hazeltine and Wheeler's neutrodyne TRF radio was the "king" of TRF radios until 1927; that was the year that General Electric introduced the tetrode vacuum tube with its new, fourth electrode, called the *screen grid*, placed between the grid and the plate. The screen grid reduced the plate-to-grid feedback capacitance by a factor of a thousand, to the *femto*farad range (fF), to the point where Hazeltine/Wheeler neutralization was not needed to avoid amplifier oscillation. Second-generation TRF radios, using tetrodes, were sold into the early 1930s (RCA had bought Armstrong's superheterodyne patents and was the sole source of superheterodyne radios, refusing until 1930—and

18 The details of the neutrodyne circuitry are in Rutland's book (note 7), pp. 37–39. There is some evidence that Hazeltine was not the sole developer of this technique and that Harold Wheeler (1903–1996) was a major contributor. Wheeler was also the inventor (in 1926) of automatic gain control (AGC) to combat signal fading in AM radio receivers. AGC first appeared in radios in 1929, and by 1932 it was a standard feature.

then only under the threat of antitrust legal action—to license other manufacturers), but by 1932 the TRF radio (with all its annoying adjustment knobs) was pretty much obsolete.[19] The modern age of the inexpensive, easy-to-use superheterodyne radio had arrived.

3.2 The Superheterodyne

Despite that last sentence, the superheterodyne principle had been known long before 1924, the year when RCA began selling the Radiola AR-812 superheterodyne, but it was so expensive that it was restricted to a high-income customer base.[20] It took another decade before mass-market, middle-class families could afford a superheterodyne radio. Armstrong is generally credited with the superheterodyne, but in fact it was an invention that was "in the air" and whose time had come. There were others working along the same line as Armstrong, men such as Lucien Levy (1892–1965) in France, Walter Schottky (1886–1976) in Germany, Henry Round (1881–1966) in England, and John Carson[21] (see note 32 in chapter 2) in the US. With or without Armstrong, the superheterodyne was the next, inevitable step in the evolution of radio.

19 See Arthur P. Harrison, "The World vs. RCA: Circumventing the Superhet," *IEEE Spectrum*, February 1983, pp. 67–71. This article has close-up photographs of the tuning mechanisms (some fiendishly clever) of TRF radio receivers. While inferior to the superheterodyne, less expensive TRF radios nevertheless sold well throughout the 1920s. During the two years 1928/29, for example, the Atwater Kent Manufacturing Company (Philadelphia) was the world's largest producer of radio receivers (outselling even the far larger RCA); those years Atwater Kent sold a total of *2 million* of its Model 40, ac-powered TRF radio for $77 each (without tubes). If you didn't already have a speaker, that was an additional $20. Atwater Kent went out of business in 1936, when its owner, Arthur Atwater Kent (1873–1949) became angry because his employees attempted to unionize; his response was to simply shut the company down.

20 The AR-812 sold for $220 *without* its six triodes, and for $286 (equivalent to over $4,300 in 2021) with tubes. It was advertised as a "portable," but it weighed 30 pounds (the battery was extra in both weight and cost). Despite those challenging numbers, RCA sold 164,000 AR-812s.

21 Carson published a paper in 1919 (written in 1918) in which he specifically discussed the demodulation of an AM signal by heterodyning it with a "locally generated wave," which is (as you'll soon see) the heart and soul of the superheterodyne. See his "A Theoretical Analysis of the Three-Element Vacuum Tube," *Proceedings of the IRE*, February 1919, pp. 187–200.

During the First World War Armstrong served in France with the American Expeditionary Forces as an officer in the US Army Signal Corps. As a well-known radio expert, he was charged with the important task of intercepting German radio transmissions. Related to that primary task was the tantalizing possibility of perhaps developing an early warning system for detecting the approach of enemy aircraft (radar was years in the future) via the electromagnetic energy emitted by ignition systems (that is, the firings of combustion engine spark plugs). Both those efforts involved detecting signals at frequencies (from 500 kHz to 3 MHz) greater than the then-existing amplifiers were able to handle. It was that difficulty that motivated Armstrong to recall (no later than 1918) Fessenden's idea of heterodyning as a way to move megahertz energy down to the frequencies at which Armstrong's amplifiers *could* perform (an idea, you'll recall from note 21, that had virtually simultaneously occurred to Carson).

To aid your understanding of the operation of the AM superheterodyne receiver, it's essential to begin with a description of the AM radio spectrum. In commercial AM broadcast radio, each station licensed to operate by the Federal Communications Commission (FCC) receives a "chunk" of spectrum 10 kHz wide. Centered on that chunk is the station's assigned carrier frequency; carrier frequencies are in the interval 540 to 1,600 kHz,[22] with a maximum allowed drift of ±20 Hz (at 1 MHz, the middle of the AM radio band, this is a maximum allowed deviation of just ±0.002%, which *is* pretty small). For example, KFI in Los Angeles operates at a carrier frequency of 640 kHz, and so its upper and lower sidebands are, respectively, 640 to 645 kHz, and 635 to 640 kHz. Transmitting a baseband spectrum of width 5 kHz is wide enough for acceptable voice and music reproduction but not for what would be considered to be high fidelity.

Carrier frequencies are spaced 10 kHz apart, but two stations operating in a common geographical area would not be assigned adjacent carrier frequencies. To do so could result in what is called *adjacent channel interference*, about which I'll say a bit more soon. Very high power stations (like KFI, which transmits at the legal maximum of 50,000 watts (W) can be received over a vast area (perhaps even the

22 In the US, carrier frequencies always end with a 0, but outside the country there are some *split-channel* stations, with carrier frequencies ending with a 5'

entire country, especially at night), and such stations originally received carrier frequency assignments that were duplicated nowhere. Such stations were called *clear-channel* stations, not because they necessarily were always received clearly but because their carrier frequencies had been "cleared" for their *exclusive* use. Today, with thousands of AM broadcast stations on the air, there simply is not enough spectrum available for all the so-called clear-channel stations to have a unique carrier frequency. However, the high-power stations that do share a common carrier frequency are at least widely separated geographically. For example, there are two clear-channel stations operating at 1,030 kHz (WBZ in Boston, and KTOW in Casper, Wyoming).

We now have all we need to understand what is going on in Figure 3.2.1, the block diagram of what Armstrong was fond of calling the "Rolls-Royce of radio receivers." We'll start at the far left (the antenna) and work our way through each stage to the far right, to the audio output of your favorite radio program. I've used a block-diagram approach here, because over the decades the specific technologies of radio receivers have undergone enormous changes. When Armstrong worked in the 1910s through the 1940s, vacuum tubes and discrete, individual components were used. Starting in the 1950s tubes gave way to discrete transistors, and then in the 1960s yet another shift to integrated circuitry (where everything in a radio receiver looked like small, featureless blobs of plastic[23]) came into play. But, whatever the evolving technology of the circuits inside each of the blocks in Figure 3.2.1 might be, the *function* of each block has remained the same.

When you tune an AM superheterodyne receiver, the knob you turn simultaneously adjusts two separate and distinct parts of the circuit. First, the antenna signal is input to a tunable rf bandpass filter. That is, you move the center frequency of the filter's passband until it aligns with the carrier frequency of the station you want to select from all the other stations whose signals are also present in

23 A *spiritual* price has been paid for this evolution in technology. As one writer observed: "Eventually the art went out of radio tinkering. Children forgot the pleasures of opening the cabinets and eviscerating their parents' old [radios]. Solid electronic blocks replaced the radio set's messy innards—so where you once could learn by tugging at soldered wires and staring into the orange glow of vacuum tubes, eventually nothing remained but featureless ready-made chips." See James Gleick's 1992 *Genius: The Life and Science of Richard Feynman*, Pantheon, 1992, p. 17.

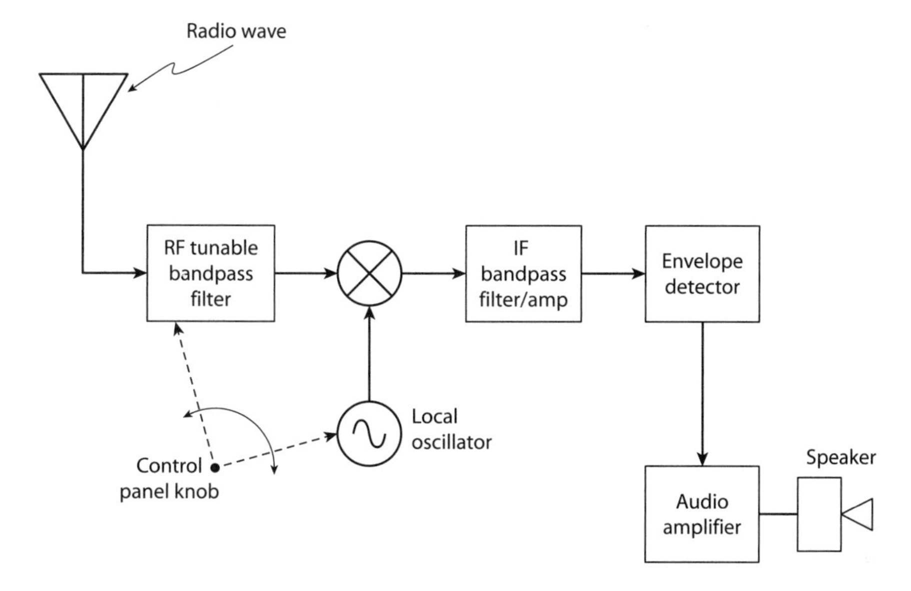

FIGURE 3.2.1. The superheterodyne.

the antenna. The passband of the antenna filter does *not* have to have anything even approaching vertical skirts; indeed, the filter's frequency response curve can be somewhat flexible in its rate of roll-off as the frequency deviates from the center frequency. (Just how flexible the filter's response can be you'll see in just a bit.) This is good, because it's very difficult (that is, expensive) to build a tunable filter with a rapid roll-off at its passband edges. Note, carefully, that this means if signals should arrive at the antenna from two stations with carrier frequencies that are not widely separated, then *both* signals will survive passage through the filter. This might appear to be the start of adjacent channel interference, but you'll soon see how the superheterodyne neatly handles this.

A second, more subtle role for the antenna filter is that it isolates the *local oscillator* (the second piece of circuitry—about which I'll tell you in just a moment—adjusted by the tuning knob) from the antenna. This is important, because the local oscillator can itself radiate rf energy into space and so interfere with nearby receivers.[24] That is,

24 A "leaky" local oscillator can be more worrisome than just merely causing interference. If you are a spy lurking in enemy territory, using a transmitter to send messages is clearly a risky business, but even just passively listening on a receiver that has a leaky

the antenna filter both lets the desired station signal move *into* the receiver while at the same time it prevents the local oscillator signal from *leaving* the receiver (that is, the antenna filter denies the local oscillator direct access to the antenna).

Next, continuing through Figure 3.2.1, the local oscillator shifts (with the aid of the multiplier, called the *mixer* in the figure) the selected chunk of spectrum that has survived passage through the antenna bandpass filter into the input of another bandpass filter with a center frequency at 455 kHz. (You'll soon see where that particular frequency comes from.) This second bandpass filter is what radio engineers call an *active filter*, because it is also an amplifier. Indeed, it is called the *IF amp* (*for intermediate-frequency amplifier*). The intermediate frequency of 455 kHz gets its name from the fact that $f_{IF}=455$ kHz is a frequency *below* the high frequency of any AM broadcast radio carrier and *above* the highest frequency in the baseband signal that modulates the carrier.

The IF amp is a high-gain bandpass filter with a bandwidth of 10 kHz (centered on and *fixed* at 455 kHz) and has, unlike the antenna filter, *very* steep skirts. The 10 kHz value allows the IF amp to pass both the upper and lower sidebands (each of width 5 kHz) of the selected station, while the sharp roll-off of the IF amp's frequency response eliminates any adjacent channel interference caused by two stations that have nearly equal carrier frequencies (both such signals, as mentioned earlier, will have gotten through the antenna filter). Since the IF amp is *not* tunable, it is relatively easy to build such a sharp cut-off bandpass filter that is stable (doesn't oscillate).

The IF amp and local oscillator are the essential parts of the superheterodyne, and so let's take a closer look at what is going on with both. Keep in mind that we'll hear (in the loudspeaker) *any* signal whose energy has been shifted into the IF amp's passband. Our starting point is the heterodyne theorem (refer to section 2.1), from which we know that if the signal out of the antenna filter is "mixed" (multiplied) with the sinusoidal signal from the local oscillator, then

local oscillator has its dangers, too. You can read about real-life occurrences of this in the 1987 book *Spy Catcher* (Viking) by Peter Wright, who was the first scientific officer of England's Security Service (MI5), the UK's domestic counterintelligence and security agency. MI5 is somewhat comparable to the US FBI (James Bond, 007, works for MI6, which is somewhat like the CIA).

the spectrum of the mixer output is the spectrum of the antenna filter signal shifted both up *and* down in frequency.

That is, if f_c and f_{LO} denote, respectively, the carrier frequency of the station we wish to tune-in, and the frequency of the local oscillator, then the mixer output spectrum will have versions of the antenna signal spectrum at both the sum *and* difference frequencies of f_c and f_{LO}; we will hear the version that gets shifted into the IF amp's passband. Thus, it should be clear that there are actually *two* different carrier frequencies (if present in the antenna) that the mixer will shift into the IF amp's passband. The first such carrier frequency is the instantly (I hope!) obvious $f_c = f_{LO} + f_{IF}$. *Obvious* because the difference frequency is $(f_{LO} + f_{IF}) - f_{LO} = f_{IF}$.

The second carrier frequency that ends up in the IF amp's passband is just slightly more subtle, as its value depends on the relative values of f_{LO} and f_{IF}. If $f_{LO} < f_{IF}$, then the carrier frequency $f_{IF} - f_{LO}$ also ends up in the IF amp's passband, because the sum frequency $(f_{IF} - f_{LO}) + f_{LO} = f_{IF}$. But, if $f_{LO} > f_{IF}$, the carrier frequency $f_{LO} - f_{IF}$ ends up in the IF amp's passband, because the difference frequency $f_{LO} - (f_{LO} - f_{IF}) = f_{IF}$.

To summarize, the two carrier frequencies that each appear in the IF amp's passband, called *image frequencies*, are

$$(1)\quad f_c = \begin{cases} f_{LO} + f_{IF} \\ f_{IF} - f_{LO} \end{cases}, \text{ if we use } f_{LO} < f_{IF},$$

$$(2)\quad f_c = \begin{cases} f_{LO} + f_{IF} \\ f_{LO} - f_{IF} \end{cases}, \text{ if we use } f_{LO} > f_{IF}.$$

This immediately motivates the question, Which situation do we choose? That is, do we use $f_{LO} < f_{IF}$, or do we use $f_{LO} > f_{IF}$? Indeed, does it *matter* which condition we choose? Yes, it does, and we'll work our way to the answer as follows.

If two stations are actually transmitting on carrier frequencies that form an image pair, that is not good, as they will end up on top of each other in the passband of the IF amp (and so in the speaker, too). The sharp cutoff of the IF amp's bandpass does not prevent this situation; rather, the sharp cutoff addresses adjacent channel interference. This is the major reason the antenna bandpass filter is part of the superheterodyne circuit. If that adjustable filter is centered on one of the image frequencies, then (obviously) it is *not* tuned to the other image. That image is therefore rejected by the antenna filter, and while some energy from the rejected image may still get through the filter, the hope is that it isn't a significant amount of energy.

The rejection by the antenna filter of the unwanted image will improve as the separation of the two image frequencies increases. So, how far apart *are* the image frequencies? In the case of $f_{LO} < f_{IF}$ (that is, for case (1)), the separation is

$$(f_{IF} + f_{LO}) - (f_{IF} - f_{LO}) = 2f_{LO},$$

a variable separation but one that is certainly always less than $2f_{IF}$, while for case (2) the image frequency separation is

$$(f_{IF} + f_{LO}) - (f_{LO} - f_{IF}) = 2f_{IF} \text{ if } f_{LO} > f_{IF}.$$

Thus, for a given value of f_{IF}, we see that the answer to our question, for maximum image frequency separation, is that we should choose $f_{LO} > f_{IF}$ to have a fixed image separation of $2f_{IF}$. If we pick a "high" value for f_{IF}, then even a "sloppy" antenna filter will provide good image frequency rejection, because it places the unwanted image frequency "far out" on the antenna filter's frequency response curve when that filter is tuned to the wanted image.

This motivates our next question: What should we use for f_{IF}? In the early days of the superheterodyne, the value of f_{IF} was much smaller than today's value (since 1938) of 455 kHz. The AR-812 sold by RCA in 1924 (see note 20), for example, used $f_{IF} = 45$ kHz, providing an image frequency separation of just 90 kHz. This low value was driven by technology limitations in building IF amps that could work at higher frequencies.[25] As the technology improved, however, it became possible to increase the value of f_{IF} to achieve ever better image rejection by the tunable antenna filter. Still, as you'll soon see, there *are* some problems with making f_{IF} arbitrarily high.

25 The biggest obstacle comes from the need for a rapid roll-off in an IF amp's frequency response (to avoid adjacent channel interference), a task that becomes ever more difficult as f_{IF} increases.

One way to approach the selection of a value for f_{IF} is to ask, What value will ensure that all possible unwanted image frequencies fall outside the AM broadcast band? *That* will certainly guarantee that no unwanted image will get into the IF amp's passband, because (by definition) there *are* no (legal) AM broadcast stations operating outside the FCC authorized AM radio band! This is an easy calculation to do. Since we have decided to use case (2), that is $f_{LO} > f_{IF}$, we know the image frequency pair consists of the two carrier frequencies

$$f_{c1} = f_{LO} + f_{IF}$$

and

$$f_{c2} = f_{LO} - f_{IF}.$$

To state the obvious, notice that $f_{c1} > f_{c2}$. Now, suppose we wish to listen to the station at the lower image frequency, f_{c2}. Then, f_{c1} is the unwanted image, and its frequency should be *greater* than 1,600 kHz. If, instead we wish to listen to the station at the higher image frequency, f_{c1}, then f_{c2} is the unwanted image frequency, and its value should be *less* than 540 kHz. So,

$$f_{LO} - f_{IF} < 540 \text{ kHz},$$
$$f_{LO} + f_{IF} > 1{,}600 \text{ kHz},$$

or multiplying through the first inequality by –1 (which reverses the direction of the inequality),

$$-f_{LO} + f_{IF} > -540 \text{ kHz},$$
$$f_{LO} + f_{IF} > 1{,}600 \text{ kHz},$$

and so, adding,

$$2f_{IF} > 1{,}060 \text{ kHz},$$

or, at last,

$$f_{IF} > 530 \text{ kHz}.$$

So, why is the actual value of $f_{IF}=455$ kHz somewhat *less* than this theoretical minimum value?

Consider if f_{IF} were actually in the AM radio band, as would be the case for $f_{IF}>530$ kHz. Then, the IF amp would be able to *directly* receive strong AM radio signals, bypassing the antenna, the antenna filter, and the mixer. To have that happen would defeat the entire concept of tuning! The value of f_{IF} that is actually used is a compromise between having f_{IF} large enough for good image rejection but not so large as to directly intrude on the AM radio band.[26]

We now have one final question to answer. For case (2), with $f_{LO}>f_{IF}$, the two carrier frequencies that put energy into the passband of the IF amp are $f_{LO}+f_{IF}$ and $f_{LO}-f_{IF}$. Which one should the antenna filter "tune" to? Does it really matter which one we choose? Theoretically, it doesn't, but from a practical, engineering point of view, it does, and here is why. In practice, when the operator of an AM superheterodyne receiver tunes the antenna filter to f_c, the local oscillator is *simultaneously* (see Figure 3.2.1 again) tuned to $f_{LO}=f_c+f_{IF}$. That is, the local oscillator is always 455 kHz *above* the desired carrier frequency. The antenna filter and the local oscillator vary together in what is called *parallel offset tracking*. The fact that the antenna filter is tunable means it's somewhat flexible in its frequency response roll-off around whatever frequency it is tuned to, and *that* means the parallel tracking of the antenna filter with the local oscillator does *not* have to be perfectly aligned.

For example, if you want to listen to KFI/Los Angeles at 640 kHz, you turn the tuning knob until the local oscillator is operating at 1,095 kHz, which you know you've accomplished when the station comes in loud and clear. But suppose the parallel tracking of the antenna filter is off just a bit, and instead of being at 640 kHz it is instead at 637 kHz. That's actually of no real concern, because KFI's sidebands at 640 ± 5 kHz will still survive passage through the antenna filter, pretty much intact, to put energy in the IF amp's passband, while the image frequency at 1,550 kHz will still be greatly suppressed and will *not* put energy (or, at least relatively little energy) in the IF amp's passband.

26 Well, you say, let's skip over the entire AM radio band and pick $f_{IF}>1{,}600$ kHz. Then we'd have *really good* image rejection! To see the objection to doing that, refer to note 25.

Now, to understand why $f_{LO}=f_c+f_{IF}$ (to understand why we tune to the carrier frequency $f_c=f_{LO}-f_{IF}$ rather than to the carrier frequency $f_c=f_{LO}+f_{IF}$), consider the following. If we tune to the carrier frequency $f_c=f_{LO}-f_{IF}$, then 540 kHz $<f_c<$ 1,600 kHz, or 540 kHz $<f_{LO}-f_{IF}<$ 1,600kHz, or 540 kHz $+f_{IF}<f_{LO}<$ 1,600 kHz $+f_{IF}$, or, since $f_{IF}=$ 455 kHz, we have 995 kHz $<f_{LO}<$ 2,055 kHz. That is, the local oscillator must be designed to operate over a 2:1 interval. However, if we tune to the carrier frequency $f_c=f_{LO}+f_{IF}$, then 540 kHz $<f_{LO}+$ 455 kHz $<$ 1,600 kHz, which says 8 5 kHz $<f_{LO}<$ 1,145 kHz, a frequency interval greater than 13:1. That is much more difficult to achieve than is 2:1, and that's why the historical decision was made to tune to $f_c=f_{LO}-f_{IF}$.

Any receiver that shifts the antenna signal frequencies to new locations in the spectrum is a heterodyne receiver. The prefix *super* is reserved for those receivers that specifically include *both* a tunable antenna filter for image rejection and a sharp-cutoff narrow-bandwidth bandpass IF amp (with a *fixed* center frequency) for adjacent channel suppression. This sort of receiver circuitry results in an easy-to-use radio that has reliably consistent tuning with a clearly audible audio output.[27] It is the gold standard of modern AM broadcast radio.

3.3 Fancy Heterodyning

We end this chapter with a couple of fancy examples of heterodyning. The first is worth looking at simply because, while it is easy to understand, it results in a way to locally (in the receiver) generate the carrier signal we *assumed* we had in our discussion of synchronous radio, which, you'll recall, doesn't transmit a carrier (refer to the discussion immediately following Figure 2.1.2). The second fancy example is used in up-upscale (expensive) AM radio receivers, as it

27 Before it was called the superheterodyne, this kind of receiver was called a *superaudible heterodyne*. Eventually, the *audible* was dropped, giving us today's term. Or so goes one explanation for the *super*. (This sounds very much like the "explanation" in note 1.) Another "explanation" I have seen is that the IF-amp center frequency (which is greater than the baseband frequencies) operates at *supersonic* frequencies, and *that's* where the *super* comes from. I find that just a bit of a stretch, but it's probably as good a tale as any other.

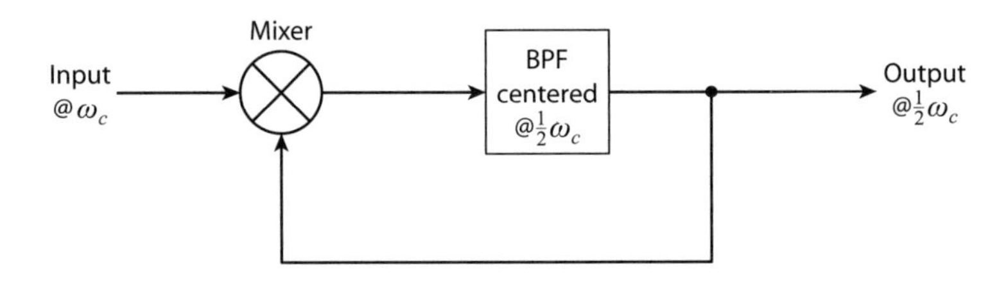

FIGURE 3.3.1. Regenerative frequency divider (÷2).

allows for excellent image rejection at *high frequencies* while retaining a standard 455 kHz IF amp.

The first example of fancy heterodyning is the *regenerative frequency divider* shown in Figure 3.3.1. With an input at frequency ω_c, the claim is that a sinusoidal output at frequency $\frac{1}{2}\omega_c$ results. To see this, we reason as follows. The bandpass filter is centered at $\omega = \frac{1}{2}\omega_c$, and so only a signal at that frequency *could* appear (let's assume the BPF has a narrow bandwidth, but, as you'll soon see, that isn't quite as restrictive as it might at first seem). But where does that output frequency of $\frac{1}{2}\omega_c$ come from? From feeding it back (hence the term *regenerative*) to the multiplier/mixer to combine with the input signal at frequency ω_c. By the heterodyne theorem, the mixer output signal has components at the sum and difference frequencies. That is, at $\omega_c + \frac{1}{2}\omega_c = \frac{3}{2}\omega_c$, and at $\omega_c - \frac{1}{2}\omega_c = \frac{1}{2}\omega_c$. The component at $\frac{3}{2}\omega_c$ is blocked by the BPF (and now we have a measure of just how narrow and sharp that filter's bandpass cutoff should be), while the component at frequency $\frac{1}{2}\omega_c$ is passed (and is just what we need to feed back to the mixer—this is an electronic version of wish fulfillment!).

But wait just a minute, you object. Where does that output signal at frequency $\frac{1}{2}\omega_c$ come from *right at the start*? Isn't this really just an electronic version of the old physics puzzle that asks, Why can't we elevate ourself off the ground by simply pulling up on our shoestrings? After all, here we get $\frac{1}{2}\omega_c$ by *assuming we already have* $\frac{1}{2}\omega_c$,

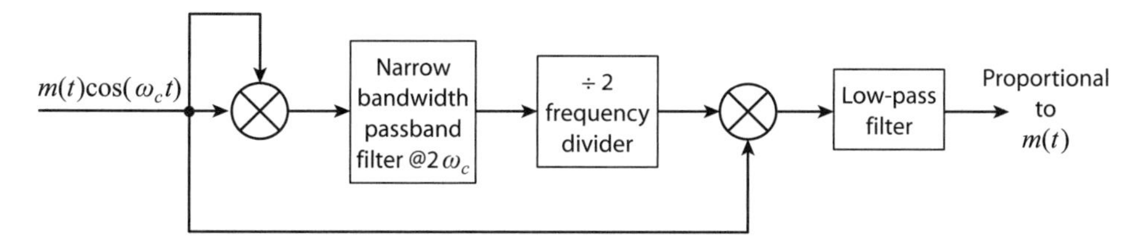

FIGURE 3.3.2. An elaboration of Figure 2.1.3.

and that argument has all the appearance of being a logical misstep. Well, no, it isn't. In the real, physical world there are always tiny thermal energy fluctuations in our circuits, resulting in rapid, random voltage fluctuations. The energy of those fluctuations is spread over a very wide spectrum, from dc to many hundreds of megahertz. It is a virtual certainty that there is always some initial energy at frequency $\frac{1}{2}\omega_c$ at the output, whatever ω_c might be.

So, of what use is such a circuit? If you refer to the synchronous demodulation receiver of Figure 2.1.3, you'll see that we assumed the receiver circuitry included a local oscillator at frequency ω_c. You might suggest we just include in the receiver a very narrow-bandwidth BPF centered on $\omega = \omega_c$, which is where the spectrum of the received signal is centered. But is there any energy there? No, because the energy at $\omega = \omega_c$, *if any*, is the dc energy of $m(t)$ that has been heterodyned up to $\omega = \omega_c$. But what *is* the dc (average) value of music or human speech over its duration. Zero. So, how *do* we obtain the assumed local oscillator signal at frequency $\omega = \omega_c$? Figure 3.3.2 shows one way to do it, using a regenerative frequency divider in an elaboration of the receiver of Figure 2.1.3.

The received signal in Figure 2.1.3, $m(t)\cos(\omega_c t)$, also shown in Figure 3.3.2, is immediately squared by the first multiplier, producing the output

$$\frac{1}{2}m^2(t) + \frac{1}{2}m^2(t)\cos(2\omega_c t).$$

As we showed in section 2.3, all the energy of $m^2(t)$ is confined to the interval $|\omega| < 2\omega_m$, where ω_m is the highest frequency in the baseband signal $m(t)$. Since $\omega_m \ll \omega_c$ in AM radio, we can reject the energy of the first term with a band-pass filter which outputs only the energy of $\frac{1}{2}m^2(t)\cos(2\omega_c t)$. And there *will be* energy in that term, because, unlike $m(t)$, $m^2(t)$ has a nonzero average value over the duration of $m(t)$. Thus, as shown in Figure 3.3.2, there is a signal at frequency $\omega = 2\omega_c$ that when run through a frequency divide-by-2 circuit, automatically gives us a signal at $\omega = \omega_c$, a signal that is mixed (heterodyned) with $m(t)\cos(\omega_c t)$ in the second multiplier. To do all this adds complexity (and cost) to a synchronous receiver, of course, which is why (as I've said before) commercial AM broadcast radio doesn't work this way but, rather, intentionally transmits a strong carrier signal. It's better to add cost to the *lone* transmitter than to each of millions of receivers.

For a second example of fancy heterodyning, consider the *dual-conversion* AM superheterodyne receiver shown in Figure 3.3.3. It uses the local oscillator/mixer/IF-amp circuitry we studied in Figure 3.2.1 to eventually shift the desired antenna signal down to 455 kHz for envelope detection (using the second *nontunable* local oscillator, the second mixer, and the second IF amp shown in Figure 3.3.3). But our new receiver circuit now also includes a preliminary stage of heterodyning (the first *tunable* local oscillator, the first mixer, and the first IF amp). To understand why one might do such a thing, suppose we want to listen to an AM transmitter with a carrier at 25 MHz. With the circuit of Figure 3.2.1, the image separation is $2f_{IF} = 910$ kHz, and a tunable antenna filter set to 25 MHz is not going to have much (if any) impact on (rejection of) an image signal less than 1 MHz away from 25 MHz. So, let's work our way through Figure 3.3.3 to see how using *two* stages of heterodyning resolves this difficulty *without* having to build a *high-frequency* narrow-bandwidth (10 kHz) IF amp.[28]

We start by assuming $f_{LO1} < f_c$, the frequency to which we wish to listen. In addition, let's assume the second IF amp is our usual 10 kHz bandwidth filter centered on 455 kHz; that is, $f_{IF2} = 455$ kHz with sharp cutoffs at 450 and 460 kHz. The first IF amp is, *for now*, left undefined in both bandwidth and center frequency. *For now*, let's just write f_{IF1} for that filter's center frequency, with an arbitrarily wide bandwidth.

28 The smaller the ratio of the IF-amp bandwidth of 10 kHz to the IF-amp center frequency, the more difficult (expensive) it is to construct the IF amp.

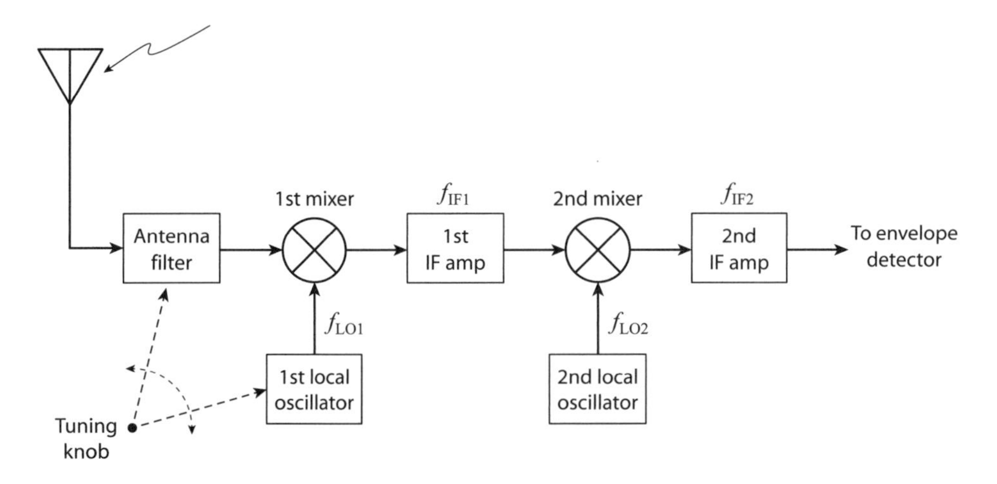

FIGURE 3.3.3. Dual-conversion AM superheterodyne receiver.

For now, all we care about is answering the question, What carrier frequencies, starting with f_c, put energy into the middle of the second IF amp's passband?' That is, what frequencies, *if present* in the antenna, would send energy to the envelope detector (and so to the loudspeaker)? Again, note *carefully* that the first local oscillator is tunable, while the second local oscillator is not.

For f_c to put energy into the second IF amp it has to first put energy into the passband of the first IF amp. To do that, let's have the tuning knob simultaneously set the first local oscillator to $f_{LO1}=f_c-f_{IF1}$ as it sets the antenna filter to f_c (in agreement with our initial assumption that $f_{LO1}<f_c$).Thus, the output of the first mixer contains a component at the difference frequency $f_c-(f_c-f_{IF1})=f_{IF1}$, which puts energy into the middle of the first IF-amp's passband. This is input to the second mixer, *and if* the second local oscillator operates at the *fixed* frequency $f_{LO2}=f_{IF1}+f_{IF2}$, then the output of the second mixer contains the difference frequency $(f_{IF1}+f_{IF2})-f_{IF1}=f_{IF2}$. So, an antenna signal at frequency f_c does indeed put energy into the middle of the passband of the second IF amp.[29] Let's call f_c "image 1."

Continuing, notice that another carrier frequency that puts energy into the middle of the second IF-amp's passband is f_c-2f_{IF1}, because

29 What if, instead, the second local oscillator operated at the fixed frequency $f_{LO2}=f_{IF1}-f_{IF2}$ (with $f_{IF1}>f_{IF2}$)? Then, the output of the second mixer would contain a component at the difference frequency $f_{IF1}-(f_{IF1}-f_{IF2})=f_{IF2}$, and so this alternative choice for f_{LO2} would also work. Challenge Problem 3.4 asks you to explore this possibility further.

then the output of the first mixer has a component at the difference frequency $(f_c - f_{IF1}) - (f_c - 2f_{IF1}) = f_{IF1}$, and so, as with the carrier frequency f_c, this second carrier also puts energy into the middle of the second IF-amp's passband, Let's call $f_c - 2f_{IF1}$ "image 2."

Next, consider the carrier at frequency $f_c - 2(f_{IF1} + f_{IF2})$. The output of the first mixer has a component at the difference frequency $(f_c - f_{IF1}) - (f_c - 2f_{IF1} - 2f_{IF2}) = f_{IF1} + 2f_{IF2}$. This is input to the first IF amp (which has, recall, *arbitrarily*—at least, for now—wide bandwidth) and so energy at frequency $f_{IF1} + 2f_{IF2}$ is input to the second mixer. The output of the second mixer has a component with the difference frequency $(f_{IF1} + 2f_{IF2}) - (f_{IF1} + f_{IF2}) = f_{IF2}$, right in the middle of the second IF-amp's passband. Let's call $f_c - 2(f_{IF1} + f_{IF2})$ "image 3."

Finally, consider the carrier at frequency $f_c + 2f_{IF2}$. The output of the first mixer has a component at the difference frequency $(f_c + 2f_{IF2}) - (f_c - f_{IF1}) = f_{IF1} + 2f_{IF2}$, and we are back to the condition we had in the case of image 3. Let's call $f_c + 2f_{IF2}$ "image 4."

Thus, we see that with dual conversion there is not an image *pair* as with the superheterodyne of Figure 3.2.1 but, rather, an image *quadruple* of frequencies: f_c, $f_c - 2f_{IF1}$, $f_c - 2(f_{IF1} + f_{IF2})$, and $f_c + 2f_{IF2}$. To insert some values to all this, suppose we wish to listen to a transmitter with a carrier frequency $f_c = 25$ MHz. As we adjust the antenna filter to 25 MHz with the tuning knob, we imagine we are simultaneously adjusting the first local oscillator to $f_{LO1} = 20$ MHz. Since mixing $f_c = 25$ MHz with the first local oscillator at 20 MHz will produce a component with the difference frequency of 5 MHz, let's set $f_{IF1} = 5$ MHz. (As stated at the start of the analysis, $f_{IF2} = 455$ kHz, with a 10 kHz bandwidth. This tells us that the second local oscillator is fixed at $f_{LO2} = f_{IF1} + f_{IF2} = 5.455$ MHz.) To have the same relative bandwidth in the first IF amp as we have in the second IF amp, we see that the bandwidth of the first IF amp should be something like 110 kHz. That is, the first IF amp easily passes energy in the passband 4.945 MHz $< f <$ 5.055 MHz and greatly attenuates energy outside that interval.

Now, our image frequency quadruple is:

image 1 = 25 MHz,
image 2 = 15 MHz,
image 3 = 14.09 MHz,
image 4 = 25.91 MHz.

If we follow each of these image frequencies through the two stages of heterodyning and IF-amp filtering, we see that image 1 (the frequency we wish to listen to) sails right through, image 2 is attenuated in some degree by the antenna filter, image 3 is attenuated by the antenna filter even more than is image 2 and is then nearly annihilated by the first IF amp, and image 4 is hardly affected by the antenna filter but is then nearly annihilated by the IF amp.

Finally, what happens with a signal that is just 10 kHz away from 25 MHz? That is, suppose we have a signal at 25.01 MHz in the antenna, along with the 25 MHz signal we wish to listen to? The 25.01 MHz signal easily passes through the antenna filter tuned to 25 MHz but mixes with the first local oscillator to give an output at 5.01 MHz, which passes right through the first IF amp. This 5.01 MHz output then mixes with the second local oscillator (at 5.455 MHz) to give an output at 0.445 MHz (445 kHz), which is greatly attenuated by the second IF amp, and so we see that we have good adjacent channel suppression.

Challenge Problem 3.1: As mentioned in note 6, the 12AX7 is a *dual* triode. So, consider the amplifier circuit in Figure CP3.1, in which both of these triodes are used. Calculate the voltage gain, $\frac{dv_{p2}}{dv_i}$.

Hint: The quiescent states of the two triodes are identical, because when $v_i = 0$, both grids are at ground potential, and so both triodes are in exactly the same state. As we did in the text, take the current in each triode (with a plate-to-cathode voltage drop of 250 volts) as $i_p = [6 + 2.5v_{gk}] \times 10^{-3}$ A. Use $v_{gk} = -1.5$ volts in the quiescent state.

Challenge Problem 3.2: Show that the high-gain differential amplifier circuit in Figure CP3.2 is a *tunable* bandpass filter (notice the two negative-feedback paths via R_3 and C_1) with the variable resistor R_2 controlling the center frequency (the frequency at which the filter output v_0 is maximum). *Hint*: Since the plus input to the differential amplifier is at zero volts (ground), then the minus input is, too. Sum currents at ① and ②, and recall there is vanishingly small input current to a high-gain differential amplifier with negative feedback. Your analysis should show that the maximum filter output is *independent* of R_2 (independent, that is, of the center frequency).

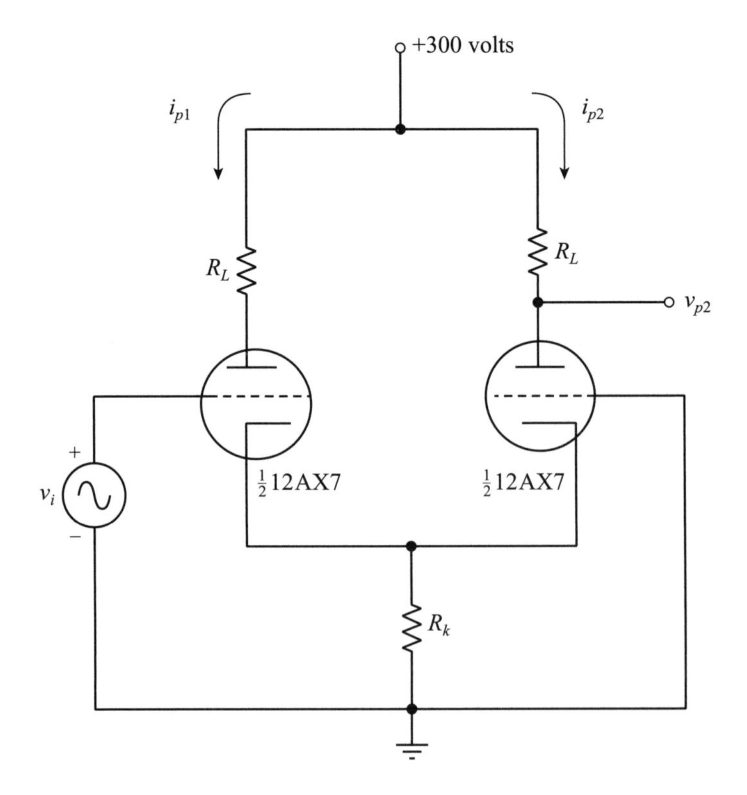

FIGURE CP3.1. Two-triode voltage amplifier.

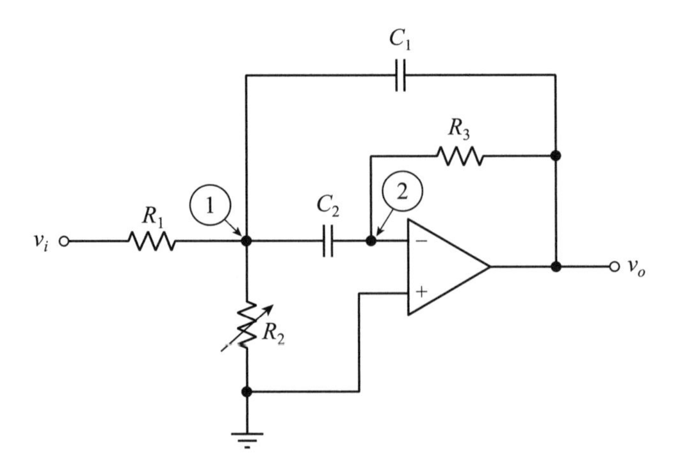

FIGURE CP3.2. A tunable bandpass filter.

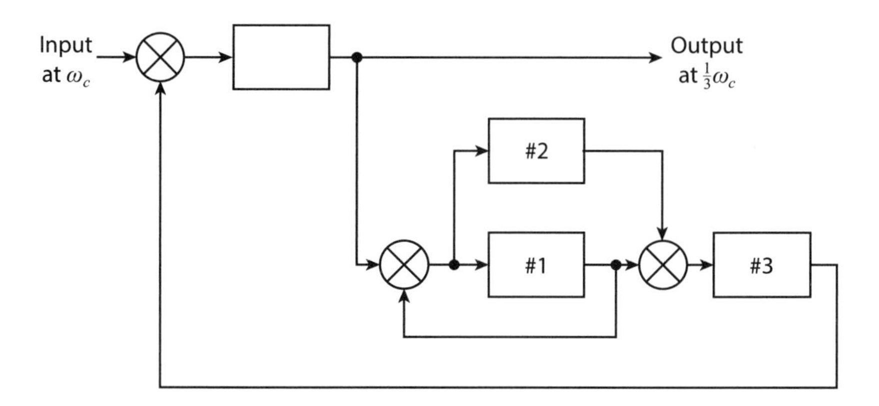

FIGURE CP3.3. Regenerative frequency divider (÷3)

Challenge Problem 3.3: The regenerative frequency divide-by-2 circuit discussed in the text is just the first example of a family of circuits that can divide frequency by an integer. For example, the circuit shown in Figure CP3.3 will, for an input at frequency ω_c, produce an output at frequency $\frac{1}{3}\omega_c$. The four boxes in the figure are narrow-bandwidth bandpass filters. Determine the center frequencies of these four filters. *Hint*: Just to get you started, be sure to not overlook the obvious, that the immediate answer for the top filter is $\frac{1}{3}\omega_c$. Next, write x and y as the center frequencies of filters #1 and #2, respectively, and then apply the heterodyne theorem (several times).

Challenge Problem 3.4: Redo, using the conditions $f_{LO1} < f_c$ and $f_{LO2} < f_{IF1}$ (see note 29), the image frequency calculations in the text for the dual-conversion AM superheterodyne radio receiver of Figure 3.3.3.

Challenge Problem 3.5: An amplifier is said to be *linear with gain* A if the output is $v_o = Av_i$, where v_i is the input. It is physically impossible for this relationship to hold for all v_i, however, as eventually $|v_i|$ will become so large that the amplifier *saturates*. That is, for any constructable electronic device (like a vacuum tube[30]) there is always a finite value V such that

$$v_o = \begin{cases} Av_i \ , |v_i| < V \\ AV, |v_i| > V \end{cases}.$$

The value of V is called the *threshold* of the amplifier. Once an amplifier has become saturated, additional increases in the input have no further effect on the amplifier output. When we discuss FM radio in chapter 5, we'll find that driving an amplifier into saturation can actually be quite useful. Now, suppose we make a *composite amplifier* by connecting n identical amplifiers in series (in a chain), with the output of each amplifier serving as the input to the next amplifier (with the exception, of course, of the last amplifier in the chain). If each amplifier in the chain has gain $A = 5$ and a saturated output of 10 volts (which means each amplifier has a threshold of 2 volts), how many amplifiers are required (what is n?) if the composite amplifier is to have a threshold of no more than 0.1 volt? What is the specific value of the composite threshold? *Hint*: The composite amplifier saturates when the *last* amplifier in the chain, saturates because it saturates before any of the others do). And remember, n has to be an integer.

Challenge Problem 3.6: Here's another circuit analysis problem that will be of great interest for us in the FM radio chapter. Show that the circuit in Figure CP3.6 simulates a *perfect* inductor with one end grounded (to electronically simulate a "floating" inductor, with *both* ends available for arbitrary connection, is a far more challenging problem). That is, show that

$$v = L_{eq} \frac{di}{dt},$$

30 The actual physical mechanism that results in saturation depends on the specific device in question, but for a vacuum tube, linear amplification ceases when the tube current either drops to zero (the tube *cuts off*) because the grid-to-cathode voltage is very negative or is the greatest it can possibly be because the space charge surrounding the cathode is already fully participating in the tube current.

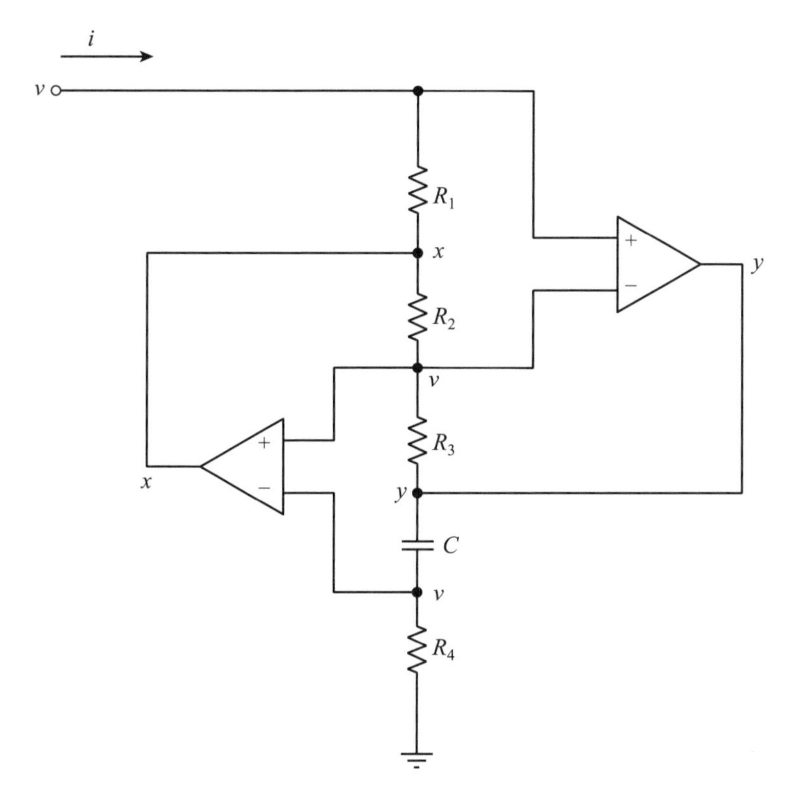

FIGURE CP3.6. A high-gain differential amplifier circuit that behaves like a perfect inductor with one end grounded.

and express L_{eq} in terms of the four resistors and the capacitor. If your analysis is correct, you should find that with $R_1 = R_2 = R_3 = 1{,}000$ ohms, $R_4 = 10{,}000$ ohms, and $C = 100\ \mu\mu F = 100\ pF = 1 \times 10^{-10} F$, then $L_{eq} = 10\ \mu H$. These are just numbers for now, but I'll remind you of them in chapter 5. *Hint*: Remember that the differential input to a high-gain differential amplifier *with negative feedback present* is, in the limit as the gain becomes arbitrarily large, zero (notice how v, because of that zero-input condition on the differential amplifiers, appears not only as the input to the circuit but also at two other places in the vertical resistor/capacitor stack). Call the outputs of the two differential amplifiers x and y, as shown, and then sum currents at all nodes *not* connected to a differential amplifier output.

Chapter 4

SSB Radio

> Single sideband super-encrypted ultra-high frequency [radio]. That's as secure as communications get.
>
> —a glowing (if today a somewhat overreaching) tribute to SSB radio from one of the characters in Tom Clancy's 1987 novel *Clear and Present Danger*.

4.1 The Origin of SSB Radio and Carson's SSB Transmitter

The preceding quote from a fictional work actually *was* pretty much on the mark during the Second World War, as SSB was the technology behind the ultra-top-secret trans-Atlantic radio/telephone link used by the English prime minister Winston Churchill and the US president Franklin Roosevelt. This system, developed by the Bell Telephone Laboratories, was one of the deepest military secrets (code named SIGSALY) of the war, perhaps second only to the atomic bomb Manhattan Project. After Roosevelt's death, in early 1945, President Harry Truman continued to use the link to discuss war-related matters with Churchill.

Given the state of electronics technology in the 1940s, SIGSALY was a physically big system, with each terminal filling several hundred cubic feet (Roosevelt's terminal was in the Pentagon, with a secure link to the White House,[1] and Churchill's was in the subbasement of a London department store, with a secure link to his underground Cabinet War Rooms). Modern SSB radio is significantly more compact![2]

1 The electrical wires connecting the White House to the Pentagon were placed inside a pressurized pipe; any attempt by an eavesdropper to penetrate the pipe (thus causing a pressure fluctuation) to get at the enclosed wires would be instantly detected by pressure-actuated switches inside the pipe.

2 You can read more about the amazing story of SIGSALY in a paper by William R. Bennett, "Secret Telephony as a Historical Example of Spread-Spectrum Communication," *IEEE Transactions on Communications*, January 1983, pp. 98–104.

By the time the SIGSALY system was built, SSB had already been around for decades. As mentioned in note 32 in chapter 2, SSB can be traced back to 1915 and the American electrical engineer and mathematician John Carson. The importance of his work can be measured by his obituary in the *New York Times*, which specifically cited SSB, alone, of all his many accomplishments (we'll encounter Carson again in the next chapter, in the discussion of FM radio). So, what did he do?

Realizing that the spectrum of a real-valued baseband signal is symmetrical (an even function about $\omega=0$; refer to Figure 2.1.2), Carson reasoned that not only did the carrier in a heterodyned signal not contain information[3] but that the spectrum itself was redundant. That is, information in the shifted spectrum for the $|\omega|> \omega_c$ half is duplicated in the $|\omega|< \omega_c$ half, and so only one of the two halves (either one) need be transmitted.[4] Carson's thinking along this line was motivated by his work at AT&T involving the transmission of information over copper wires. The bandwidth of such wires, compared with that of modern fiber-optic cables and wireless microwave links, is like comparing a canoe to an aircraft carrier, and so Carson was searching for a way to "compress" more message-carrying capability into the limited bandwidth of copper wires. Here's what he came up with.

To transmit numerous baseband signals (each a long-distance telephone call) over a wire, each signal was heterodyned with a carrier whose frequency (ω_c) was different for each baseband signal (each of these individual carrier frequencies was called a *subcarrier*), which separated the baseband signals. The result is shown in the top and middle portions of Figure 4.1.1. If the bandwidth of each baseband signal is 4 kHz, the bandwidth of the heterodyned signal is then 8 kHz (it is double-sided about its subcarrier). Carson's idea was to take each heterodyned signal and use a bandpass filter to remove the redundant part. The bottom portion of Figure 4.1.1 shows the result when the BPF rejects the energy in the frequencies $|\omega|< \omega_c$. That is, the output of the BPF is, in Figure 4.1.1, the *upper sideband*

3 Recall that the only reason a large carrier is present in broadcast AM radio is to make the envelope detection process *at the receiver* easy (that is, inexpensive) to do (refer to section 2.2).

4 Refer to note 10 in chapter 2.

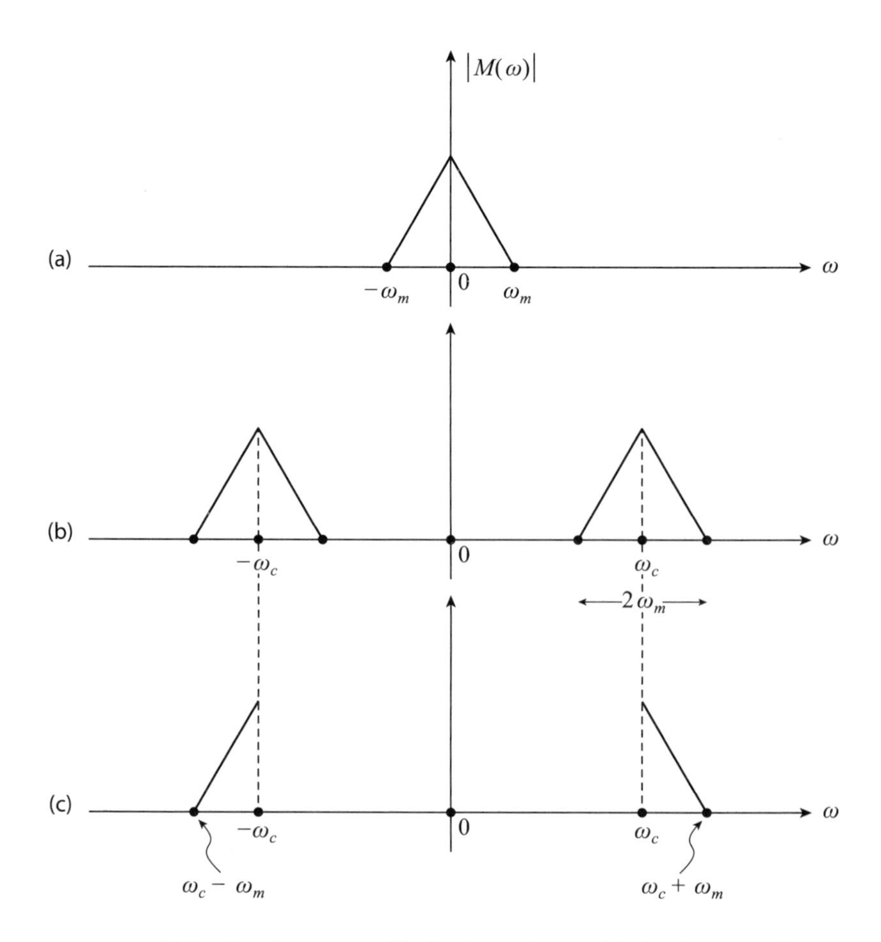

FIGURE 4.1.1. (a) Baseband spectrum; (b) after heterodyning; (c) after high-pass filtering, to produce an upper sideband signal.

of the heterodyned baseband signal, and this filtered version of the double-sided signal has a bandwidth of just 4 kHz, not the 8 kHz of the double-sided heterodyned signal. The immediate result was a doubling of the number of telephone calls that could be transmitted over wires (see the following box for a bit more detailed discussion). The subcarrier was *not* present in the final, filtered SSB signal.

The first use of SSB in telephone circuits was in 1918, on a line connecting Baltimore and Pittsburgh, and it was just four years later that SSB made the journey from telephone wires to transoceanic radio.[5] A

5 The extension of SSB from telephone circuitry to radio seems to have first been suggested in 1922 by Lloyd Espenschied (1889–1986), an electrical engineer at Bell Telephone

successful demonstration showing the feasibility of trans-Atlantic SSB radio was made by the Western Electric engineer Raymond Heising (1888–1965) in January 1923. This was followed in 1927 with the start of commercial SSB radio service between New York City and London. That service transmitted a single sideband signal (the *lower* sideband, selected using rejection filtering of the upper sideband) with a bandwidth of 2.7 kHz, using a carrier frequency that could be anywhere in the interval 41 to 71 kHz.

The baseband signal spectrum for a telephone call over AT&T's system was limited to frequencies from 300 to 3,400 Hz, and so a single call could easily be transmitted using a 4 kHz chunk of spectrum. Starting with that, AT&T heterodyned each of 12 such calls with a subcarrier individually selected from 12 subcarriers distributed evenly over the interval 60 to 108 kHz ($60\text{ kHz}+12\times4\text{ kHz}=108\text{ kHz}$). Each call was heterodyned using diode ring modulators (see Figure 2.5.3). The filtered output of each modulator was the baseband voice spectrum of a call, shifted by a subcarrier frequency. At this point, the modulator outputs were double-sided, and filtering rejected the upper sidebands. The result was 12 SSB spectrums that when combined formed what was called a *basic group*, occupying a chunk of spectrum 48 kHz wide. Five such basic groups were then individually heterodyned with yet higher frequency subcarriers, spaced 48 kHz apart over the interval 312 to 552 kHz ($312\text{ kHz}+5\times48\text{ kHz}=552\text{ kHz}$). The filtered outputs of the heterodyning modulators were, again, double-sided, and then more filtering rejected the lower sidebands. Combining five basic groups created what was called a *super group*. In the same way, additional stages of heterodyning, filtering, and combining continued (10 super groups made a *master group*, and 6 master groups made a *jumbo group*). The final stage combined 3 jumbo groups to create a *frequency division multiplexed* (FDM) signal that allowed a single wire to simultaneously carry $12\times5\times10\times6\times3=10{,}800$ calls. FDM was

Laboratories, in his paper "Application to Radio of Wire Transmission Engineering," *Proceedings of the IRE*, October 1922, pp. 344–368.

retired from modern telephone applications with the advent of digital technology, but FDM in telephone circuits was the origin of SSB radio.

Using filters to reject one of the sidebands in a double-sideband signal, to get a single-sideband signal, is the obvious, brute-force approach. It works, yes, but how *much more* elegant it would be to design a transmitter circuit that *directly* generates a signal with a spectrum like the one in the bottom sketch of Figure 4.1.1. This was accomplished by Carson's colleague at AT&T, Ralph Hartley (see Challenge Problem 2.7), who filed a patent application in 1925 (granted in 1928) for what is known as the *phasing method*. In the next two sections you'll see how all the Fourier theory we developed in the last two sections of chapter 1 and the end of chapter 2 explains how Hartley directly generated an SSB signal *without* the use of sideband filtering.[6] This is *not* to say, however, that we are no longer going to see filters in SSB radio. Indeed, you'll see by the end of this chapter a new kind of filter—the *all-pass*—and to prepare for that, look at Challenge Problem 4.1.

4.2 The Hilbert Transform and Hartley's SSB Transmitter

To start our analysis, let's review two Fourier transform results, involving the step and impulse functions, we developed in section 1.7:

$$\text{(A)} \qquad \delta(t) \leftrightarrow 1,$$

$$\text{(B)} \qquad \frac{1}{2}\delta(t) - j\frac{1}{2\pi t} \leftrightarrow u(-\omega).$$

We can use (A) and (B) to answer the question, What time function pairs with $u(\omega)$? (Recall I asked you to think about this in Challenge Problem 1.8.) To answer the question, notice that $u(-\omega)+u(\omega)=1$,

6 Fourier theory, however, was *not* the mathematical approach of the original theoretical discussions of SSB radio. In Raymond Heising's classic paper, "Production of Single Sideband for Trans-Atlantic Radio Telephony," *Proceedings of the IRE*, June 1925, pp. 291–312, you'll find only trigonometric identities. The early radio pioneers were very clever analysts, and they could "see" how things work without using formal Fourier theory; I think, though, that you'll find Fourier theory makes what they did easy to understand.

and so $u(\omega)=1-u(-\omega)$. Thus, the answer to our question is, The time function that pairs with $u(\omega)$ is the time function that has the Fourier transform $1-u(-\omega)$. Since (A) tells us that $\delta(t)$ pairs with 1, then combining that result with (B) says the time function that pairs with $u(\omega)$ is

$$\delta(t)-\left[\frac{1}{2}\delta(t)-j\frac{1}{2\pi t}\right]=\frac{1}{2}\delta(t)+j\frac{1}{2\pi t};$$

that is,

$$\frac{1}{2}\delta(t)+j\frac{1}{2\pi t}\leftrightarrow u(\omega).$$

Next, let's review one last result from Fourier theory, the *time convolution theorem*, which you were asked to derive in Challenge Problem 2.6 (if you didn't do it then, do it now, or at least look in the solutions section at the back of the book):

$$\text{(C)} \qquad m(t)*g(t)=\int_{-\infty}^{\infty}m(u)g(t-u)du\leftrightarrow M(\omega)G(\omega).$$

In words, convolution in the time domain pairs with multiplication in the frequency domain. *Now* we are all set to see how Hartley put an SSB transmitter circuit together without using sideband filtering.

With $m(t)$ denoting a baseband signal (the output of a microphone), we have the pair $m(t)\leftrightarrow M(\omega)$. We then define a new signal $z_+(t)$, paired with the transform $Z_+(\omega)$. That is, $z_+(t)\leftrightarrow Z_+(\omega)$, where $z_+(t)$ is such that $Z_+(\omega)=M(\omega)u(\omega)$. Since $u(\omega)$ is the step function in the frequency domain, then the spectrum of the time signal $z_+(t)$ is zero for negative frequencies and equal to the positive-frequency half of $M(\omega)$. The top sketch of Figure 4.2.1 shows $Z_+(\omega)$. Clearly, $Z_+(\omega)$ is not a symmetrical spectrum, and so we know $z_+(t)$ is not a real-valued time function (see the discussion immediately following Figure 1.6.1). This will not be of any concern to us because we will not need to actually generate $z_+(t)$—it is strictly a mathematical object that we'll use along our way to a circuit that we *can* build. Using the time convolution theorem of (C), with $g(t)=z_+(t)$, we see that $Z_+(\omega)=M(\omega)u(\omega)$ says that

$$z_+(t)=m(t)*\left[\frac{1}{2}\delta(t)+j\frac{1}{2\pi t}\right]=\frac{1}{2}m(t)*\delta(t)+j\frac{1}{2\pi}m(t)*\frac{1}{t}.$$

Recalling that $m(t) * \delta(t) = m(t)$—refer to Challenge Problem 2.6[7]—we have

$$z_+(t) = \frac{1}{2}\left[m(t) + j\frac{1}{\pi}\int_{-\infty}^{\infty} \frac{m(u)}{t-u}du\right].$$

The expression in the square brackets is called an *analytic signal*, and the imaginary part of that expression is called the *Hilbert transform*, $\bar{m}(t)$, of $m(t)$; that is,

$$\bar{m}(t) = \frac{1}{\pi}\int_{-\infty}^{\infty} \frac{m(u)}{t-u}du,$$

and

$$z_+(t) = \frac{1}{2}\left[m(t) + j\bar{m}(t)\right].$$

Notice that, unlike the Fourier transform, the Hilbert transform does not change domains: the Hilbert transform of a time signal is another time signal. Any time signal of the form $x(t) \pm j\bar{x}(t)$, where $\bar{x}(t) = \frac{1}{\pi}\int_{-\infty}^{\infty} \frac{x(u)}{t-u}du$ denotes the Hilbert transform of $x(t)$, is called *analytic*, a name due to the Hungarian-born electrical engineer Dennis Gabor (1900–1971), who coined the term in a 1946 paper on communication theory. (Gabor received the 1971 Nobel Prize in Physics.) "Hilbert transform" is a name due to Hardy, who published the integral in an analysis quite different from the one I just took you through, and so, much to a resurrected Hardy's almost certain astonishment, his mathematics appears—*in a central way*—in the theory of SSB radio! He initially thought the integral was original with him but later learned that the German mathematician David Hilbert (1862–1943) had, in a nonradio context, known of it for years, and so Hardy named it the Hilbert transform (in 1928). But even Hilbert was not the first, as the integral had appeared decades earlier in the 1873 doctoral

7 If you haven't been doing the challenge problems, you now should see the error of your ways! At the least, study the solutions at the back of the book.

dissertation of the Polish-born Russian mathematician Yulian-Karl Vasilievich Sokhotsky (1842–1927).

To shift the spectrum of $z_+(t)$ up to rf for transmission, at carrier frequency ω_c, we simply multiply $z_+(t)$ by $e^{j\omega_c t}$, to arrive at

$$z_+(t)e^{j\omega_c t} = \frac{1}{2}[m(t) + j\overline{m}(t)][\cos(\omega_c t) + j\sin(\omega_c t)]$$
$$= \frac{1}{2}[m(t)\cos(\omega_c t) - \overline{m}(t)\sin(\omega_c t)]$$
$$+ j\frac{1}{2}[\overline{m}(t)\cos(\omega_c t) + m(t)\sin(\omega_c t)].$$

The spectrum of this complex time function is shown in the second sketch of Figure 4.2.1; it represents the upper sideband portion of an SSB signal. To get a real-valued time signal that we can actually generate, we of course need, as mentioned earlier, a symmetrical spectrum.

We can achieve that by repeating what we've just done, but this time we put together a time signal that gives us the *negative*-frequency part of an SSB signal. So, let's write the pair $z_-(t) \leftrightarrow Z_-(\omega)$, where $Z_-(\omega) = M(\omega)u(-\omega)$, as shown in the third sketch of Figure 4.2.1. That is, $z_-(t)$ is the time signal that has a spectrum that is zero for positive frequencies and is equal to the negative-frequency portion of $M(\omega)$. From (B), then, we have the analytic signal

$$z_-(t) = m(t) * \left\{\frac{1}{2}\delta(t) - j\frac{1}{2\pi t}\right\} = \frac{1}{2}[m(t) - j\overline{m}(t)].$$

Next, we multiply $z_-(t)$ by $e^{-j\omega_c t}$ to left-shift the spectrum of $z_-(t)$ by ω_c on the frequency axis (see the fourth sketch in Figure 4.2.1), and we arrive (you should do the easy algebra) at

$$z_-(t)e^{-j\omega_c t} = \frac{1}{2}[m(t)\cos(\omega_c t) - \overline{m}(t)\sin(\omega_c t)]$$
$$- j\frac{1}{2}[\overline{m}(t)\cos(\omega_c t) + m(t)\sin(\omega_c t)].$$

Adding the two spectrums in the second and fourth sketches of Figure 4.2.1 gives us the *symmetrical* spectrum (a spectrum we can

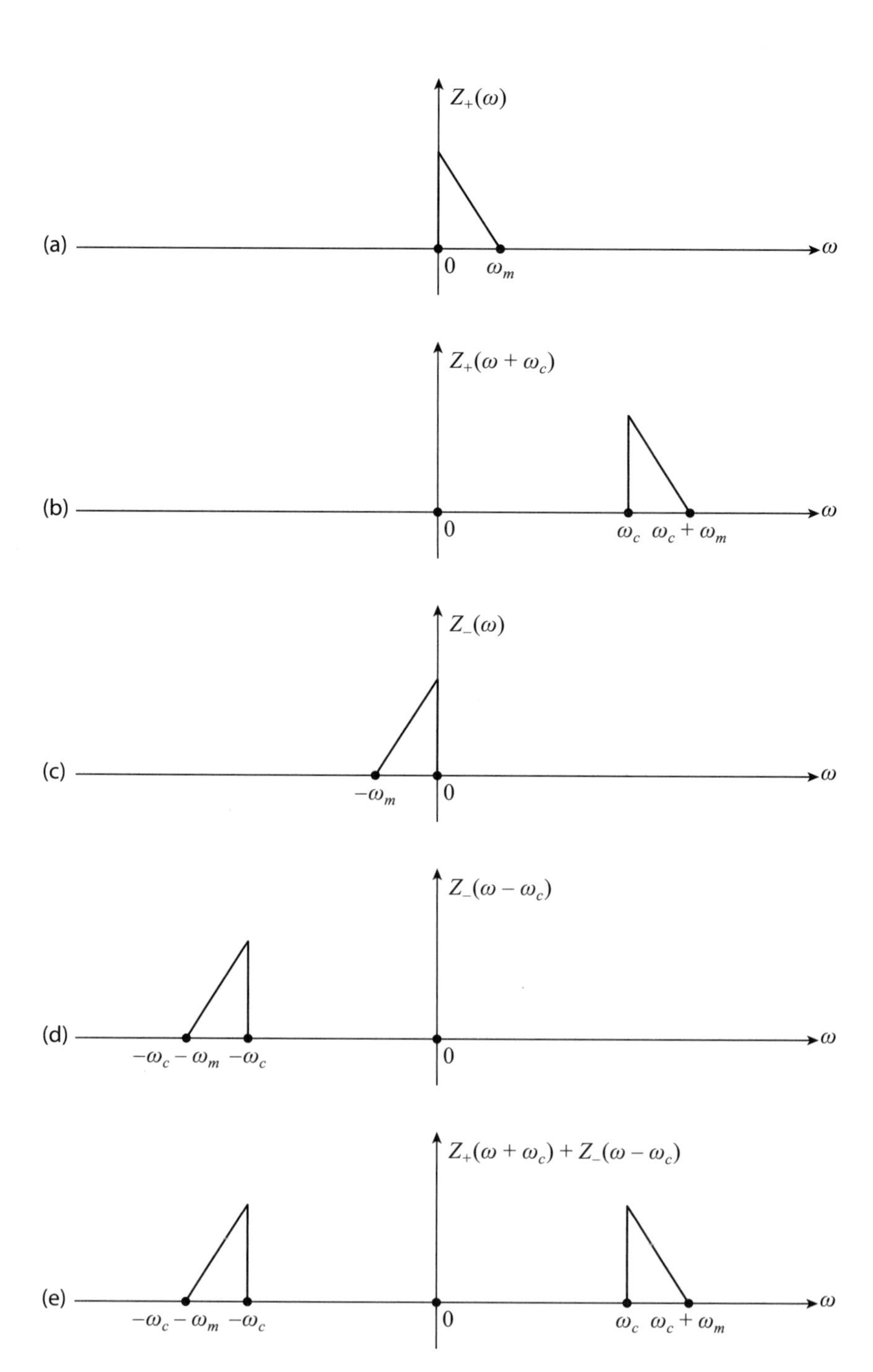

FIGURE 4.2.1. Constructing an upper-sideband SSB signal.

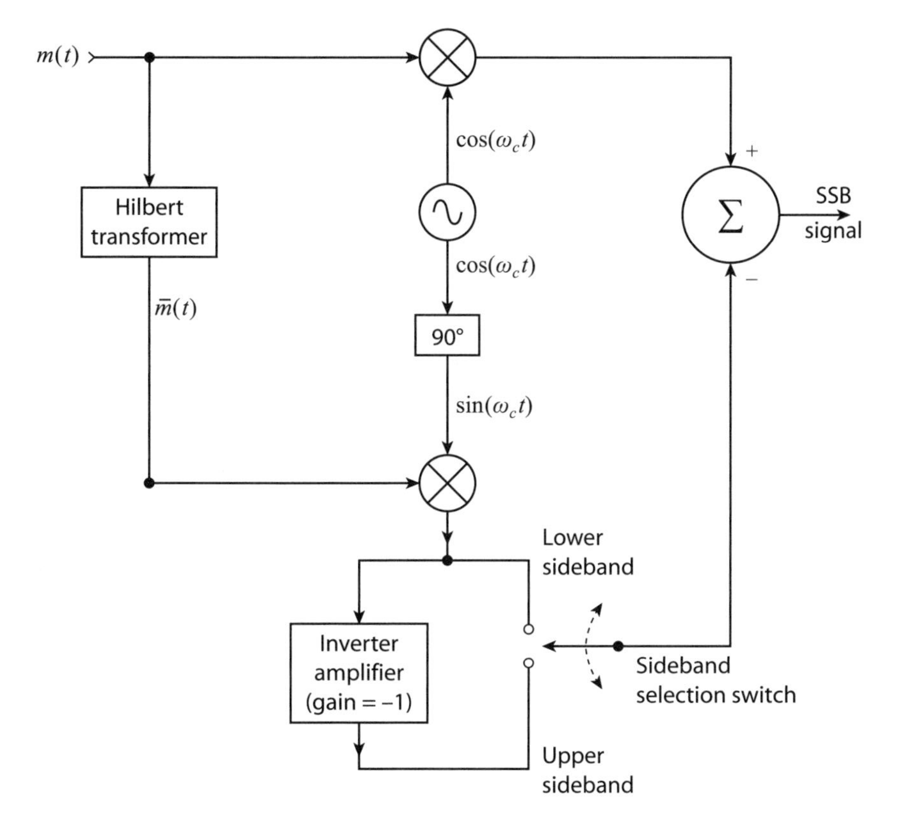

FIGURE 4.2.2. Hartley's SSB radio transmitter circuit.

generate!) in the fifth (bottom) sketch of Figure 4.2.1, which is the upper-sideband SSB spectrum of the *real-valued* rf time signal

$$z_+(t)e^{j\omega_c t} + z_-(t)e^{-j\omega_c t} = m(t)\cos(\omega_c t) - \overline{m}(t)\sin(\omega_c t).$$

If you repeat this entire analysis for the lower-sideband SSB case (and I encourage you to do that) you'll find the resulting time signal is $m(t)\cos(\omega_c t) + \overline{m}(t)\sin(\omega_c t)$. This is the same as for the upper-sideband case except that the minus sign has been replaced with a plus sign. Thus, by adding an inverting amplifier and a switch at the output of the multiplier in the Hilbert transformer signal path, we can select either sideband—upper or lower—with the flip of that switch. Figure 4.2.2 shows the transmitter circuit—Hartley's invention—that does all this, where we *assume* we have a box (a circuit called a *Hilbert transformer*) that for an input of the baseband signal $m(t)$ generates an output of $\bar{m}(t)$.

4.3 The Mathematics of the Hilbert Transformer

It's one thing to draw a box and label it "'Hilbert transformer," as in Figure 4.2.2, but it's entirely different to give a detailed description of the circuitry that is supposedly inside that box. Let me now show you that, first, it is impossible to give such a description (that's the bad news), but it *is* possible to sidestep that perhaps shocking result (that's the good news).

An important theoretical concept in electrical engineering is the *impulse response* of a circuit, defined as the output signal resulting in response to an input of $\delta(t)$. As discussed in section 1.4 (in particular, see note 26 in chapter 1), a circuit is impossible to build if it is not causal, that is, if it generates an anticipatory output before the input appears. The impulse response $h(t)$ of the Hilbert transformer is, with $m(t)=\delta(t)$, given by

$$h(t)=\frac{1}{\pi}\int_{-\infty}^{\infty}\frac{\delta(u)}{t-u}du=\frac{1}{\pi t},\ |t|<\infty,$$

where the sampling property (see section 1.7) of the impulse (which occurs at $u=0$) has been used to evaluate the integral. Since $h(t)$ represents a nonzero output at $t<0$ to an impulse applied at $t=0$, we see that the Hilbert transformer is noncausal, and so it simply does not exist (except on paper)!

So, is Hartley's SSB transmitter circuit in Figure 4.2.2 the radio engineering version of a shaggy dog story? Is all this chapter simply an outrageous joke? Well, of course, you suspect that's not the case, and mathematics again shows us how things are not nearly so grim. To understand what follows, we need to develop a *physical* interpretation of the Hilbert transform.

We start the mathematical development of such an interpretation by observing that if $m(t)=\delta(t)$, then $M(\omega)=1$, and so the Fourier transform of the Hilbert transformer output is the product

$M(\omega)F[h(t)] = F[h(t)] = F\left[\frac{1}{\pi t}\right]$, which describes how the Hilbert transformer modifies the input-frequency spectrum $M(\omega)$ to arrive at the output-frequency spectrum. You might ask, How do we know that the *product* $M(\omega)F[h(t)]$ is the Fourier transform of the output? We know that because a theoretical result from electrical engineering (which I have *not* proven here) is that the output of a linear, time-invariant circuit[8] is the time convolution of the input with the circuit's impulse response,[9] and we know that convolution in the time domain pairs with multiplication in the frequency domain.

So, in other words, $F\left[\frac{1}{\pi t}\right]$ is the transfer function $H(\omega)$ of the Hilbert transformer. The calculation of $H(\omega)$ immediately leads us to a historically famous problem in mathematics, that of evaluating the integral

$$\int_{-\infty}^{\infty} h(t)e^{-j\omega t}dt = \frac{1}{\pi}\int_{-\infty}^{\infty} \frac{e^{-j\omega t}}{t}dt$$
$$= \frac{1}{\pi}\left[\int_{-\infty}^{\infty} \frac{\cos(\omega t)}{t}dt - j\int_{-\infty}^{\infty} \frac{\sin(\omega t)}{t}dt\right].$$

The first integral is zero, because the integrand is odd, and so we are left with

$$H(\omega) = -\frac{j}{\pi}\int_{-\infty}^{\infty} \frac{\sin(\omega t)}{t}dt.$$

This integral is the famous *Dirichlet's discontinuous integral* (you'll see where the *discontinuous* comes from in just a moment), named after the German mathematician Gustav Peter Lejeune Dirichlet

8 The linearity of the Hilbert transformer immediately follows from its integral definition. That is, the transform of a sum of time signals is the sum of the individual transforms, and the transform of $km(t)$ is k times the transform of $m(t)$ The time invariance of the Hilbert transformer (which means that a shift in the input results in the same shift in the output) is easily found from the integral definition, as well—see Challenge Problem 4.2.

9 You can find a proof of this in any undergraduate linear systems text. See, for example, David K. Cheng, *Analysis of Linear Systems*, Addison-Wesley, 1959, pp. 233–234.

(1805–1859). Its evaluation is just within the scope of sophomore calculus (see Challenge Problem 4.3) that shows

$$\int_0^\infty \frac{\sin(ax)}{x}dx = \begin{cases} \frac{\pi}{2}, & a>0 \\ 0, & a=0 \\ -\frac{\pi}{2}, & a<0 \end{cases},$$

or since the integrand is even (and writing ω for a and t for x)[10],

$$\int_{-\infty}^\infty \frac{\sin(\omega t)}{t}dt = \begin{cases} \pi, & \omega>0 \\ 0, & \omega=0 \\ -\pi, & \omega<0 \end{cases}.$$

So,

$$H(\omega) = \begin{cases} -j, & \omega>0 \\ 0, & \omega=0 \\ j, & \omega<0 \end{cases}$$

is the transfer function of the Hilbert transformer. This is often written as $H(\omega)=-j\,\text{sgn}(\omega)$, where sgn (ω) is the *sign* function. That is, sgn $(\omega)=+1$ when $\omega>0$, and $\text{sgn}(\omega)=-1$ when $\omega<0$. The sgn (ω) function is obviously an odd function.

Since $\bar{m}(t)$ comes out of a Hilbert transformer when $m(t)$ goes in, we see that the spectrum of $\bar{m}(t)$ is given by

10 This integral, for the $\omega=1$ case, was actually first evaluated by Euler, sometime just a few years before his death (long before Dirichlet was born). To go from the $\omega=1$ case to the case of arbitrary ω is borderline trivial (try it and see). The integral appears all through communication and information theory, and so it should be no surprise to see it in radio engineering. If somehow given the opportunity to make my case directly to a resurrected Hardy, that radio offers a wonderful example of the intersection of "real" mathematics and practical technology, I would be quite sure to direct his attention to how Dirichlet's integral arises naturally in the theory of SSB radio. That's because Hardy was fascinated with this integral (see, for example, the discussion in *The G. H. Hardy Reader*, note 7 in the preface, pp. 311–321). And I think that fascination might well have sparked his interest in (if not learning how to solder) at least learning how radio works.

$$\bar{M}(\omega) = H(\omega)M(\omega) = -j\ \mathrm{sgn}(\omega)M(\omega).$$

From this it immediately follows that

$$|\bar{M}(\omega)|^2 = |M(\omega)|^2 \text{ for all } \omega,$$

which says the energy (see section 1.6) of $\bar{m}(t)$ is precisely equal to the energy of $m(t)$. If $m(t)$ is a baseband signal, then so, too, is $\bar{m}(t)$.

You can use the transfer function of the Hilbert transformer to also quickly conclude that the Hilbert transform of the Hilbert transform of $m(t)$ is $-m(t)$. That is, $\bar{\bar{m}}(t) = -m(t)$. The proof of this is straightforward. Let $y(t) = \bar{m}(t)$. Then, $Y(\omega) = \bar{M}(\omega)$, and so, as we just found,

$$\bar{M}(\omega) = Y(\omega) = \begin{cases} -j\ M(\omega),\ \omega > 0 \\ +j\ M(\omega),\ \omega < 0 \end{cases}.$$

In the same way,

$$\bar{Y}(\omega) = \begin{cases} -j\ Y(\omega),\ \omega > 0 \\ +j\ Y(\omega),\ \omega < 0 \end{cases} = \begin{cases} -j\ \{-j\ M(\omega)\},\ \omega > 0 \\ +j\ \{+j\ M(\omega)\},\ \omega < 0 \end{cases}.$$

But $\bar{Y}(\omega) = \bar{\bar{M}}(\omega)$, and so (as $j = \sqrt{-1}$),

$$\bar{\bar{M}}(\omega) = \begin{cases} -M(\omega),\ \omega > 0 \\ -M(\omega),\ \omega < 0 \end{cases} = -M(\omega),$$

and this holds for *any* ω. That is, the Fourier transform of $\bar{\bar{m}}(t)$ is $-M(\omega)$, the negative of the Fourier transform of $m(t)$. So, $\bar{\bar{m}}(t) = -m(t)$.

To end this section, here's one more interesting mathematical property of the Hilbert transform. For $m(t)$ any real time signal, $m(t)$ and its transform are *orthogonal*; that is,

$$\int_{-\infty}^{\infty} m(t)\bar{m}(t)dt = 0.$$ [11]

11 As an example of this, suppose the input to a Hilbert transformer is $m(t) = \delta(t)$. By definition, the output of the Hilbert transformer is $\bar{\delta}(t)$, as well as being the impulse response of the transformer, $\frac{1}{\pi t}$. So, $\bar{\delta}(t) = \frac{1}{\pi t}$, and the orthogonality result says

This is only a bit more involved to show if you recall an earlier result from section 2.3: if $m(t)$ and $g(t)$ are any two real-time functions, then we have the Fourier pair

$$m(t)g(t) \leftrightarrow \frac{1}{2\pi}\int_{-\infty}^{\infty} G(u)M(\omega-u)du,$$

and so, if we let $g(t)=\bar{m}(t)$,

$$m(t)\bar{m}(t) \leftrightarrow \frac{1}{2\pi}\int_{-\infty}^{\infty} \bar{M}(u)M(\omega-u)\,du.$$

Now, from earlier in this section we have

$$\bar{M}(\omega) = -j\ \mathrm{sgn}(\omega)M(\omega),$$

and thus

$$m(t)\bar{m}(t) \leftrightarrow \frac{1}{2\pi}\int_{-\infty}^{\infty} -j\ \mathrm{sgn}(u)M(u)M(\omega-u)\,du.$$

That is, the Fourier transform of $m(t)\bar{m}(t)$ is the right-hand side of that pair, and so

$$\int_{-\infty}^{\infty} m(t)\bar{m}(t)e^{-j\omega t}dt = \frac{1}{2\pi}\int_{-\infty}^{\infty} -j\ \mathrm{sgn}(u)M(u)M(\omega-u)\,du.$$

This holds for all values of ω and, in particular, for $\omega=0$. Setting $\omega=0$, we get

$$\int_{-\infty}^{\infty} m(t)\bar{m}(t)dt = \frac{1}{2\pi}\int_{-\infty}^{\infty} -j\ \mathrm{sgn}(u)M(u)M(-u)\,du\ .$$

Since

$\int_{-\infty}^{\infty} \delta(t)\frac{1}{\pi t}dt = 0.$ That is, $\int_{-\infty}^{\infty} \frac{1}{t}\delta(t)dt = 0$, a result you may have already showed in Challenge Problem 1.9.

$$M(\omega) = \int_{-\infty}^{\infty} m(t) e^{-j\omega t} dt,$$

then, as $m(t)$ is real,

$$M(-\omega) = \int_{-\infty}^{\infty} m(t) e^{j\omega t} dt = M^*(\omega),$$

and therefore

$$\begin{aligned}\int_{-\infty}^{\infty} m(t)\bar{m}(t)dt &= \frac{-j}{2\pi}\int_{-\infty}^{\infty} \text{sgn}(u)M(u)M^*(u)du \\ &= \frac{-j}{2\pi}\int_{-\infty}^{\infty} \text{sgn}(u)|M(u)|^2 du.\end{aligned}$$

Since $m(t)$ is real, we know $|M(u)|^2$ is even, and as $\text{sgn}(u)$ is odd, we can conclude that the integrand on the right is odd, and so the integral is zero. That is,

$$\int_{-\infty}^{\infty} m(t)\bar{m}(t)dt = 0.$$

Okay, that's enough (for now) of the mathematics of the Hilbert transformer (other important mathematical properties of the transform are left for you to ponder in Challenge Problem 4.4). Let's now move on to fulfill my promise to give you a *physical* interpretation of what a Hilbert transformer actually does.

4.4 The Physics of the Hilbert Transformer

I'll start by simply telling you what the Hilbert transformer does *physically*, and then I'll show you the mathematical justification for the physics. *The Hilbert transformer takes each frequency component of m(t) and delays it in time by one-fourth of a period.* That's it! What could be simpler? It's *so* simple, in fact, that you should ask yourself, Where's the fly in the ointment? And, after all, there *has* to be a fly somewhere, because the Hilbert transformer is impossible to construct. The answer is easy, once seen: delaying any *particular* frequency component by $90^\circ = \frac{\pi}{2}$ radians (one-fourth of a period) *is* easy, but that delay is different for each frequency component, and that broad requirement results in the physical impossibility of the Hilbert transformer. (But in just a moment I'll show you how radio engineers get around this apparent mathematical obstacle.)

But first, where does this physical interpretation come from? Let $f(t)$ be the sinusoidal signal $f(t) = \cos(\omega_0 t)$ with period $T = \frac{2\pi}{\omega_0}$, and so $\omega_0 T = 2\pi$. This is a representative frequency component of the many frequencies that are present in any "interesting" $m(t)$. Then,

$$f\left(t - \frac{T}{4}\right) = \cos\left[\omega_0\left(t - \frac{T}{4}\right)\right] = \cos\left(\omega_0 t - \frac{\omega_0 T}{4}\right) = \cos\left(\omega_0 t - \frac{\pi}{2}\right).$$

Recalling Euler's identity, we have

$$f\left(t - \frac{T}{4}\right) = \frac{e^{j\left(\omega_0 t - \frac{\pi}{2}\right)} + e^{-j\left(\omega_0 t - \frac{\pi}{2}\right)}}{2} = \frac{1}{2}\left[e^{j\left(\omega_0 t - \frac{\pi}{2}\right)} + e^{j\left(-\omega_0 t + \frac{\pi}{2}\right)}\right]$$

$$= \frac{1}{2}\left[e^{-j\frac{\pi}{2}} e^{j\omega_0 t} + e^{j\frac{\pi}{2}} e^{-j\omega_0 t}\right] = \frac{1}{2}[-je^{j\omega_0 t} + je^{-j\omega_0 t}].$$

Referring to Figure 1.2.1, we see that the two exponential terms in the last, rightmost pair of brackets are counterrotating vectors in the complex plane. That is, with increasing time, $e^{j\omega_0 t}$ rotates counterclockwise (CCW) while $e^{-j\omega_0 t}$ rotates clockwise (CW). You'll recall we associate CCW rotation with positive frequency (ω_0), and CW rotation with negative frequency ($-\omega_0$). So, in the rightmost brackets we see the positive-frequency CCW rotating vector $e^{j\omega_0 t}$ multiplied by $-j$, and the negative-frequency CW rotating vector $e^{-j\omega_0 t}$ multiplied by $+j$. But that's just a word description of the transfer function of the Hilbert transformer.

Now, how do radio engineers get around the "difficulty" of the physical nonexistence of the Hilbert transformer—a difficulty that says it's impossible to build a circuit with one input which generates two constant, equal-amplitude outputs that, at any frequency, are 90° out of phase? (In radio engineering jargon, this is called a *phase splitter*.) We do that with a three-step argument. First, we observe that we need our circuit to work over only a *finite* interval of frequencies, not over *all* frequencies. For acceptable SSB voice transmission, this interval is typically taken to be 300 to 3,000 Hz. Second, we demand that the two output signals have to be only "almost" 90° out of phase over the frequency interval 300 to 3,000 Hz. And finally, referring to Hartley's circuit in Figure 4.2.2, we additionally don't require a *fixed* 0° phase shift in the upper path to the top mixer and a *fixed* 90° phase shift in the lower path to the bottom mixer, but, rather, we'll accept a *variable* phase shift of $\varphi(\omega)$ in the upper path *if* it is accompanied by a phase shift of $\varphi(\omega)+90°$ in the lower path. If we can arrange for these three conditions, then Hartley's SSB transmitter circuit still works.

I'll show you the mathematics that supports that last sentence in just a bit, but let me first elaborate a bit more on the last paragraph. To replace the Hilbert transformer in Figure 4.2.2, we'll instead use the configuration shown in Figure 4.4.1, where $m(t)$ is the output of a bandpass filter that guarantees $m(t)$ is limited to the audio frequencies of 300 to 3,000 Hz. The outputs of the two all-pass filters, A and B, over the frequency interval 300 to 3,000 Hz are supposed to be (1) constant, with equal amplitudes, and (2) have a phase difference of 90°. The obvious question now, of course, is, What is an all-pass filter?

If you do Challenge Problem 4.1, you'll see that the circuit in Figure CP4.1 has a transfer function such that $|H(\omega)|^2=1$ for *all* ω (hence the name *all-pass* filter), which might make you wonder, What good is a filter that lets *everything* through, no matter the frequency? After all, a bare wire can do that! The answer is, the *phase shift* that occurs *is* a function of frequency (a bare wire introduces no phase shift). Indeed, when you work through the analysis of that problem you should arrive at the following expression for the phase shift:

$$\varphi(\omega)=-2\tan^{-1}\left(\frac{\omega}{\omega_0}\right),\ \omega_0=\frac{1}{RC}.$$

A paper[12] in the late 1920s by the Bell Laboratories electrical engineer Otto Julius Zobel (1887–1970)precipitated great interest in discovering practical circuitry that could go into making the filter

12 O. J. Zobel, "Distortion Correction in Electrical Circuits," *Bell System Technical Journal*, July 1928, pp. 438–534. Zobel's name is commonly associated with lattice filters (see Figure CP1.5).

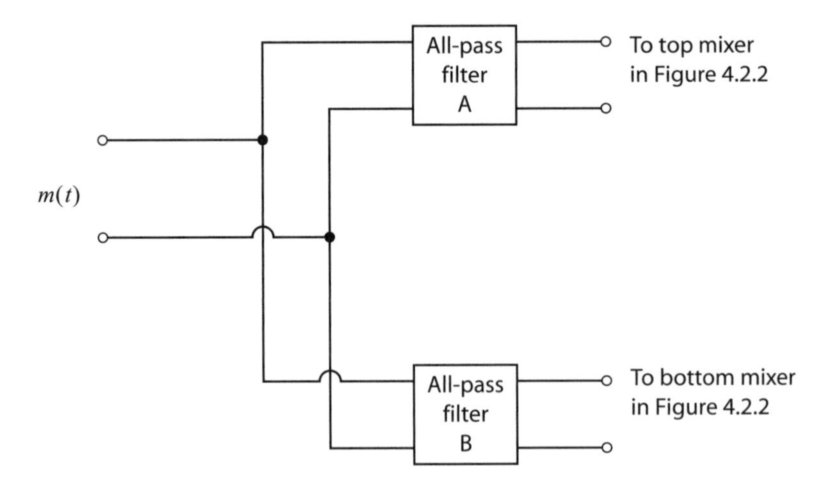

FIGURE 4.4.1. Making a 90° phase-*difference* circuit from two all-pass filters.

boxes A and B of Figure 4.4.1, to give constant, equal-amplitude signals that have a phase *difference* of 90° over the passband of the filters. That search culminated in the early 1950s with the invention by Donald K. Weaver Jr. (1924–1998), an electrical engineer at the Stanford Research Institute, of astonishingly simple circuitry (using just a handful of resistors and capacitors).[13]

Weaver's circuitry is shown in Figure 4.4.2, with two copies of the same circuit (using, of course, different component values; hence, for example, Z_{1A} and Z_{1B} in the figure distinguish the two versions of Z_1) forming the A and B filters. Notice, in particular, that the input to each filter box is from the output of a center-tapped transformer (refer to Figure 2.4.3), where each of the impedances Z_1 and Z_3 is a resistor in parallel with a capacitor, and Z_2 is a resistor in series with a capacitor. That is, the expressions for Z_1 and Z_3 have the forms (you should verify this)

$$Z_1 = \frac{R_1}{1+(\omega R_1 C_1)^2} - j\frac{\omega R_1^2 C_1}{1+(\omega R_1 C_1)^2},$$

13 Donald K. Weaver Jr., "Design of RC Wide-Band 90-Degree Phase-Difference Network," *Proceedings of the IRE*, April 1954, pp. 671–676.

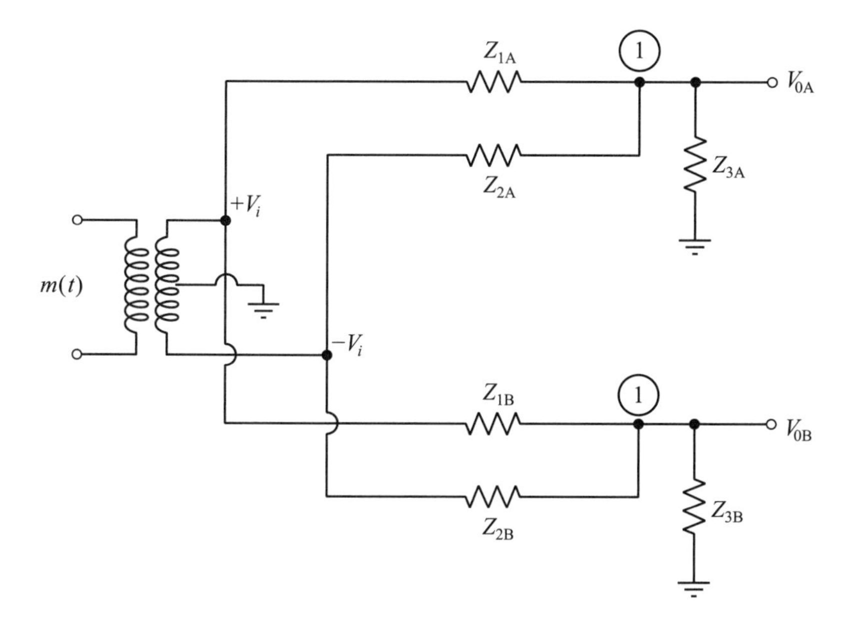

FIGURE 4.4.2. A detailed version of Figure 4.4.1.

and

$$Z_3 = \frac{R_3}{1+(\omega R_3 C_3)^2} - j\frac{\omega R_3^2 C_3}{1+(\omega R_3 C_3)^2},$$

while the expression for Z_2 has the form (this should be obvious by inspection)

$$Z_2 = R_2 - j\frac{1}{\omega C_2}.$$

So, besides the center-tapped transformer, Weaver's circuit has just six resistors and six capacitors.

In his paper, Weaver presented the computational steps used to arrive at the values of the circuit components: "'cookbook' fashion without any need for an understanding of the underlying theory [which, in fact, is simply undergraduate circuit theory]." Starting with just three parameters, (1) f_1, the low-frequency start of the all-pass passband; (2) f_2, the high-frequency end of the all-pass passband; and (3) ε, the maximum allowable deviation from 90° of the phase difference, the component values are calculated with some routine

Table 4.4.1. Weaver's circuit values

A Filter	B Filter
$R_1 = 562.7 \times 10^3$	$R_1 = 562.7 \times 10^3$
$R_2 = 100 \times 10^3$	$R_2 = 100 \times 10^3$
$R_3 = 228.9 \times 10^3$	$R_3 = 228.9 \times 10^3$
$C_1 = 592.1 \times 10^{-12}$	$C_1 = 150.1 \times 10^{-12}$
$C_2 = 3{,}331 \times 19^{-12}$	$C_2 = 844.8 \times 10^{-12}$
$C_3 = 1{,}456 \times 10^{-12}$	$C_3 = 369.2 \times 10^{-12}$

algebra. Weaver's paper concludes with a detailed example, using $f_1 = 300$ Hz, $f_2 = 3{,}000$ Hz, and $\varepsilon = 1.1°$ The resulting component values (in ohms and farads) are shown in Table 4.4.1.

We can confirm that Weaver's circuit and component values do indeed work, with the following analysis (coupled with some extensive computer calculations that I'll soon describe). If we apply Kirchhoff's current law at either of the points marked ① in Figure 4.4.2, we have

$$\frac{V_i - V_o}{Z_1} + \frac{-V_i - V_o}{Z_2} = \frac{V_o}{Z_3},$$

which quickly reduces to give the transfer function

$$\frac{V_o}{V_i} = \frac{Z_2 Z_3 - Z_1 Z_3}{Z_1 Z_2 + Z_2 Z_3 + Z_1 Z_3}.$$

The transfer function for each of the A and B filters has this form.

Now, imagine that we evaluate the numerical value of these two transfer functions in 1 Hz steps, from f_1 to f_2. At each step, each of the six Z's is calculated, and from them the transfer function for each filter,

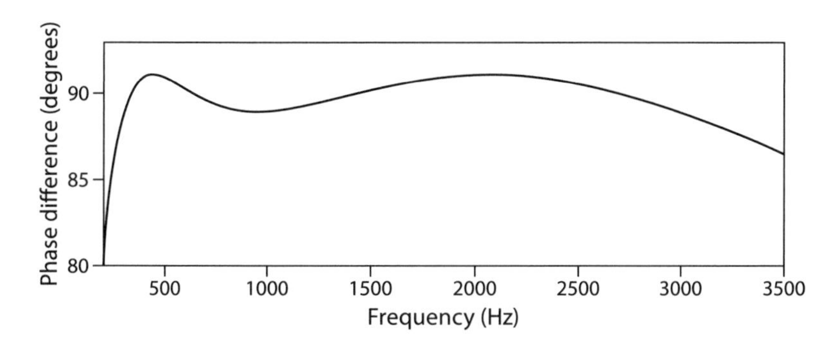

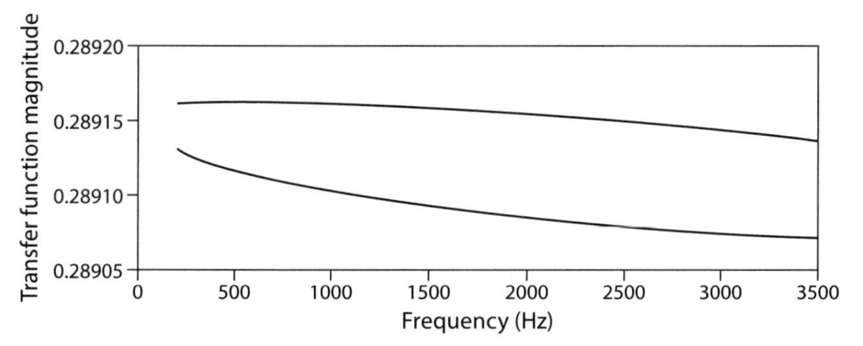

FIGURE 4.4.3. The phase difference (top) and the filter transfer function magnitudes (bottom) of Weaver's circuit in Figure 4.4.2 with $\epsilon = 1.1°$.

at each frequency step, follows with just a few more multiplications and a division. (I'm not counting mere additions and subtractions.) Next, the phase difference between each filter's output, and the amplitude of each filter's output, can easily be found. With nearly 3,000 frequency steps from 300 to 3,000 Hz, this complex-valued arithmetic requires less than five seconds on an ordinary laptop running modern software (like MATLAB). A resurrected Hardy would, I believe, despite his suspicious view of technology, be utterly fascinated by such computational wizardry (for more on this point, see Challenge Problem 4.5). The result is displayed in Figure 4.4.3.[14]

14 The computational power of my laptop is itself left in the dust by that of modern video game consoles, which perform *trillions* of floating-point operations per second. The basic version of the PS5 (PlayStation 5), for example, sold by SONY for less than $400 and now in millions of homes, is benchmarked at more than 10 teraflops (10^{13} floating-point operations per second), while my seven-year-old PS4 runs at just under 2 teraflops. Hardy would, of course, be appalled at the violent games in the *Hitman* and *Sniper Elite* franchises (of which your author admits being an ardent fan, something I think it best

The top plot shows that the phase difference does indeed stay within 1.1° of 90° over the entire interval from 300 to 3,000 Hz (the plot actually extends from 200 to 3,500 Hz, to show how the phase performance of the filters quickly degrades at frequencies outside the design interval). This is pretty impressive, but Weaver mentions at the end of his paper that he had constructed filters that worked over the same passband with a maximum phase deviation of just 0.2°. That's *very* impressive! The bottom plot in Figure 4.4.3 shows that while the magnitudes of the filter transfer function are not equal, they differ only in the fourth decimal place.

The perfect Hilbert transformer is impossible to build, yes, but extremely good approximations to it *can* be built (over a finite passband) that are arbitrarily close to perfection.

Now, to finish this section, let me clean up one last loose end, namely, the claim I made earlier that if the two paths in Figure 4.2.2 (Hartley's SSB transmitter circuit) leading to the upper and lower path mixers have a phase difference of 90°, then we still have an SSB transmitter. To see this, suppose the phase shift through the Weaver all-pass filter that feeds into the upper mixer is $\varphi(\omega)$, and the phase shift through the all-pass filter that feeds into the lower mixer is $\varphi(\omega)+\frac{\pi}{2}$. (The case of $\varphi(\omega)=0$ is that of the Hilbert transformer.) Let's now explore how $\varphi(\omega)$ shows up in the output of Hartley's transmitter.

Fix your attention in Figure 4.4.1 on one (any) frequency component of $m(t)$, and call that frequency ω_p. If we write that component as $\cos(\omega_p t)$ then, with $\varphi(\omega)$ as the phase shift through filter A, we have the signal $\cos\{\omega_p t-\varphi(\omega)\}$ going to the upper mixer. This means the signal $\cos\left\{\omega_p t-\left[\varphi(\omega)+\frac{\pi}{2}\right]\right\}=\cos\left\{\omega_p t-\varphi(\omega)-\frac{\pi}{2}\right\}$ is the signal going to the lower mixer.[15] So, the signals out of the two mixers are (written out in detail to avoid dropping one or more terms):

---

to perhaps not tell Hardy), but I believe he would have enjoyed games based on puzzle-solving, like *Myst* and *Portal*.

15 The phase shifts through the two filters A and B are *delays*, and so we *subtract* the phase shifts from $\omega_p t$ to get the arguments of the two cosine signals going to the mixers.

$$\text{upper mixer: } \cos\{\omega_p t - \varphi(\omega)\}\cos(\omega_c t)$$
$$\text{lower mixer: } \cos\left\{\omega_p t - \varphi(\omega) - \frac{\pi}{2}\right\}\sin(\omega_c t).$$

Expanding using well-known trigonometric identities, for the upper mixer we have

$$\begin{aligned}
&\cos\{\omega_p t - \varphi(\omega)\}\cos(\omega_c t)\\
&= \left[\cos(\omega_p t)\cos\{\varphi(\omega)\} + \sin(\omega_p t)\sin\{\varphi(\omega)\}\right]\cos(\omega_c t)\\
&= \cos(\omega_p t)\cos(\omega_c t)\cos\{\varphi(\omega)\} + \sin(\omega_p t)\cos(\omega_c t)\sin\{\varphi(\omega)\}\\
&= \frac{1}{2}\{\cos(\omega_c t - \omega_p t) + \cos(\omega_c t + \omega_p t)\}\cos\{\varphi(\omega)\}\\
&\quad + \frac{1}{2}\{\sin(\omega_p t - \omega_c t) + \sin(\omega_c t + \omega_p t)\}\sin\{\varphi(\omega)\}\\
&= \frac{1}{2}\left[\cos\{(\omega_c - \omega_p)t\} + \cos\{(\omega_c + \omega_p)t\}\right]\cos\{\varphi(\omega)\}\\
&\quad + \frac{1}{2}\left[-\sin\{(\omega_c - \omega_p)t\} + \sin\{(\omega_c + \omega_p)t\}\right]\sin\{\varphi(\omega)\},
\end{aligned}$$

or, finally, the output of the upper mixer is

$$\begin{aligned}
&\frac{1}{2}\left[\cos\{(\omega_c - \omega_p)t\}\cos\{\varphi(\omega)\} - \sin\{(\omega_c - \omega_p)t\}\sin\{\varphi(\omega)\}\right]\\
&+ \frac{1}{2}\left\{\cos\{(\omega_c + \omega_p)t\}\cos\{\varphi(\omega)\} + \sin\{(\omega_c + \omega_p)t\}\sin\{\varphi(\omega)\}\right\}.
\end{aligned}$$

If we repeat this procedure for the output of the lower mixer, we have

$$\begin{aligned}
\cos\left\{\omega_p t - \varphi(\omega) - \frac{\pi}{2}\right\}\sin(\omega_c t) &= \cos\left\{\omega_p t - \left[\varphi(\omega) + \frac{\pi}{2}\right]\right\}\sin(\omega_c t)\\
&= \cos(\omega_p t)\cos\left\{\varphi(\omega) + \frac{\pi}{2}\right\}\sin(\omega_c t)\\
&\quad + \sin(\omega_p t)\sin\left\{\varphi(\omega) + \frac{\pi}{2}\right\}\sin(\omega_c t),
\end{aligned}$$

or as $\cos\left\{\varphi(\omega) + \frac{\pi}{2}\right\} = -\sin\{\varphi(\omega)\}$, and as $\sin\left\{\varphi(\omega) + \frac{\pi}{2}\right\} = \cos\{\varphi(\omega)\}$, the output of the lower mixer is

$$-\cos(\omega_p t)\sin(\omega_c t)\sin\{\varphi(\omega)\}+\sin(\omega_p t)\sin(\omega_c t)\cos\{\varphi(\omega)\}$$
$$=-\frac{1}{2}\left\{\sin(\omega_c t-\omega_p t)+\sin(\omega_c t+\omega_p t)\right\}\sin\{\varphi(\omega)\}$$
$$+\frac{1}{2}\left\{\cos(\omega_c t-\omega_p t)-\cos(\omega_c t+\omega_p t)\right\}\cos\{\varphi(\omega)\},$$

or, finally, the output of the lower mixer is

$$\frac{1}{2}\left[\cos\left\{(\omega_c-\omega_p)t\right\}\cos\{\varphi(\omega)\}-\sin\left\{(\omega_c-\omega_p)t\right\}\sin\{\varphi(\omega)\}\right]$$
$$-\frac{1}{2}\left[\cos\left\{(\omega_c+\omega_p)t\right\}\cos\{\varphi(\omega)\}+\sin\left\{(\omega_c+\omega_p)t\right\}\sin\{\varphi(\omega)\}\right].$$

Now, if we *add* the upper-path and the lower-path mixer outputs, we see that the result is

$$\cos\left\{(\omega_c-\omega_p)t\right\}\cos\{\varphi(\omega)\}-\sin\left\{(\omega_c-\omega_p)t\right\}\sin\{\varphi(\omega)\}$$
$$=\cos\left\{(\omega_c-\omega_p)t+\varphi(\omega)\right\},$$

which is, indeed, a signal at only the *lower* sidetone frequency, in agreement with Figure 4.2.2. Alternatively, if we *subtract* the lower-path mixer output from the upper-path mixer output, we see that the result is

$$\cos\left\{(\omega_c+\omega_p)t\right\}\cos\{\varphi(\omega)\}+\sin\left\{(\omega_c+\omega_p)t\right\}\sin\{\varphi(\omega)\}$$
$$=\cos\left\{(\omega_c+\omega_p)t-\varphi(\omega)\right\},$$

which is, indeed, a signal at *only* the *upper* sidetone frequency, again in agreement with Figure 4.2.2.

But what, you are no doubt wondering, *about the $\varphi(\omega)$ that appears in our final expressions?* After all, $\varphi(\omega)$ *is* a function of frequency, and so each frequency component of $m(t)$ will experience a different phase shift—won't that "scramble up" $m(t)$ at the receiver? In theory, yes, but in fact, no, because for an audio $m(t)$, that is, a signal meant to be *heard*, the human ear is experimentally found to be insensitive to phase. The actual $m(t)$ and the "phase-scrambled" $m(t)$ sound pretty much the same.

4.5 Receiving an SSB Signal

At this point we have launched an SSB signal into space. How does a distant receiver recover $m(t)$ from that signal? To keep the mathematics straightforward, let's return to Hartley's SSB transmitter circuit of Figure 4.2.2 incorporating a perfect Hilbert transformer (which we know from Weaver can be approximated as closely as we wish), with the antenna signal at the receiver[16] being

$$r(t) = m(t)\cos(\omega_c t) \pm \bar{m}(t)\sin(\omega_c t).$$

The $\pm$ is, of course, either + or – (not both) depending on which sideband we are using. Be quite clear, too, on the value of ω_c: (1) it is generated at the *transmitter*, and (2) there is no carrier at that frequency in $r(t)$. This may remind you of the situation in chapter 2 (see Figure 2.1.3) with synchronous demodulation, and to recover $m(t)$ from our single-sideband $r(t)$ we'll do exactly the same thing we did then. That is, let's multiply $r(t)$ with a locally (at the receiver) generated signal that is as close as possible to the transmitter oscillator's frequency. That is, let's imagine our receiver (somehow) generates the signal $\cos\{(\omega_c + \Delta\omega)t + \theta\}$, where $\Delta\omega$ and θ are both zero *if we have achieved perfect synchrony*. So, forming

$$r(t)\cos\{(\omega_c + \Delta\omega)\,t + \theta\},$$

we have the signal

$$\begin{aligned}&\left[m(t)\cos(\omega_c t) \pm \bar{m}(t)\sin(\omega_c t)\right]\cos\{(\omega_c + \Delta\omega)t + \theta\}\\ &= \frac{1}{2}m(t)\left[\cos\{(2\omega_c + \Delta\omega)t + \theta\} + \cos\{(\Delta\omega)t + \theta\}\right]\\ &\quad \pm \frac{1}{2}\bar{m}(t)\left[\sin\{(2\omega_c + \Delta\omega)t + \theta\} + \sin\{(\Delta\omega)t + \theta\}\right],\end{aligned}$$

16 The $r(t)$ expression was derived in section 4.2 as the transmitted signal, and so there would, of course, be some time delay (phase shift) when it is received at a distant location. Since we are now working strictly inside the receiver, however, we can (for mathematical ease) take that delay as zero.

which, after low-pass filtering, gives the detected signal as

$$\frac{1}{2}m(t)\cos\{(\Delta\omega)t+\theta\}\pm\frac{1}{2}\bar{m}(t)\sin\{(\Delta\omega)t+\theta\}.$$

If $\Delta\omega$ and θ are both zero, then this detected signal reduces to $\frac{1}{2}m(t)$ and so, as claimed, the original baseband signal has been successfully recovered.

But what happens if we don't have perfect synchronization? Suppose, for instance, that $\Delta\omega=0$, but $\theta\neq 0$. Then, the detected signal is $\frac{1}{2}m(t)\cos\{\theta\}\pm\frac{1}{2}\bar{m}(t)\sin\{\theta\}$. The first term should remind you, as I mentioned earlier, of the result from section 2.1 for phase mismatch in the synchronous demodulation of double-sideband, suppressed-carrier AM, but now we also have a second term involving the Hilbert transform of $m(t)$. What impact does that extra term have?

Focus your attention on an arbitrary, particular frequency component of $m(t)$, which we'll write as $\cos(\omega_p t)$. Then, for that component we know (because you are going to show it in Challenge Problem 4.4) $\bar{m}(t)=\sin(\omega_p t)$, and so the detected signal is

$$\frac{1}{2}\{\cos(\omega_p t)\cos(\theta)+\sin(\omega_p t)\sin(\theta)\}.$$

Recalling the identity you derived (right?) in Challenge Problem 1.11,

$$A\cos(\alpha t)+B\sin(\alpha t)=\sqrt{A^2+B^2}\cos\left\{\alpha t-\tan^{-1}\left(\frac{B}{A}\right)\right\}$$

where A, B, and α are constants, then with $A=\cos(\theta)$, $B=\sin(\theta)$, and $\alpha=\omega_p$, the detected signal is

$$\begin{aligned}\frac{1}{2}\left[\sqrt{\cos^2(\theta)+\sin^2(\theta)}\right]\cos\left\{\omega_p t-\tan^{-1}\left(\frac{\sin(\theta)}{\cos(\theta)}\right)\right\}\\ =\frac{1}{2}\cos\left[\omega_p t-\tan^{-1}\{\tan(\theta)\}\right]\\ =\frac{1}{2}\cos(\omega_p t-\theta),\end{aligned}$$

which is just a nonattenuated, phase-shifted version of $m(t)$. Since this is so for every frequency component of $m(t)$—recall that ω_p is arbitrary—then unlike in the case of synchronous demodulation of a double-sideband, suppressed carrier AM signal, there is *no* phase mismatch that affects the amplitude of the detected SSB signal. Rather, each frequency component of the detected $m(t)$ simply experiences the same phase shift, and since the human ear is relatively insensitive to phase, the intelligibility of the SSB signal survives.

Our next question is the obvious one: What if $\theta = 0$ but now $\Delta\omega \neq 0$? In this case, of frequency mismatch at the receiver, the detected signal after low-pass filtering of

$$r(t)\cos\{(\omega_c + \Delta\omega)t + \theta\}$$

is (with $\theta = 0$),

$$\frac{1}{2}m(t)\cos\{(\Delta\omega)t\} \pm \frac{1}{2}\bar{m}(t)\sin\{(\Delta\omega)t\}.$$

The first term represents a time-varying amplitude effect, which can be quite noticeable, even with $\Delta\omega$ "small," but now we again also have a second term involving the Hilbert transform of $m(t)$. And so, again, we ask: What impact does that term have? As before, we concentrate on an arbitrary frequency component of $m(t)$, and so $m(t) = \cos(\omega_p t)$ and $\bar{m}(t) = \sin(\omega_p t)$. Thus, the detected signal is

$$\begin{aligned}\frac{1}{2}\cos(\omega_p t)\cos\{(\Delta\omega)t\} \pm \frac{1}{2}\sin(\omega_p t)\sin\{(\Delta\omega)t\} \\ = \frac{1}{2}\left[\cos\{(\omega_p + \Delta\omega)t\} + \cos\{(\omega_p - \Delta\omega)t\}\right] \\ \pm \frac{1}{2}\left[\cos\{(\omega_p - \Delta\omega)t\} - \cos\{(\omega_p + \Delta\omega)t\}\right].\end{aligned}$$

Therefore, depending on which sideband we are using (refer to Figure 4.2.2)—recall that we use the minus sign for the upper sideband and the plus sign for the lower sideband—the detected signal is

$\cos\{(\omega_p + \Delta\omega)t\}$ if we are listening to the upper sideband

or

$$\cos\{(\omega_p - \Delta\omega)t\} \text{ if we are listening to the lower sideband.}$$

In either case, we see that the Hilbert transform term has resulted in shifting ω_p by the frequency mismatch.

Since ω_p is arbitrary, then *every* frequency component of $m(t)$ is shifted by the same amount, and that destroys the harmonic relationships that exist among the frequency components in the original $m(t)$, but unless $\Delta\omega$ is "large," the intelligibility of the received signal is not lost. Indeed, if the frequency mismatch can be held to no more than 30 Hz or so, speech reproduction is generally good.[17] For frequency mismatches greater than that people do begin to sound a bit like Donald Duck, but even then, speech is not usually rendered unintelligible. It's important to understand that this effect of frequency mismatch is *not* the same as changing the *pitch* of speech. A pitch change results when, for example, a recording of an audio signal is played back either too slowly or too fast. In that case each frequency component is shifted *proportionally*, which leaves harmonic relationships unchanged.

4.6 Weaver's SSB Transmitter

Two years after publishing his work on phase-difference circuitry, Weaver (now an electrical engineering professor at Montana State College, later University) published an astonishing new paper that essentially rendered his phase-difference work moot. He had discovered yet a new way, a *third* way (an alternative to Carson's sideband rejection filtering, and to Hartley's Hilbert transformer) to build an SSB transmitter,[18] shown in Figure 4.6.1. To see how it works is actually not that difficult.

The vertical skirts on the two LPFs in Figure 4.6.1 may seem as restrictive as is Carson's brute-force sideband filtering approach, but

17 Refer to section 3.3 for one way to get ω_c from a signal without a large carrier.

18 Donald K. Weaver Jr., "A Third Method of Generation and Detection of Single-Sideband Signals," *Proceedings of the IRE*, December 1956, pp. 1703–1705.

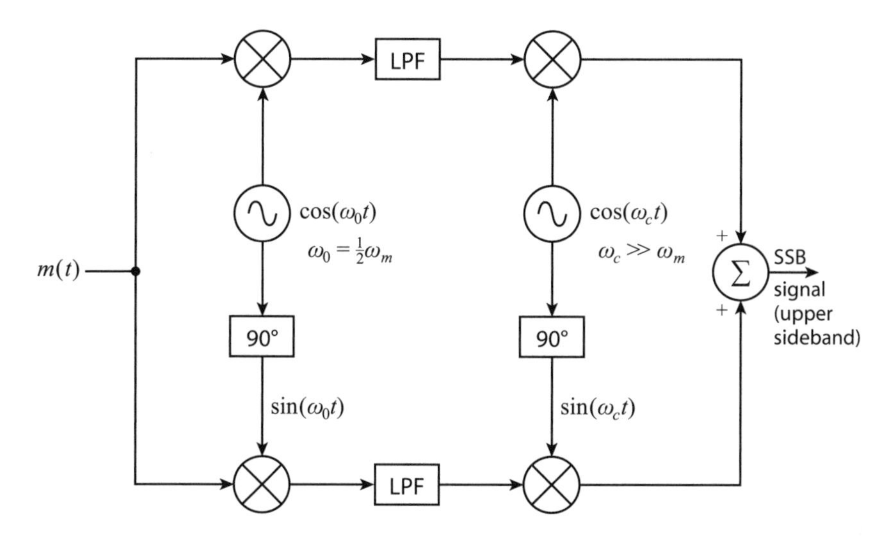

FIGURE 4.6.1. Weaver's SSB transmitter (the low-pass filters have vertical skirts at $\omega = \omega_0$).

Weaver's "third way" has a "trick" to get around that—to be explained at the end of this discussion.

As Figure 4.6.1 shows, a total of *four* heterodyne operations take place in Weaver's transmitter. To understand them we'll simply follow signal spectrums through the circuit with the aid of Fourier theory, complex algebra, and the heterodyne theorem. In the analysis that follows, I'll follow Weaver's notation, and write ω_0 as $\frac{1}{2}\omega_m$ (ω_m is, as before, the highest frequency present in $m(t)$); ω_0 is the heterodyne frequency that appears in the initial mixing operations in both the upper and lower paths of Figure 4.6.1.

Figure 4.6.2a shows the input spectrum of $m(t)$, and we'll follows what happens as that spectrum moves through the *upper* path of Weaver's circuit (what happens in the *lower* path is *almost*, but not quite, the same). The first mixer shifts the spectrum up *and* down the frequency axis by $\omega_0 = \frac{1}{2}\omega_m$, as shown in Figure 4.6.2b. That's because when we multiply $m(t)$ by $\cos\left(\frac{1}{2}\omega_m t\right)$ the Fourier transform of the mixer output is (refer to section 1.6)

$$F\{m(t)\cos(\omega_0 t)\} = F\left\{m(t)\frac{e^{j\omega_0 t} + e^{-j\omega_0 t}}{2}\right\}$$
$$= \frac{1}{2}F\left\{m(t)e^{j\frac{1}{2}\omega_m t}\right\} + \frac{1}{2}F\left\{m(t)e^{-j\frac{1}{2}\omega_m t}\right\},$$

which says the mixer output transform is the input transform shifted *up* the frequency axis by $\frac{1}{2}\omega_m$ and multiplied by $\frac{1}{2}$, as well as shifted *down* the frequency axis by $\frac{1}{2}\omega_m$ and multiplied by $\frac{1}{2}$. This is shown in Figure 4.6.2b, where a $\frac{1}{2}$ beside each shifted spectrum indicates the multiplicative factor.

Figure 4.2.6c shows the result of low-pass filtering Figure 4.6.2b, using the assumption that the upper-path LPF has a vertical skirt at $\omega = \omega_0 = \frac{1}{2}\omega_m$. Figure 4.6.2d shows the up- and downshifted transforms in the upper path that result from the second mixer operation, which multiplies the output of the LPF by $\cos(\omega_c t)$, where ω_c is the radio frequency at which efficient radiation of energy from an antenna occurs. Again, there is a multiplication by a factor of $\frac{1}{2}$, and so in Figure 4.6.2d there is a $\frac{1}{4}$ beside each shifted transform $\left(\left(\frac{1}{2}\right)\left(\frac{1}{2}\right) = \frac{1}{4}\right)$.

Starting in Figure 4.6.2e, we repeat this entire process in the lower path, but now with multiplications using $\sin(\omega_0 t)$ in the first mixer and $\sin(\omega_c t)$ in the second mixer. For the first mixer, Euler's identity tells us that this results in multiplicative factors of $\frac{1}{2j}$ for the upshifted transform and $-\frac{1}{2j}$ for the downshifted transform, which are written next to the associated transform. Figure 4.6.2f shows the transform after passing through the LPF, and Figure 4.6.2g shows the second mixer output, the up- and downshifted transforms resulting after multiplication with $\sin(\omega_c t)$; this multiplication again introduces multiplicative factors of $\frac{1}{2j}$ for the upshifted transform and $-\frac{1}{2j}$ for the downshifted transform. Those new factors should explain to you

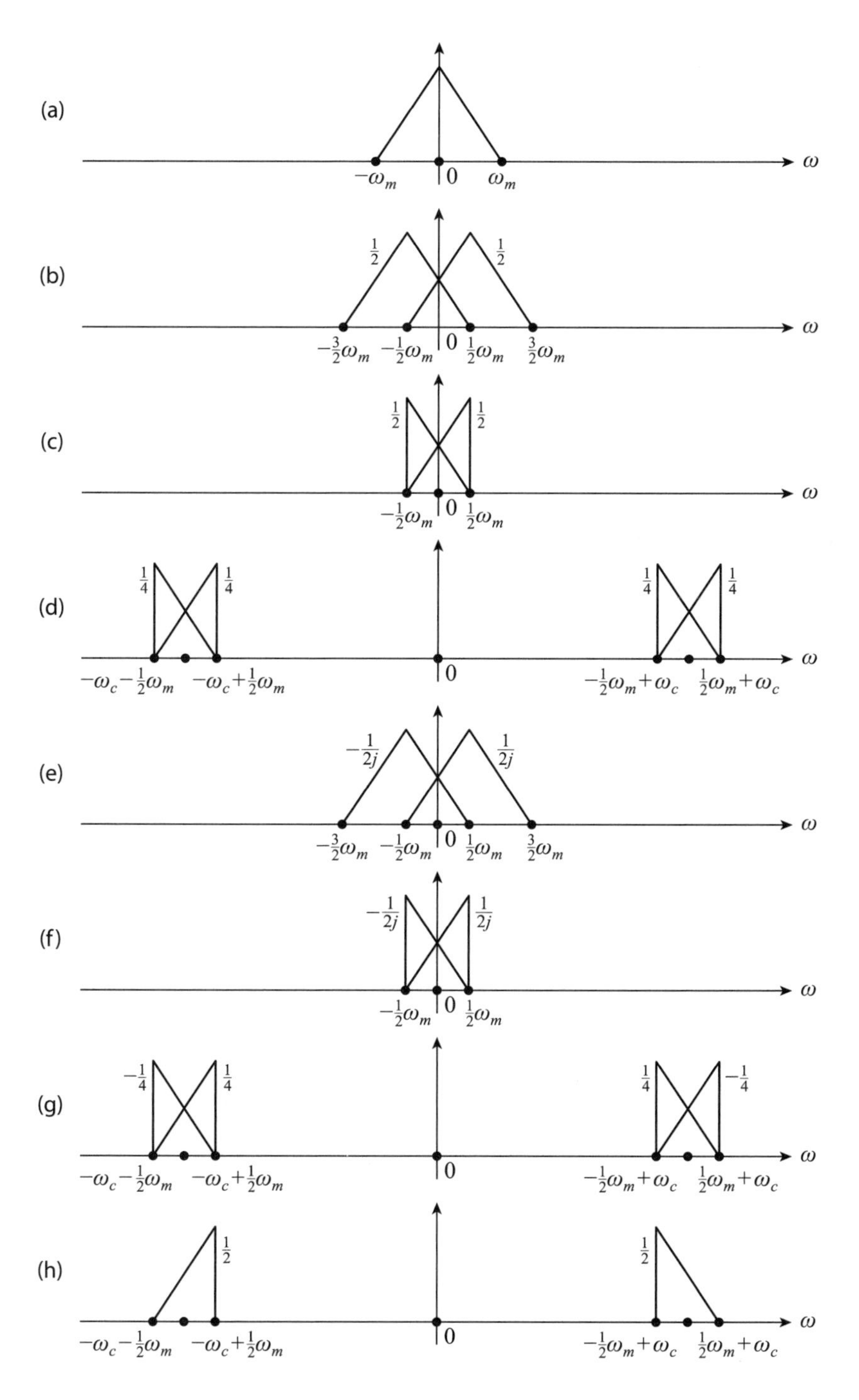

FIGURE 4.6.2. Signal spectrums in Weaver's SSB transmitter.

why $-\frac{1}{4}$ and $\frac{1}{4}$ appear next to the various transforms in Figure 4.6.2g:

$$\left(-\frac{1}{2j}\right)\left(-\frac{1}{2j}\right)=-\frac{1}{4},\ \left(\frac{1}{2j}\right)\left(-\frac{1}{2j}\right)=\frac{1}{4},\text{ and }\left(\frac{1}{2j}\right)\left(\frac{1}{2j}\right)=-\frac{1}{4}.$$

As shown in Figure 4.6.1, if we *add* the final (rightmost) signals in the upper and lower paths (see Figures 4.6.2d and 4.6.2g), we get the transform of the Weaver SSB transmitter output. The $\frac{1}{4}$ and $-\frac{1}{4}$ portions cancel each other, and the $\frac{1}{4}$ and $\frac{1}{4}$ portions add. The result is that only the *upper* sideband of $m(t)$ has survived passage through Weaver's circuit—the lower sideband has self-canceled. If, instead of *adding*, we had *subtracted* the lower-path output from the upper-path output, then the final output would be the *lower* sideband. In practice, the flip of a switch determines whether one adds or subtracts, just as with Hartley's SSB transmitter of Figure 4.2.2.

Finally, to complete this chapter, we discuss the "trick" mentioned earlier that Weaver's "third way" uses to avoid the need for the low-pass filters in Figure 4.6,1 to have vertical skirts. It's based on the fact that the spectrum of $m(t)$ in Weaver's circuit doesn't extend all the way down to dc. Rather, $m(t)$ is assumed to be prefiltered so as to have a lowest-frequency component of 300 Hz. That is, the spectrum of $m(t)$ has a gap of finite width centered on $\omega=0$, a gap that contains no energy. The lack of such low-frequency energy results in very little degradation of intelligibility, but it is in that gap that the LPFs can have a nonvertical skirt without allowing any undesired energy through (because there *is* no energy in the gap).

Weaver's circuit was so novel and unexpected that it prompted a great deal of interest not only among radio professionals but also among radio amateurs. One was so intrigued he built a working transmitter the very next year, directly from Weaver's paper. That's a golden endorsement![19]

19 Howard F. Wright, "The Third Method of S.S.B.: How It Works in Theory and Practice," *QST*, September 1957, pp. 11–15. Wright's construction project was truly a home-basement effort; he wrote: "[T]he parts were largely 'scrounged' from interested bystanders and any part that would work, regardless of size or shape, was used."

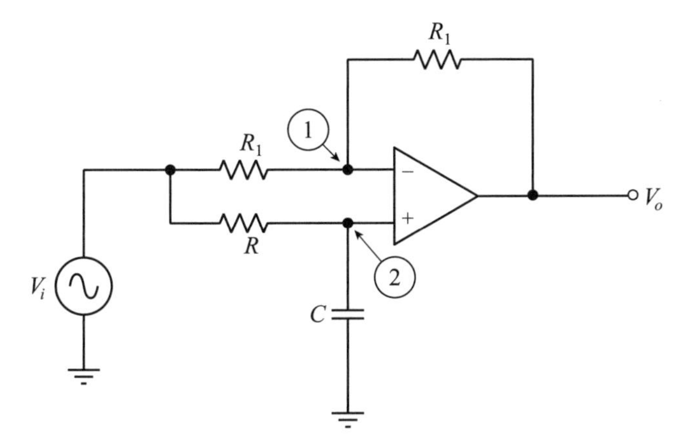

FIGURE CP4.1. An all-pass filter.

Challenge Problem 4.1: Find the transfer function, $H(\omega)=\dfrac{V_o(\omega)}{V_i(\omega)}$, for the high-gain differential amplifier circuit of Fig CP4.1. Does your result provide a hint as to why this circuit is called an *all-pass filter*? *Hint*: Sum currents at ① and ②, and recall that with negative feedback present the input differential voltage is (as the gain increases without bound) zero, as are the input currents to the differential amplifier.

Challenge Problem 4.2: The definition of the Hilbert transformer is that the input signal $m(t)$ produces the Hilbert transform output signal $\frac{1}{\pi}\int_{-\infty}^{\infty}\frac{m(u)}{t-u}du$, which is a time function; let's call it $g(t)$. *Time invariance* means that if a is any constant, then the input $m(t-a)$ produces the output $g(t-a)=\frac{1}{\pi}\int_{-\infty}^{\infty}\frac{m(u)}{t-a-u}du$. Show that the Hilbert transform is, indeed, time invariant. *Hint*: Write $m(t-a)=y(t)$ as the input, which by definition produces the output $\frac{1}{\pi}\int_{-\infty}^{\infty}\frac{y(u)}{t-u}du=\frac{1}{\pi}\int_{-\infty}^{\infty}\frac{m(u-a)}{t-u}du$. Then, change variable to $z=u-a$, and so $dz=du$ and $u=z+a$. Thus, (now you finish the argument).

Challenge Problem 4.3: An elegant derivation of Dirichlet's discontinuous integral (which was central to the derivation of the transfer function of the Hilbert transformer) can be found in my book *Inside Interesting Integrals* (2d ed.), Springer, 2020, pp. 110–112. That derivation does, however, use Leibniz's rule for differentiating an integral, which might be just outside the reach of freshman calculus. So, try this alternative approach which, while a bit longer, *is* within the scope of freshman calculus. (I suspect, however, that Hardy wouldn't care for the rather casual nature of this analysis.)

(1) Suppose p and q are two real quantities (with $p > 0$) that are each independent of x. Evaluate the integral $\int_0^\infty e^{(-p+jq)x}dx$ to arrive at $\frac{p+jq}{p^2+q^2}$.

(2) Use Euler's identity on e^{jqx} and equate real parts to arrive at

$$\int_0^\infty e^{-px}\cos(qx)dx = \frac{p}{p^2+q^2}.$$

(3) Integrate (2) with respect to q from a to b, reverse the order of integration (do the q-integral first, *then* the x-integral), and arrive at $\int_0^\infty e^{-px}\left\{\frac{\sin(bx)-\sin(ax)}{x}\right\}dx = \frac{1}{p}\int_a^b \frac{dq}{1+\left(\frac{q}{p}\right)^2}$.

(4) Change the variable to $x = \frac{q}{p}$ in that last integral to arrive at $\frac{1}{p}\int_{a/p}^{b/p}\frac{p\,dx}{1+x^2} = \tan^{-1}\left(\frac{b}{p}\right) - \tan^{-1}\left(\frac{a}{p}\right)$.

(5) Set $b=0$ in (3) and (4) to arrive at $\int_0^\infty e^{-px}\frac{\sin(ax)}{x}dx = \tan^{-1}\left(\frac{a}{p}\right)$.

(6) Let $p \to 0$, and now complete the final step of this derivation.

Challenge Problem 4.4: (a) Show that the Hilbert transform of $\cos(\omega t)$ is $\sin(\omega t)$. Thus, Euler's identity $e^{j\omega t} = \cos(\omega t) + j\sin(\omega t)$ is an analytic signal. *Hint*: Trying to directly evaluate the Hilbert transform integral is *not* the way to do this. Rather, use the idea that the Hilbert transform subtracts 90° of phase from positive frequencies and adds 90° of phase to negative frequencies. Then, use Euler's identity. (b) Prove that the Hilbert transform of any constant is zero. *Hint*: In this case, actually do the integral, but be sure *not* to integrate across the discontinuity. That is, write the integral with limits that approach the singularity from both sides. (c) What is the Hilbert transform of $\frac{1}{t}$? *Hint*: Recall that if the Hilbert transform of $x(t)$ is $\overline{x}(t)$, then the Hilbert transform of $\overline{x}(t)$ is $-x(t)$. Then, recall the impulse response of the Hilbert transformer.

Challenge Problem 4.5: Here's another example of the tremendous aid computers can provide modern analysts, aid that surely even Hardy would admit is admirable. You'll recall that in chapter 1 (section 1.3) I showed you how MATLAB could solve differential equations, a task Hardy found to often be "soul-destroying." And in this chapter we used MATLAB again to do all the mind-numbing complex arithmetic associated with Weaver's 90° phase-difference circuitry. Similarly, MATLAB can make short work of calculating Hilbert transforms, a task made particularly challenging because the integrand of the transform integral has a discontinuity. For example, suppose $m(t)=\begin{cases}1,\ 0<t<1\\ 0,\ \text{otherwise}\end{cases}$. A MATLAB code (which I'll not pain you with) quickly shows that $\overline{m}(t)$ is given by the plot in Figure CP4.5. (The shaded rectangle is $m(t)$.) See if you can do the actual integration. That is, evaluate $\overline{m}(t)=\int_0^1 \frac{du}{t-u}$, and confirm your result is consistent with the MATLAB plot. *Hint*: Again, be sure to avoid the discontinuity.

Challenge Problem 4.6: Show that the Hilbert transform of $\frac{\sin(t)}{t}$ is $\frac{1-\cos(t)}{t}$. I'll remind you of this result when we discuss the concept of *instantaneous frequency* in the next chapter, on FM. Hint: This can be a bit of a tricky problem, so here's a start for you. The formal calculation is the evaluation of the integral

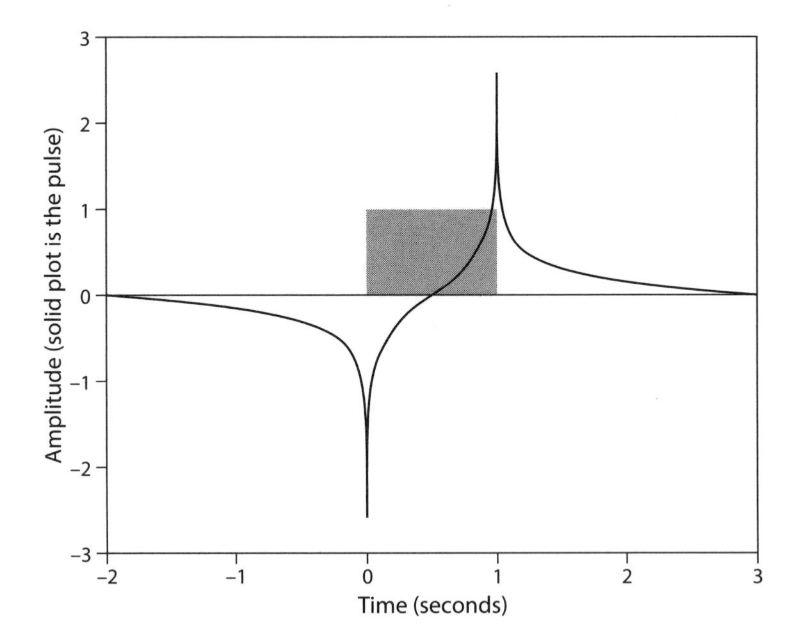

FIGURE CP4.5. Hilbert transform of a pulse.

$$\frac{1}{\pi}\int_{-\infty}^{\infty}\frac{\frac{\sin(u)}{u}}{t-u}du=\frac{1}{\pi}\int_{-\infty}^{\infty}\frac{\sin(u)}{u(t-u)}du=\frac{1}{\pi}\int_{-\infty}^{\infty}\frac{1}{t}\left\{\frac{\sin(u)}{(t-u)}+\frac{\sin(u)}{u}\right\}du$$

$$=\frac{1}{t}\left[\frac{1}{\pi}\int_{-\infty}^{\infty}\frac{\sin(u)}{t-u}du+\frac{1}{\pi}\int_{-\infty}^{\infty}\frac{\sin(u)}{u}du\right].$$

The first integral is the Hilbert transform of $\sin(t)$, which you know from Challenge Problem 4.4a, and the second integral is something you also know because you immediately recognize it as a special case (first done by Euler) of Dirichlet's discontinuous integral. Now, complete the final steps of this analysis.

Chapter 5

FM Radio

> You don't make inventions by fancy mathematics. You make it by jackassing storage batteries around the laboratory.
>
> —angry words attributed[1] to Edwin Armstrong after listening to a January 1937 talk on frequency modulation by radio theoretician Murray Crosby (1903–1974). Hardy would have been shocked, and so would be most modern engineers.

5.1 Why FM?

Along with asking *why* FM? we need to be clear in our minds *what* FM is. That will, in fact, be the most direct way to understanding what prompted Armstrong to utter his harsh words in response to Crosby. Two defining parameters of the signal generated by a sinusoidal oscillator are the amplitude and the frequency of the signal. Information can be impressed on this signal by modulating either one of these parameters, and, of course, *amplitude* modulation (AM) is what this book has been discussing so far. In this chapter we'll take a look at the second possibility, that of frequency modulation (FM).

AM was the practice from the earliest days of broadcast radio, but the (perhaps) surprising fact is that FM was already being thought about *decades* before the start of the 1920s. This is surprising to many because the traditional story of FM radio is that it was invented by Armstrong at the end of 1933, with Armstrong as a lone-wolf genius who did it all in the face of powerful interests (RCA, Westinghouse, and General Electric) who, fearing FM would destroy the value of their AM radio investments, stiff-armed Armstrong at every turn. In the end (so goes this tale), Armstrong, driven to despair by the relentless obstruction, plunged to his death January 1954 from a window in his 13th-floor New York City apartment. In other words,

1 From a March 1992 interview (archived at the IEEE History Center, Rutgers University, New Brunswick, NJ) with radio pioneer Harold Beverage (1893–1993).

the development of FM radio has been imagined as a cosmic battle raging over 20 years between a solitary, heroic, almost saintlike but ultimately tragic figure and a band of greedy, ruthless, unethical, amoral corporate thugs. What could be a more "made-for-television" scenario than that?[2] Well, it *is* a good tale, but it is simply not true.

Quite specific descriptions of FM itself, and of one way to achieve it, were made in two US patent applications filed in 1902 (granted in 1905) when Armstrong was just 12 years old. The applicant, in both cases, was Cornelius Dalzell Ehret (1874–1955), a patent attorney in Philadelphia.[3] There is no evidence that Ehret's patents had even the slightest impact on future events, but the point is that FM itself was "in the air" *long* before Armstrong began his radio career. Much FM research had been conducted by RCA and others well before 1933 (research that Armstrong had full knowledge of and access to[4]), and the traditional Armstrong/FM story fails to explain the huge investments made by RCA, Westinghouse, and GE in television, a development that certainly *did* doom "old-time radio."

Tucker (note 3) tells us nothing more about Ehret, but in the half century since he wrote, the internet has come into existence, and we can easily do today historical research that would have been much more difficult in 1970. The website *Find a Grave* shows Ehret's birth/death dates on his gravestone, and Ancestry.com reveals that he graduated from Cornell in 1896 with a degree in electrical engineering. He then took a position as an examiner in the electrical division of the US Patent Office—along the way he earned a law degree from the National University Law School

2 If the ultraconservative philosopher/writer Ayn Rand (1905–1982) had used a background setting of radio, rather than of architecture, Armstrong could have been the hero in her 1943 bestselling novel *The Fountainhead* that advocated the intrinsic superiority of the lone genius over the undeniable (in Rand's view) mediocrity of the collective.

3 D. G. Tucker, "The Invention of Frequency Modulation in 1902," *The Radio and Electronic Engineer*, July 1970, pp. 33–37.

4 The most dramatic example of this was a mathematical paper on FM written by the AT&T electrical engineer John Carson (developer of sideband-rejection filtering for SSB), "Notes on the Theory of Modulation," *Proceedings of the IRE*, February 1922, pp. 57–64. I'll say a *lot* more about Carson's paper later in this chapter.

in Washington, DC—a position he resigned from in 1901 to enter private legal practice. (And yet, when he sailed to New York from London in 1906, he listed his occupation on the passenger manifest as *engineer*.) Over the next 30 years Ehret received 35 patents on various electrical and mechanical gadgets, even though there is no evidence he actually constructed any of them! This is quite puzzling and, all in all, Ehret is a real mystery, one that would be a good topic for a historian of technology.

Long before the modern age of vacuum tube AM broadcast radio, the earlier technologies of radio-wave transmission (spark-gap, high-speed alternator, and arc) had made it clear that the radio spectrum is a *finite* resource: as the number of radio transmitters using those technologies grew, the remaining spectrum available for future use shrank. Then, with the invention of AM broadcast radio and the resulting further increase in the number of transmitters on the air, the need for conserving the spectrum became even more urgent, and FM offered what appeared to many to be a way to accomplish that. Varying the frequency of a carrier by "just a little bit" around its nominal value (resulting in what is known today as *narrowband FM*) was intuitively (but falsely) thought to result in a signal occupying only a "sliver of spectrum."

An understanding of the sideband structure of an FM signal was, however, lacking. (Such spectrums are vastly more complicated than are the spectrums of AM signals; recall Fleming's confusion on the much less complicated issue of AM sidebands in section 2.1.) The seminal contribution of Carson's paper (note 4) was the development of a mathematical description of the nature of FM sidebands. His work showed that spectrum is *not* saved with narrowband FM. What Armstrong developed, however, falls outside of much of Carson's paper (a point not understood at the time), as Armstrong's work was based on what is now called *wideband FM*, which *does* use a lot of the spectrum in an absolute sense but not in a relative sense.[5]

5 The FM frequency band is 88 to 108 MHz. A 10 kHz chunk of AM radio spectrum at 1 MHz is 1% of the carrier frequency, while the 200 kHz of spectrum allotted to an FM station (at 100 MHz) is 0.2% of the carrier frequency, five times *smaller* than in the AM case.

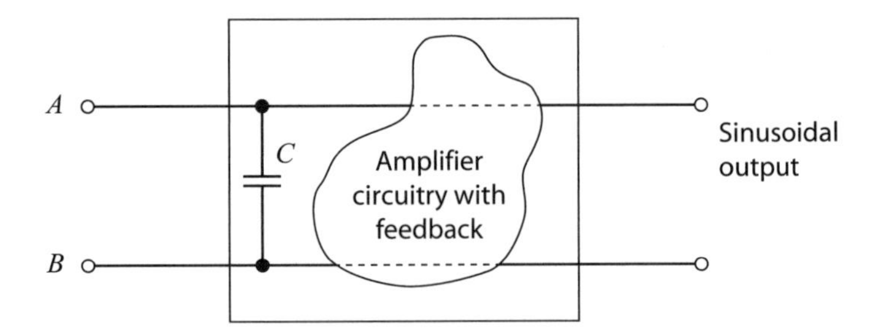

FIGURE 5.1.1. A sinusoidal oscillator with A and B *not* connected.

How, then, do we generate an FM signal? Imagine we have an oscillator circuit whose frequency depends on the value of some tuning capacitor C. We have seen such circuits before (the phase-shift oscillator of Figure 1.5.2, and the tank oscillator of Figure CP2.7). Now, consider the general situation shown in Figure 5.1.1, where points A and B are *not* connected (yet), and the oscillator produces a sinusoidal output of angular frequency ω_0.

Next, suppose we connect what is called a *capacitor microphone* to points A and B in Figure 5.1.1. Such a microphone is simply a parallel-plate capacitor, with one of its plates rigidly fixed and the other a flexible diaphragm that is sensitive to the sound-wave pressure on it, as shown in Figure 5.1.2. As the flexible plate moves back and forth in response to the impinging sound waves, a varying capacitance (*varying* because the distance between the flexible plate and the rigid plate changes as the flexible plate vibrates in and out) is in parallel with C, and so the *total* tuning capacitance varies with the amplitude of the sound waves and thus so, too, does the *instantaneous frequency*[6] of the oscillator output.

Specifically, if $C=C_0$ results in an oscillator angular frequency of ω_0, we can imagine that with the capacitor microphone connected across C the *total* tuning capacitance is $C=C_0\,[1-2h\sin(pt)]$, where

6 The concept of *instantaneous frequency* is a tricky one, all too easy to dismiss as "obvious" until closer inspection reveals it to be not nearly so obvious at all. I'll discuss this in more detail in the next section.

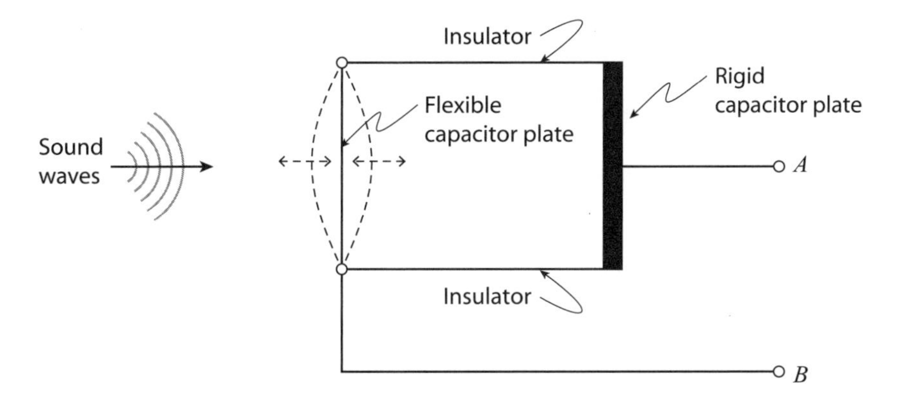

FIGURE 5.1.2. A capacitor microphone.

p is the angular frequency of an audio-tone sound wave incident on the microphone, and h is a dimensionless factor directly proportional to the sound wave *amplitude*. (I am following Carson's notation here, and you'll soon see why he wrote $2h$ instead of just h.) Thus, C varies from a minimum of $C_0(1-2h)$ to a maximum of $C_0(1+2h)$. All the oscillators discussed in this book have oscillation frequencies that vary inversely with the square root of C: the Colpitts oscillator of Figure CP2.7, for example, has the approximate angular frequency $\omega = \frac{1}{\sqrt{LC}}$, which for the present discussion is

$$\frac{1}{\sqrt{L\left[C_0\ \{1-2h\ \sin(pt)\}\right]}} = \frac{1}{\sqrt{LC_0}}\frac{1}{\sqrt{1-2h\ \sin(pt)}} = \frac{\omega_0}{\sqrt{1-2h\ \sin(pt)}}.$$

If we assume h is "small," then

$$\omega \approx \omega_0\{1+h\ \sin(pt)\},$$

an approximation that gets better as h gets smaller (and now you can see why Carson included that factor 2 where he did—to avoid having a factor of $\frac{1}{2}$ in the expression for the oscillation frequency ω). At this point Carson wrote the following:

> From the foregoing reasoning it has frequently been concluded that the oscillation circuit generates a continuously varying frequency of instantaneous value [in hertz] $\frac{\omega_0}{2\pi}\{1 + h\sin(pt)\}$ so that the generated frequency [in hertz] varies between the limits $\frac{\omega_0}{2\pi}(1-h)$ and $\frac{\omega_0}{2\pi}(1-h)$."

That is, the frequency variation (in hertz) defines a chunk of spectrum of width $\frac{2\omega_0 h}{2\pi}$. Now, you'll recall from all our previous discussion of AM, that the sidebands of AM (actually, side*tones* for the discussion here) are separated by $\frac{2p}{2\pi}$ Hz.

Having earlier in his paper observed that with SSB technology the successful transmission of an AM signal requires only half this spectrum width, that is $\frac{p}{2\pi}$ Hz, Carson continued:

> According to this theory, if $2\omega_0 h$ is made less than p, the range of frequencies transmitted [in FM] will be smaller than $\frac{p}{2\pi}$, which is the minimum range required in amplitude modulation.

Carson called this conclusion "ingenious and plausible" because it seems to say that by making h arbitrarily small an FM signal can require an arbitrarily small chunk (a "sliver") of spectrum. This was the seductive lure of FM for radio engineers who were eager to conserve spectrum—but then Carson went on in his paper to show the sideband structure of an FM signal reveals the lure to be a false one, as I'll show you later in this chapter.

To finish this section, let's examine what an FM signal looks like. The amount of frequency variation (the *frequency deviation*) in the signal depends on the distance the flexible plate in the capacitor microphone of Figure 5.1.2 moves. That is, the deviation from f_0 (the oscillator frequency with no sound waves present) is a measure of the *amplitude* of the sound waves. If the sound wave is a single-frequency tone, the *rate* of the frequency deviation is at the fre-

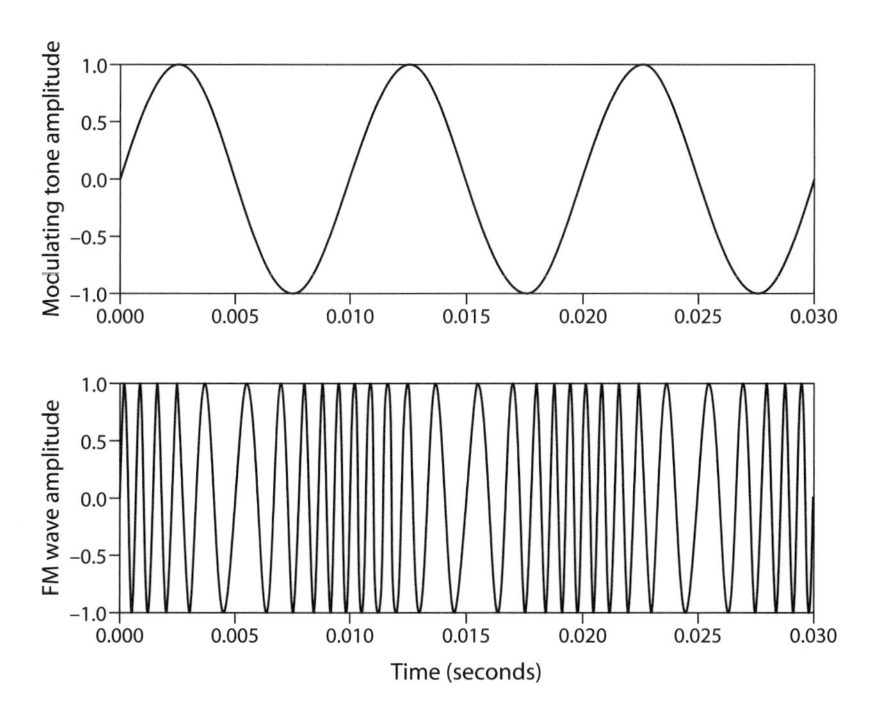

FIGURE 5.1.3. An FM signal modulated by a single frequency tone.

quency of the tone, as Figure 5.1.3 shows, where you'll particularly notice that the *amplitude* of the FM wave is *constant*. Unlike the case with AM, frequency modulation does *not* require the transmitter to vary its power from moment to moment. All the information present in an FM signal is in the instantaneous frequency of the signal and not in the amplitude. This was one of the features of FM that was considered to be a big advantage over AM, as the typical "noise" heard on AM radio is due to extraneous amplitude variations caused by atmospherics, such as weather (lightning strikes), and signal fading (fluctuations in the ionosphere). As was soon discovered, however, many of the early FM detectors (circuitry that transforms the received instantaneous frequency into amplitude variations that are then sent to an ordinary envelope detector, just as in an AM radio receiver) were also subject to such distractions (called *additive noise*), as well—until a new circuit called the *limiter* was introduced. The limiter, to be discussed later, allowed FM radio to enjoy its theoretical freedom from external amplitude variations, and received FM signals are generally *much* less noisy because of atmospherics than are AM signals.[7]

The upper plot of Figure 5.1.3 shows three cycles of a 100 Hz audio tone over the time interval $0 \le t \le 0.03$ second. The lower plot shows the resulting FM signal for a carrier frequency of 1,000 Hz modulated by the audio tone. (The values of 100 Hz and 1,000 Hz were picked strictly for mathematical convenience; actual FM broadcast radio frequencies are quite different.) These plots are for the FM signal $\sin\{\omega_c(t) + M \sin(\omega_a t)\}$, where $\omega_c = 2\pi$ (1,000), $\omega_a = 2\pi$ (100) and M (the so-called *modulation index*) is 5. Later in this chapter, you'll see the origin of the sine-in-a-sine function and the M.

5.2 Instantaneous Frequency

A central concept in FM radio theory is that of "instantaneous frequency," which Carson clearly understood. Indeed, we find those words first used in his published 1922 paper, a theoretical analysis of FM, in which he devoted a significant portion to a discussion of the concept. To summarize what he concluded:

> Suppose we have a function $\sin\{\Omega(t)\}$ where $\Omega(t)$ is any specific function of time. . . . We define the *generalized* [instantaneous] *frequency* as equal [in hertz] to $\left(\frac{1}{2\pi}\right)\Omega'(t)$ [where $\Omega'(t) = \frac{d}{dt}\Omega(t)$]. . . . In the case where $\Omega(t) = \omega t$ [as in $\sin(\omega t)$] it agrees with the usual definition of frequency [in hertz of] $\omega/2\pi$.

In the decades that followed, analysts went far beyond Carson, and today it is common to find the following for the definition of instanta-

7 In the July 1928 issue of the *Proceedings of the IRE* Carson published a paper ("The Reduction of Atmospheric Disturbances," pp. 966–975) where he wrote: "We are unavoidably forced to the conclusion that static, like the poor, will always be with us." Years later Carson's assertion found its way into one of Raymond Chandler's novels featuring private eye Philip Marlowe (the 1944 *The Lady in the Lake*), where, at one point, we read of someone in a "low-rent diner" ("the manager of the joint [was] a low budget tough guy in shirt sleeves and a mangled cigar") fiddling with a radio "that was as full of static as the mashed potatoes were full of water."

neous frequency: If $x(t)$ is any given time function, define the analytic signal $x(t)+j\bar{x}(t)$, with the phase function $\varphi(t)=\tan^{-1}\left\{\frac{\bar{x}(t)}{x(t)}\right\}$. The instantaneous frequency (in hertz) of $x(t)$ is then given by

$$\left(\frac{1}{2\pi}\right)\frac{d}{dt}\varphi(t)=\left(\frac{1}{2\pi}\right)\frac{d}{dt}\tan^{-1}\left\{\frac{\bar{x}(t)}{x(t)}\right\}.$$

For example, if $x(t)=\cos(\omega t)$, then (as you showed in Challenge Problem 4.4a) the Hilbert transform $\bar{x}(t)$ is $\sin(\omega t)$, and so the instantaneous frequency of $x(t)$ is

$$\left(\frac{1}{2\pi}\right)\frac{d}{dt}\tan^{-1}\left\{\frac{\sin(\omega t)}{\cos(\omega t)}\right\}=\left(\frac{1}{2\pi}\right)\frac{d}{dt}\tan^{-1}\{\tan(\omega t)\}$$
$$=\left(\frac{1}{2\pi}\right)\frac{d}{dt}(\omega t)=\frac{\omega}{2\pi}.$$

Even easier (indeed, trivial) is the calculation of the instantaneous frequency of a constant (which we would intuitively assume to be zero). And so it is, because as you showed in Challenge Problem 4.4b, the Hilbert transform of any constant c is zero, and so the instantaneous frequency is, as we expect,

$$\left(\frac{1}{2\pi}\right)\frac{d}{dt}\tan^{-1}\left(\frac{0}{c}\right)=\left(\frac{1}{2\pi}\right)\frac{d}{dt}\tan^{-1}(0)=0.$$

This probably seems pretty straightforward, but defining the instantaneous frequency in this way with the analytic signal has some surprises for us. For example, suppose we have the decaying oscillatory function $x(t)=\frac{\sin(t)}{t}$ with an "obvious" constant frequency of $\frac{1}{2\pi}$ hertz (*obvious* because the zeros of $\sin(t)$ are at twice that frequency). Now, as you showed in Challenge Problem 4.6, $\bar{x}(t)=\frac{1-\cos(t)}{t}$ and so the instantaneous frequency is

$$\left(\frac{1}{2\pi}\right)\frac{d}{dt}\tan^{-1}\left\{\frac{1-\cos(t)}{\sin(t)}\right\}.$$

Recalling the differentiation formula

$$\frac{d}{dt}\tan^{-1}\{u(t)\}=\left(\frac{1}{1+u^2}\right)\frac{du}{dt},$$

with $u(t)=\dfrac{1-\cos(t)}{\sin(t)}$, we have the instantaneous frequency as

$$\left(\frac{1}{2\pi}\right)\frac{1}{1+\left\{\dfrac{1-\cos(t)}{\sin(t)}\right\}^2}\left[\frac{\sin^2(t)-\{1-\cos(t)\}\cos(t)}{\sin^2(t)}\right]$$

$$=\left(\frac{1}{2\pi}\right)\left\{\frac{\sin^2(t)}{\sin^2(t)+1-2\cos(t)+\cos^2(t)}\right\}\left\{\frac{\sin^2(t)-\cos(t)+\cos^2(t)}{\sin^2(t)}\right\}$$

$$=\left(\frac{1}{2\pi}\right)\frac{1-\cos(t)}{2-2\cos(t)}$$

$$=\frac{1}{4\pi}.$$

That is, the definition of instantaneous frequency, in terms of the analytic signal, correctly gives a constant value but one that is *half* of what we intuitively expect. The fact that the oscillations decay with time seems to have a (nonobvious) effect on the instantaneous frequency.

Here's an even more (perhaps) perplexing calculation. It is not difficult to use the results of Challenge Problem 1.13 to show that if $x(t)=\dfrac{1}{1+t^2}$, then $\overline{x}(t)=\dfrac{t}{1+t^2}$ (see Challenge Problem 5.1). Notice, as we start, that $x(t)$ *doesn't oscillate*. In any case, the instantaneous frequency is

$$\left(\frac{1}{2\pi}\right)\frac{d}{dt}\tan^{-1}(t)=\left(\frac{1}{2\pi}\right)\frac{1}{1+t^2}.$$

This does go to zero as $t\to\pm\infty$, but just *what is actually oscillating* for any $|t|<\infty$?

Questions like this have led to a lot of controversy over just *what* "instantaneous frequency" actually means. Some have argued it is impossible to measure a frequency *instantaneously*, that in fact a time function has to be observed over a nonzero time interval to estimate the frequency (for example, the time interval between consecutive zero crossings after the dc level of the signal has been removed). Another objection is that to use the analytic signal definition assumes we know the true behavior of the signal for all time, something we can achieve only in a theoretical analysis.

So, what do we do now, in our discussion here of FM radio? The direct answer is: follow Carson. We all *think* we have an intuitive feeling for what we are talking about when we use the words *instantaneous frequency* as we modulate a carrier frequency, but keep in mind that perhaps our "feeling" might not be quite so obvious as we'd like to believe! Still, for all that follows, we will use Carson's definition of the instantaneous frequency of a signal as the time derivative of the signal's phase angle.

To see how what we've done so far leads us to the plot in Figure 5.1.3, let's write the FM signal produced by using an audio tone of frequency ω_a (that has amplitude A) to modulate a carrier of frequency ω_c as

$$e(t) = \sin\{\omega_c t + kA\sin(\omega_a t)\},$$

where k is some positive constant. We do this because, following Carson, we then have the instantaneous frequency

$$\frac{d}{dt}\{\omega_c t + kA\sin(\omega_a t)\} = \omega_c + kA\omega_a\cos(\omega_a t),$$

which describes a frequency varying above and below ω_c at the tone frequency ω_a. The amplitude factor $kA\omega_a$ defines the maximum frequency deviation around ω_c, which we'll write as ΔF. That is, $kA\omega_a = \Delta F$ (thus, $kA = \frac{\Delta F}{\omega_a}$), and so our FM signal is

$$e(t) = \sin\left\{\omega_c t + \frac{\Delta F}{\omega_a}\sin(\omega_a t)\right\}.$$

It is usual to write $M = \frac{\Delta F}{\omega_a}$, called the *modulation index*,[8] and so we arrive at the FM signal of Figure 5.1.3:

$$e(t) = \sin\{\omega_c t + M \sin(\omega_a t)\}.$$

5.3 FM Sidebands

The major contribution of Carson's 1922 paper was the discovery of the true nature of the sideband structure of an FM signal. The details of his analysis are, however, borderline obscure (in my opinion) and certainly must have gone right over the heads of 99.9% of the radio engineers of 1922. With only a couple of offhand references to "Mathieu's equation" and "Hill's equation" (neither of which he explained), referring readers only to the famous book *Modern Analysis* by Whittaker and Watson,[9] Carson then simply declared that the sideband structure of an FM signal is, contrary to the then popular belief, infinite in bandwidth and that the amplitudes of the individual sideband terms are expressible as Bessel functions.[10] In this section I'll show you a way, using no more than a couple of trigonometric identities and Euler's identity, to completely derive Carson's results.

8 The modulation index clearly is not a constant but instead continuously varies with time. For example, since M varies with A, then the modulation index varies as one speaks more or less loudly into a microphone. In commercial broadcast FM radio ΔF is a maximum of 75 kHz, while f_a is a minimum of 50 Hz. So, M can vary from zero (with no audio input $A = 0$ and so $\Delta F = 0$) up to 1,500 ($\Delta F = 75{,}000$ and $f_a = 50$).

9 Hardy would have instantly appreciated this reference. The English mathematician Edmund Whittaker (1873–1956) had been one of Hardy's teachers at Cambridge when Whittaker was giving lectures that would later become his famous 1902 book *A Course on Modern Analysis*. Hardy was also well acquainted with George Watson (1886–1965), who collaborated with Whittaker to write the 1915 second edition of the book (the edition referenced by Carson).

10 Carson's paper was reprinted in the June 1963 issue of the *Proceedings of the IEEE* as part of a series of important papers in the history of electronics. As the editorial introduction put it: "Once upon a time, not really so long ago, . . . band width was apt to imply Sousa's marching formation, Bessel's functions might have been considered more organic than mathematical [what a gentle way to conjure up what might otherwise be a most disturbing image!] . . . At such a time the broadcast band was theoretically all at [a wavelength of] 360 meters, amplitude modulation was the vogue, frequency space was even then at a premium and a means of compressing signal band width was eagerly sought."

The FM signal we arrived at in the previous section has unit amplitude, and so, to be just a bit more general, let's assume here that our FM signal has amplitude E_0. Also, for historical reasons, I'll (for now) write x instead of M to emphasize the point made in note 8 that the modulation index is a variable, not a constant. So, we start with the FM signal at carrier frequency ω_c, with a modulating audio tone of frequency ω_a, written as

$$e(t) = E_0 \sin\{\omega_c t + x \sin(\omega_a t)\},$$

or, temporarily writing $\theta = \omega_a t$, we have

$$e(t) = E_0 \sin\{\omega_c t + x \sin(\theta)\}.$$

From the trigonometric identity $\sin(A+B) = \sin(A)\cos(B) + \cos(A)\sin(B)$ we arrive at

$$e(t) = E_0[\sin(\omega_c t)\cos\{x \sin(\theta)\} + \cos\{\omega_c t\}\sin\{x \sin(\theta)\}].$$

The problem at hand is now clear—if we develop series expansions of $\cos\{x\ sin(\theta)\}$ and $\sin\{x\ sin(\theta)\}$, then each of the terms in those expansions will heterodyne with the FM carrier of frequency ω_c (via multiplication with $\sin(\omega_c t)$ and $\cos(\omega_c t)$). The result will give us the sideband structure of the FM signal.

We start with the function $e^{\frac{x}{2}\left(t-\frac{1}{t}\right)}$—you'll soon see why this curious function is useful—and imagine it written as a power series in t. That is, writing (for historical reasons) the series coefficients as $J_n(x)$, we have

$$e^{\frac{x}{2}\left(t-\frac{1}{t}\right)} = \sum_{n=-\infty}^{\infty} J_n(x) t^n.$$

Our next step is to find $J_n(x)$. Since

$$e^{\frac{x}{2}\left(t-\frac{1}{t}\right)} = e^{\frac{x}{2}t} e^{-\frac{x}{2t}} = e^{\left(\frac{x}{2}\right)t} e^{\left(-\frac{x}{2}\right)t^{-1}},$$

then, from the well-known power series expansion of the exponential function, we have

$$\sum_{n=-\infty}^{\infty} J_n(x)t^n = \left\{\sum_{r=0}^{\infty} \frac{\left(\frac{x}{2}\right)^r}{r!} t^r\right\}\left\{\sum_{s=0}^{\infty} \frac{\left(-\frac{x}{2}\right)^s}{s!} t^{-s}\right\}.$$

Now, fix your attention on any particular value of s in the second sum on the right. Then, for the particular value of $r=n+s$ in the first sum on the right, the product of these two gives a t^n term. That is, for the particular value of s that we picked, we get the product term

$$\frac{\left(\frac{x}{2}\right)^{n+s}}{(n+s)!} \frac{\left(-\frac{x}{2}\right)^s}{s!} t^n = \frac{\left(\frac{x}{2}\right)^{n+s}\left(\frac{x}{2}\right)^s (-1)^s}{s!(n+s)!} t^n = \frac{(-1)^s\left(\frac{x}{2}\right)^{n+2s}}{s!(n+s)!} t^n.$$

So, summing over all possible values of s, we have

$$J_n(x) = \sum_{s=0}^{\infty} (-1)^s \frac{1}{s!(n+s)!}\left(\frac{x}{2}\right)^{n+2s},$$

or

$$J_n(x) = \sum_{s=0}^{\infty} \frac{(-1)^s}{s!(n+s)!2^{n+2s}} x^{n+2s}.$$

The $J_n(x)$ are called *Bessel functions of the first kind of order n*, after the German astronomer Friedrich Wilhelm Bessel (1784–1846), because they are solutions to a nonlinear differential equation (*Bessel's equation*, used in the physics of planetary motion) that he studied (an equation we will *not* need to pursue in our discussion of FM radio!). We will not discuss the other Bessel functions. The general nature of $J_n(x)$ is illustrated in Figure 5.3.1, which shows plots of $J_0(x)$, $J_1(x)$, and $J_2(x)$, all of which look somewhat sinusoidal but are not (in general, the amplitude of $J_n(x)$ decreases as x increases, and the zero crossings of $J_n(x)$ are not evenly spaced). Notice that $J_n(0)=0$ *except* for the $n=0$ case; $J_0(0)=1$. As you'll soon see, this has physically important implications for FM. As the index n increases, the first peak of $J_n(x)$, and all subsequent peaks, move to the right Second only to the familiar trigonometric functions, Bessel functions appear everywhere in advanced radio engineering[11]—Hardy would feel right at home with them.

Now, let $t=e^{j\theta}$ in our original power series expansion of $e^{\frac{x}{2}\left(t-\frac{1}{t}\right)}$. We then have

$$e^{\frac{x}{2}\left(t-\frac{1}{t}\right)}=e^{\frac{x}{2}(e^{j\theta}-e^{-j\theta})}=e^{jx\sin(\theta)}=\cos\{x\sin(\theta)\}+j\sin\{x\sin(\theta)\}$$
$$=\sum_{n=-\infty}^{\infty}J_n(x)e^{jn\theta}=J_0(x)+\left\{J_1(x)e^{j\theta}+J_{-1}(x)e^{-j\theta}\right\}$$
$$+\left\{J_2(x)e^{j2\theta}+J_{-2}(x)e^{-j2\theta}\right\}+\cdots.$$

To evaluate $J_{-1}(x)$, $J_{-2}(x)$, and so on, we replace n with $-n$ in our $J_n(x)$ expression. That is, for n a *positive* integer (and so $-n<0$), we have

$$J_{-n}(x)=\sum_{s=0}^{\infty}\frac{(-1)^s}{s!\,(s-n)!2^{2s-n}}x^{2s-n}.$$

For $s=0,1,2,\ldots,n-1$, we observe that $(s-n)!$ is the factorial of a negative integer and so *blows up*,[12] which means the first n terms in the sum are zero, and so we can start the sum from $s=n$. Equiv-

11 Just as with the trigonometric functions, any good collection of math formulas will have tables of numerical values for the Bessel functions. Just as a note for the curious, in MATLAB to get the numerical value of $J_n(x)$ to more decimal places than you'll probably need, type the command *besselj(n,x)*. For example, *besselj* (2,4) $=J_2(4)=0.36412814585207\ldots$ (compare with Figure 5.3.1).

12 The fact that the factorial $k!$ blows up to either $+\infty$ or to $-\infty$ if k is a negative integer is probably not anything new for a mathematician, but others reading this may wonder how that happens. For k a *positive* integer we write $(k+1)!=(k+1)(k)(k-1)(k-2)\ldots(2)(1)=(k+1)k!$ That is, $k!=\frac{(k+1)!}{k+1}$. We can use this form for $k!$ to extend the calculation of factorials backward into the negative integers. To start, if $k=0$, then $0!=\frac{1!}{1}=1$. Thus, if $k=-1$, then $(-1)!=\frac{0!}{0}=\frac{1}{0}=\infty$. If $k=-2$, then $(-2)!=\frac{(-1)!}{-1}=\frac{\infty}{-1}=-\infty$. Continuing in this way, you can quickly see that $|k!|=\infty$ for k any negative integer.

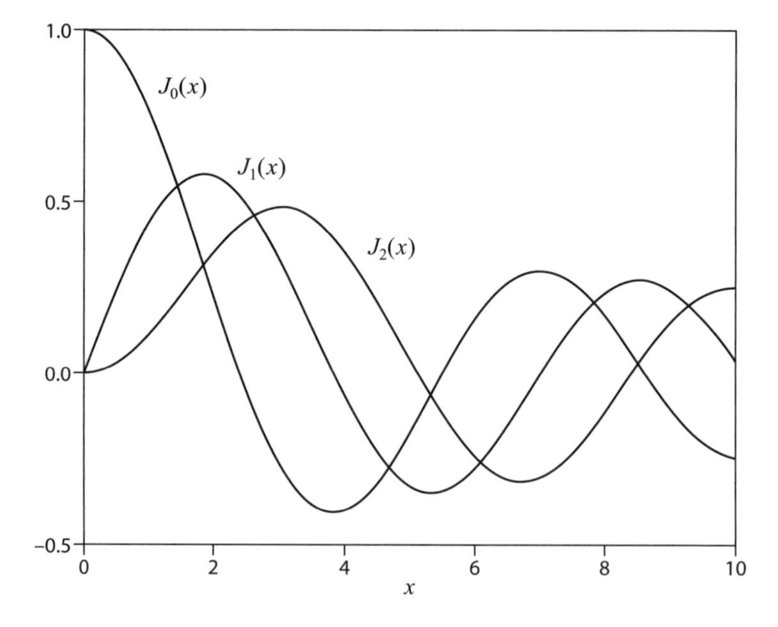

FIGURE 5.3.1. The first three Bessel function of the first kind for $n \geq 0$.

alently, we can still start the sum at $s = 0$ *if* we replace s with $s + n$, and so we have

$$J_{-n}(x) = \sum_{s=0}^{\infty} \frac{(-1)^{s+n}}{(s+n)!\ s!2^{2(s+n)-n}} x^{2(s+n)-n},$$

or

$$J_{-n}(x) = \sum_{s=0}^{\infty} \frac{(-1)^{s+n}}{s!(n+s)!2^{n+2s}} x^{n+2s}.$$

Comparing this equation with $J_n(x)$, we see that

$$J_{-n}(x) = (-1)^n J_n(x).$$

That is,

$$J_{-1}(x) = -J_1(x),\ J_{-2}(x) = J_2(x),\ J_{-3}(x) = -J_3(x),\ \ldots$$

Thus,

$$\begin{aligned}
\cos\{x \sin(\theta)\} + j \sin\{x \sin(\theta)\} &= J_0(x) + \{J_1(x)e^{j\theta} - J_1(x)e^{-j\theta}\} \\
&\quad + \{J_2(x)e^{j2\theta} + J_2(x)e^{-j2\theta}\} + \{J_3(x)e^{j3\theta} - J_3(x)e^{-j3\theta}\} \\
&\quad + \{J_4(x)e^{j4\theta} + J_4(x)e^{-j4\theta}\} + \cdots \\
&= J_0(x) + j2J_1(x)\sin(\theta) + 2J_2(x)\cos(2\theta) + j2J_3(x)\sin(3\theta) \\
&\quad + 2J_4(x)\cos(4\theta) + \cdots
\end{aligned}$$

or, equating real and imaginary parts, we at last have the series expansions we need to determine the sideband structure of our FM signal, $e(t)$:

$$\cos(x)\,\sin(\theta) = J_0(x) + 2\sum_{n=1}^{\infty} J_{2n}(x)\cos(2n\theta)$$

$$\sin(x)\,\sin(\theta) = 2\sum_{n=1}^{\infty} J_{2n-1}(x)\sin\{(2n-1)\theta\}.$$

So, remembering that x is a stand-in for the modulation index, M, and that θ is a stand-in for $\omega_a t$, we have

$$e(t) = E_0 \left[\begin{array}{c}
\sin(\omega_c t)\left(J_0(M) + 2\sum\limits_{n=1}^{\infty} J_{2n}(M)\cos(2n\omega_a t)\right) \\
+ \\
\cos(\omega_c t)\left(2\sum\limits_{n=1}^{\infty} J_{2n-1}(M)\sin\{(2n-1)\,\omega_a t\}\right)
\end{array}\right]$$

$$= E_0 \left[\begin{array}{c}
\sin(\omega_c t)J_0(M) \\
+ \\
2\,\sin(\omega_c t)\left\{\begin{array}{l} J_2(M)\cos(2\omega_a t) + J_4(M)\cos(4\omega_a t) \\ + J_6(M)\cos(6\omega_a t) + \cdots \end{array}\right\} \\
+ \\
2\,\cos(\omega_c t)\left\{\begin{array}{l} J_1(M)\sin(\omega_a t) + J_3(M)\sin(3\omega_a t) \\ + J_5(M)\sin(5\omega_a t) + \cdots \end{array}\right\}
\end{array}\right]$$

or, using the identity $\sin(A)\cos(B) = \frac{1}{2}\{\sin(A-B) + \sin(A+B)\}$ with $A = \omega_c t$, $B = \omega_a t$, and $\omega_c > \omega_a$, we have $e(t) =$

$$E_0\begin{bmatrix} J_0(M)\sin(\omega_c t)+J_2(M)\{\sin(\omega_c t-2\omega_a t)+\sin(\omega_c t+2\omega_a t)\} \\ + \\ J_4(M)\{\sin(\omega_c t-4\omega_a t)+\sin(\omega_c t+4\omega_a t)\} \\ + \\ J_6(M)\{\sin(\omega_c t-6\omega_a t)+\sin(\omega_c t+6\omega_a t)\} \\ +\cdots \\ J_1(M)\{-\sin(\omega_c t-\omega_a t)+\sin(\omega_c t+\omega_a t)\} \\ + \\ J_3(M)\{-\sin(\omega_c t-3\omega_a t)+\sin(\omega_c t+3\omega_a t)\} \\ + \\ J_5(M)\{-\sin(\omega_c t-5\omega_a t)+\sin(\omega_c t+5\omega_a t)\}+\cdots, \end{bmatrix}$$

or, with a little rearrangement,

$$e(t)=E_0\begin{bmatrix} J_0(M)\sin(\omega_c t)+J_1(M)\{\sin[(\omega_c+\omega_a)t]-\sin[(\omega_c-\omega_a)t]\} \\ + \\ J_2(M)\{\sin[(\omega_c+2\omega_a)t]+\sin[(\omega_c-2\omega_a)t]\} \\ + \\ J_3(M)\{\sin[(\omega_c+3\omega_a)t]-\sin[(\omega_c-3\omega_a)t]\} \\ + \\ J_4(M)\{\sin[(\omega_c+4\omega_a)t]+\sin[(\omega_c-4\omega_a)t]\} \\ + \\ J_5(M)\{\sin[(\omega_c+5\omega_a)t]-\sin[(\omega_c-5\omega_a)t]\} \\ + \\ J_6(M)\{\{\sin[(\omega_c+6\omega_a)t]+\sin[(\omega_c-6\omega_a)t]\}\}+\cdots \end{bmatrix}$$

As discovered by Carson in 1922, we see that frequency modulating a carrier with a *single* tone produces an *infinity* of upper and lower sidebands, a situation very much different from that in the AM case (where a single tone generates *one* upper-sideband frequency and *one* lower-sideband frequency). As in the AM case, however, FM sidebands are uniformly spaced by the modulating tone frequency. It's no doubt obvious that we aren't going to transmit all those sidebands (remember, FM was originally imagined as a spectrum-*conserving* technique!) What has become known as *Carson's rule*[13] will tell us how much finite bandwidth we need to transmit to have an intelligible FM signal, and as a prelude to the origin of that rule, consider the following.

The FM wave of Figure 5.1.3 is a sinusoid (of varying frequency) with constant peak amplitude, E_0. From Challenge Problem 1.3 we know the average power of a sinusoid *of any frequency* is the peak value divided by $\sqrt{2}$. So, the average power of our FM wave is the constant

$$\left(\frac{E_0}{\sqrt{2}}\right)^2 = \frac{1}{2}E_0^2.$$

In the same way, the average power at the carrier frequency is

$$\left(E_0\frac{J_0(M)}{\sqrt{2}}\right)^2 = \frac{1}{2}E_0^2 J_0^2(M).$$

In the same way, the average power at the first *upper*-sideband frequency is

$$\left(E_0\frac{J_1(M)}{\sqrt{2}}\right)^2 = \frac{1}{2}E_0^2 J_1^2(M).$$

In the same way, the average power at the first *lower*-sideband frequency is

$$\left(E_0\frac{J_1(M)}{\sqrt{2}}\right)^2 = \frac{1}{2}E_0^2 J_1^2(M).$$

13 In modern texts and Web blogs that discuss Carson's rule, the usual statement is that the rule comes from Carson's 1922 paper. Well, in a way that's true, but only in the same way as saying the Pythagorean theorem follows from geometrical arguments. In both cases, there is a *lot more* that needs to be said! What is called "Carson's rule," in fact, does not appear anywhere in Carson's 1922 paper. We will finish this section with a complete derivation of it.

In the same way, the average power at the second upper-sideband frequency is

$$\left(E_0 \frac{J_2(M)}{\sqrt{2}}\right)^2 = \frac{1}{2} E_0^2 J_2^2(M),$$

and so on.

Since the average total power of the FM wave is the sum of the average power of the carrier plus the average powers of all the sidebands, we must have

$$\frac{1}{2}E_0^2 = \frac{1}{2}E_0^2 J_0^2(M) + \frac{1}{2}E_0^2 J_1^2(M) + \frac{1}{2}E_0^2 J_1^2(M) + \frac{1}{2}E_0^2 J_2^2(M) + \cdots,$$

or, finally, we arrive at the purely mathematical result (see challenge Problem 5.2)

$$1 = J_0^2(\mathrm{M}) + 2\sum_{n=1}^{\infty} J_n^2(M),$$

an identity that is trivially true for $M=0$ but *not* so obviously true when $M \neq 0$. (Recall that $J_0(0)=1$ while $J_n(0)=0$ for $n \geq 1$.) The physical interpretation of this identity is that as M varies, the distribution of the FM wave power moves back and forth between the carrier and the sidebands, and the total power in the signal remains unchanged.

Now, suppose we agree to increase the bandwidth of our transmitted FM wave until at least 98% of the total sideband power is included (the 98% is an arbitrary value that we hope is large enough that we will not affect the intelligibility of the signal because we failed to send that last 2% of sideband power—see Challenge Problem 5.3). As before, since the average power of the FM wave is $\frac{1}{2}E_0^2$, and since the average power of the carrier is $\frac{1}{2}E_0^2 J_0^2(\mathrm{M})$, then the total average power in *all* the sidebands is $\frac{1}{2}E_0^2\left[1 - J_0^2(M)\right]$. Now, the total average power in the first N sidebands (upper and lower) is

$$E_0^2 2\sum_{n=1}^{N} \frac{1}{2} J_n^2(M) = E_0^2 \sum_{n=1}^{N} J_n^2(M),$$

and so, what we are after is the *smallest* integer N such that

$$E_0^2 \sum_{n=1}^{N} J_n^2(M) \geq 0.98 \frac{1}{2} E_0^2 \left[1 - J_0^2(M)\right].$$

That is, we are asking for the smallest integer N such that

$$\sum_{n=1}^{N} J_n^2(M) \geq 0.49 \left[1 - J_0^2(M)\right].$$

This inequality is very easy to code, for any value of M, using software like MATLAB; for M varying into the hundreds, the smallest N that works is numerically found to always be equal to $M+1$. That is, for an FM wave with modulation index M, we transmit at least 98% of the total sideband power if we transmit the first $M+1$ sidebands. Since the sidebands are uniformly spaced in frequency by f_a (the modulation tone frequency), the bandwidth required is $2(M+1)f_a$, where the factor of 2 accounts for the fact that we are transmitting both the upper and the lower sidebands. Recalling that $M = \frac{\Delta f}{f_a}$, where Δf is the maximum frequency deviation around the carrier, we see the required maximum FM bandwidth is

$$2\left(\frac{\Delta f}{f_a} + 1\right) f_a = 2(\Delta f + f_a),$$

a result popularly known today as *Carson's rule*. In commercial FM broadcast radio, $\Delta f = 75$ kHz and $f_a = 15$ kHz, and so Carson's rule says the maximum required bandwidth is $2(75+15) = 180$ kHz. This is why commercial FM stations are licensed for a 200 kHz bandwidth (see note 5).

These comments are for the *wideband* FM that Armstrong developed, not the narrowband FM studied by Carson (recall his "*h* is small" assumption, from just after Figure 5.1.2).[14] It is this distinction that explains Armstrong's angry response to Murray Crosby that opened this chapter. At the end of his 1922 paper, Carson wrote that for

14 Narrowband FM typically has $\Delta f = 5$ kHz, and so, with $f_a = 5$ kHz (amateurs and private business operators are typical users of narrowband FM with these specifications), the required bandwidth would be just 20 kHz.

the narrowband FM that he had studied, "[t]his system of modulation, therefore, discriminates against high frequencies and therefore inherently introduces distortion." This is indeed true for narrowband FM but not for the wideband FM that was Armstrong's interest. Many (if not all) the radio engineers of 1922 failed to appreciate this distinction, and so FM in general got a poor reputation. Armstrong incorrectly blamed Carson for that, but there is little doubt that Carson's rather terse paper would have been *vastly* improved with the inclusion of just a bit more explanatory detail.

5.4 The FM Receiver

At this point we have generated and transmitted an FM signal. Our question now is, How does a remote receiver recover the original modulating tone for a listener to hear? The basic idea is simple: let's just build circuitry that transforms frequency variations into amplitude variations and then proceed just as we did with an AM superheterodyne receiver. That's pretty simple, all right and, in fact, building actual circuitry that does the job is *conceptually* almost as simple. The block diagram in Figure 5.4.1 of an FM receiver—with the exception of the two blocks labeled *limiter* and *FM detector—does* look a lot like the superheterodyne AM receiver of Figure 3.2.1.

The *intermediate frequency* (IF) for commercial FM radio is $f_{IF}=10.7$ MHz (see Challenge Problem 5.4); that is, the IF amplifier is a bandpass filter centered on 10.7 MHz with a fixed bandwidth of 200 kHz (extending from 10.6 to 10.8 MHz.[15] The assigned FM carrier frequencies (in the US) are 88.1 to 107.9 MHz, in steps of 0.2 MHz. The local oscillator tunes 10.7 MHz *above* the desired station's carrier (just as in AM radio) to heterodyne the received signal into the IF amplifier/bandpass filter, and so tunes from 98.8 to 118.6 MHz. Even though the limiter is encountered first in Figure 5.4.1 as the received signal makes its way through the receiver, we'll discuss the FM detector first. That's because it is a characteristic of the FM detector that

15 The IF bandwidth–to–IF center-frequency ratio in FM radio is $\frac{200\times10^3}{10.7\times10^6}=0.0187$, which is not very different from the ratio of $\frac{10\times10^3}{455\times10^3}=0.022$ for AM broadcast radio.

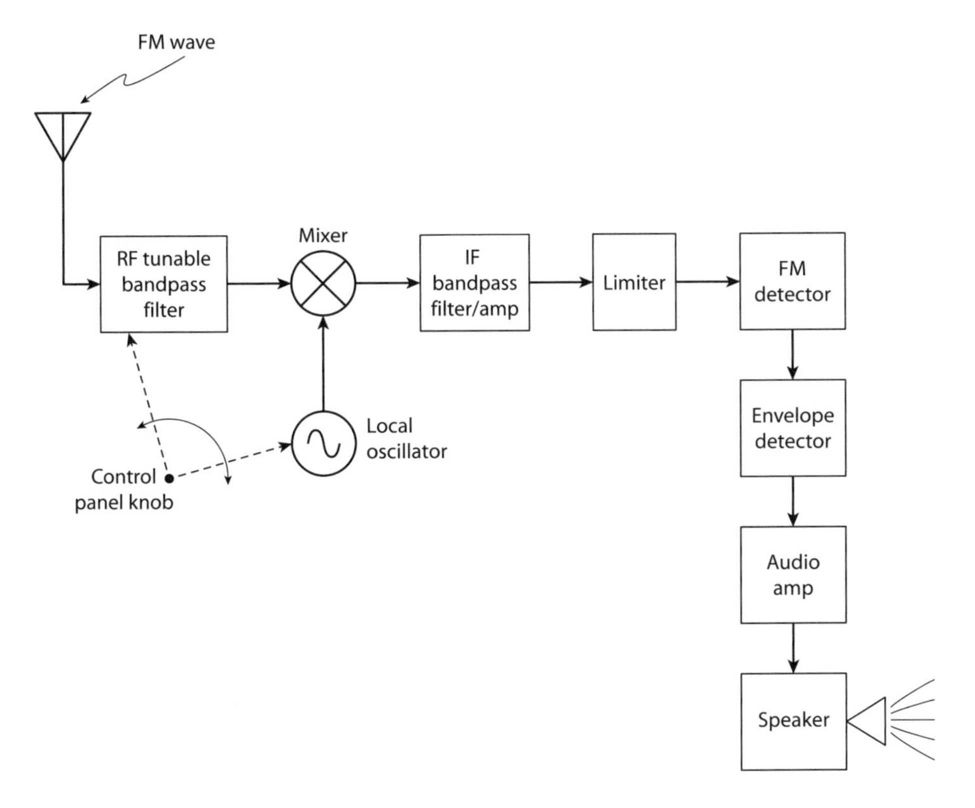

FIGURE 5.4.1. FM superheterodyne receiver.

motivates why the limiter is in the receiver, and you'll better appreciate a discussion of the limiter when you understand *why* it's there.

The simplest FM detector is the *slope detector*, which functionally appeared in the 1902 patent by Ehret (refer to the first section of this chapter). To understand the slope detector, look at Figure 5.4.2, which you'll recognize as being nothing more than a simple high-pass filter (see Figure 1.4.1). A high-gain differential amplifier provides the output—connected as a voltage follower (refer to Figure CP2.1a)—to give a buffered output that guarantees the circuitry following the slope detector has no influence on the operation of the slope detector.

Figure 5.4.3 shows the transfer function of the HPF, which, since

$$H(\omega)=\frac{R}{R+\frac{1}{j\omega C}}=\frac{j\omega RC}{1+j\omega RC},$$

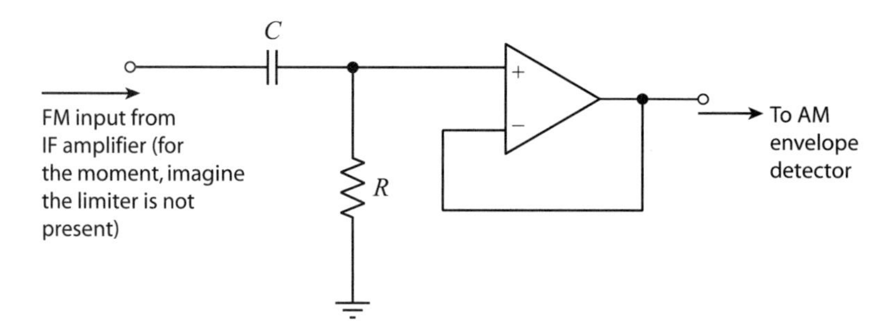

FIGURE 5.4.2. A simple FM detector (slope detector).

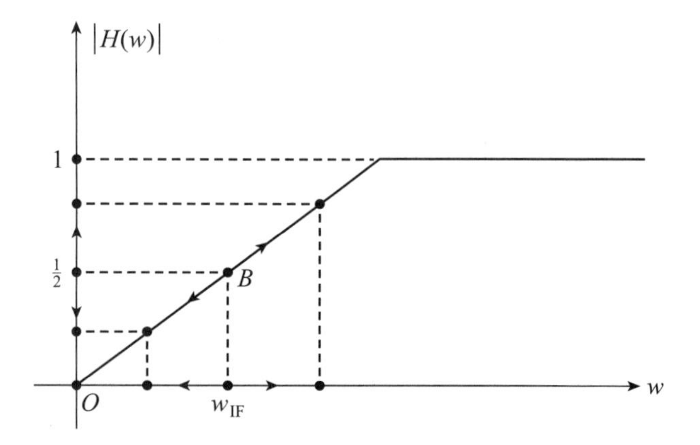

FIGURE 5.4.3. How the slope detector works.

says

$$|H(\omega)| = \frac{\omega RC}{\sqrt{1+(\omega RC)^2}}.$$

In Figure 5.4.3 the IF frequency ($\omega_{IF}=2\pi f_{IF}$) is set to the frequency where $|H(\omega)| = \frac{1}{2}$. This assignment is somewhat arbitrary, with the only consideration being to put ω_{IF} at what appears to be the *middle* of what looks like a *linear* portion of the transfer function. As the instantaneous frequency of the heterodyned FM wave varies around ω_{IF} (the frequency of the FM carrier heterodyned into the middle of the IF amplifier's passband), we see that the response of the HPF maps those frequency variations along the horizontal axis into amplitude variations along the vertical axis.

By definition,

$$\frac{\omega_{IF}RC}{\sqrt{1+(\omega_{IF}RC)^2}}=\frac{1}{2},$$

from which it easily follows that

$$RC=\frac{1}{\omega_{IF}\sqrt{3}}.$$

Thus,

$$|H(\omega)|=\frac{\omega\dfrac{1}{\omega_{IF}\sqrt{3}}}{\sqrt{1+\left(\omega\dfrac{1}{\omega_{IF}\sqrt{3}}\right)^2}}=\frac{\left(\dfrac{\omega}{\omega_{IF}}\right)}{\sqrt{3+\left(\dfrac{\omega}{\omega_{IF}}\right)^2}}.$$

Since the instantaneous frequency of a commercial FM radio signal can swing 75 kHz both above and below the IF center frequency, the value of the instantaneous frequency of the slope detector input can vary from 10.625 to 10.775 MHz. The value of $\frac{\omega}{\omega_{IF}}$ can therefore vary from 10.625/10.7 to 10.775/10.7, that is, from 0.993 to 1.007. So, $|H(\omega)|$ varies from $\frac{0.993}{\sqrt{3+(0.993)^2}}$ to $\frac{1.007}{\sqrt{3+(1.007)^2}}$, that is, from 0.4974 to 0.5026. These numbers immediately reveal a serious problem with the slope detector—a *very* low sensitivity to changes in the instantaneous frequency of its input.

Be sure you understand what the output of the slope detector *is*: with an input of an FM wave of *constant* amplitude E_0 and instantaneous *frequency* ω, the output is an FM wave with the same instantaneous frequency *and* instantaneous *amplitude* $|H(\omega)|E_0$. This signal of varying amplitude is the input to an envelope detector (refer to section 2.6), just as in an AM radio receiver. A simple way to visualize what Figure 5.4.3 means, in general, is to imagine a bead, B, threaded on a wire (the linear portion of the slope detector's transfer function). With no modulation present, the bead is at rest, positioned as shown in Figure 5.4.3. *With* modulation present, the bead slides back and forth along the wire: the maximum distance of that slide along the wire, in either direction, is a measure of the *amplitude* of the modulation, while the *rate* at which the bead passes through the "at rest" position is the (audio) frequency of the modulation. In particular, with no carrier modulation present, we have $\frac{\omega}{\omega_{IF}} = 1$, and the slope detector output is an FM wave with constant amplitude $\frac{1}{2}E_0$. The output of the envelope detector is the *constant* $\frac{1}{2}E_0$. If this constant signal is the input to the primary winding of a transformer, then, because the primary signal is *unchanging*, the signal induced in the secondary winding of the transformer (which serves as the input to an audio amplifier) is zero, and so no sound is heard.

A big improvement on the basic slope detector was the 1935 invention of the *Travis detector*.[16] Interestingly, in a paper[17] published that year, Travis presented his work in the context of a problem that has no apparent connection with FM demodulation. Rather, his concern was that of automatically correcting for the drifting of the frequency of the local oscillator in an AM superheterodyne receiver; a sufficiently large drift would heterodyne the received signal to frequencies outside the passband of the IF bandpass filter/amplifier.[18] As Travis wrote of this situation: "Oscillator drift, if not corrected by . . . readjustment, is capable of mistuning the [received] signal [by a lot] in the course of a few hours." Travis's answer to this problem was to develop a detector circuit that could generate a voltage linearly proportional

16 After Charles Travis (1882–1941). Travis is a bit of a mysterious figure in radio history. He had technical degrees from the University of Pennsylvania (BS, 1902; PhD, 1906), but they were not in electrical engineering or physics but, rather, in *civil engineering*! Before 1926, when he became a radio engineer for the Atwater Kent Company, Travis worked as a geologist. He must have been one of those turn-of-the-century self-taught enthusiasts who became caught up in the "romance of early radio."

17 Charles Travis, "Automatic Frequency Control," *Proceedings of the IRE*, October 1935, pp. 1125–1141. The work reported in this paper was considered to be sufficiently important to be cited in Travis's *New York Times* obituary notice.

18 Once a receiver is turned on, the temperature of its circuitry can gradually increase and so alter the electrical properties of the components. This can result in slowly changing the local oscillator frequency.

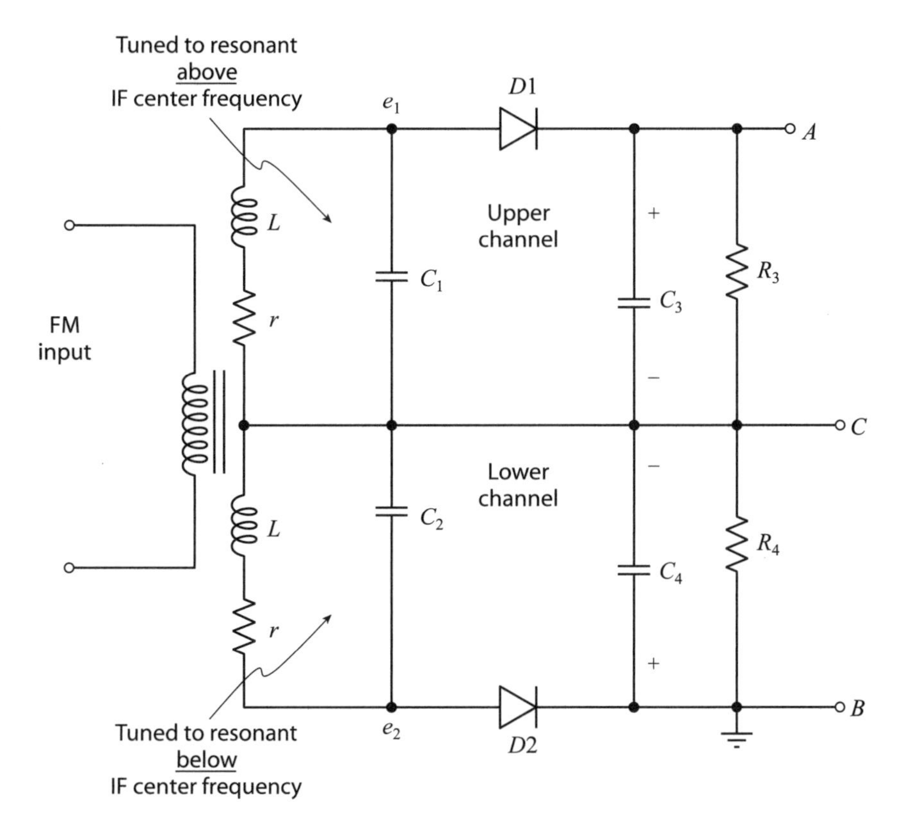

FIGURE 5.4.4. The Travis FM-to-AM circuit.

to the amount of frequency drift. This voltage would then serve as a negative-feedback signal to drive the oscillator frequency back to its original value. While useful for solving Travis's original problem, his circuit really came into its own when applied to FM radio. Figure 5.4.4 shows its basic form.

We'll begin the discussion of Figure 5.4.4 with an explanation of it and then follow up with a simple mathematical analysis. The Travis circuit is built around a center-tapped transformer, with its primary winding receiving the IF filter/amplifier FM output (10.6 to 10.8 MHz). Again, as we did for the slope detector in Figure 5.4.2, imagine (for the moment) that the limiter is not present. The two secondary windings of the transformer, each with the same value of inductance L and the same value of coil resistance r—and therefore the same Q (refer to the discussion of Figure 2.6.4)—are the start of what are essentially two *independent* slope detector circuits. Each of these circuits consists of a tuned front end that has a frequency-dependent response (e_1 and e_2 in Figure 5.4.4), which is the input to the diode of an envelope detector. We'll call these two circuits the *upper channel* (L, r, C_1, $D1$, C_3, and R_3) and the *lower channel* (L, r, C_2, $D2$, C_4, and R_4). For the upper channel, L, r, and C_1 are the frequency-dependent front end, and $D1$, C_3, and R_3 form an envelope detector. Similarly, for the lower channel, L, r, and C_2 are the frequency-dependent front end, and $D2$, C_4, and R_4 form an envelope detector.

Travis's key idea was to tune the two frequency-dependent front ends to *different* frequencies, one somewhat *above* the IF frequency of 10.7 MHz (we'll arbitrarily pick this to be the upper channel, with resonant frequency $\alpha\omega_{IF}$, $\alpha>1$) and the other to somewhat *below* 10.7 MHz (the lower channel, with resonant frequency $\beta\omega_{IF}$, $\beta<1$). One way to do the tuning in Figure 5.4.4 is to have C_1 and C_2 as variable capacitors that can be adjusted *at the radio factory and then sealed* (for a casual user to fiddle with the tuning circuits in an FM radio demodulator is to risk disaster!). Alternatively, the center-tapped transformer might have threaded ferrite cores that can be positioned (with a screwdriver) inside the transformer's secondary windings to change the inductances of the secondaries. In this case, too, the adjustments are *factory* adjustments. This second approach does introduce the complication that the two resonant circuits have different Q-values (refer to section 2.6 and, in particular, Figure 2.6.4). From here on we'll assume we tune via C_1 and C_2. Nominal values for the resonant L's and C's are 4.5 μH and 33 pF (4.5×10^{-6} H and 33×10^{-12} F), allowing for adjustments of the two resonant frequencies to above and below the 10.7 MHz IF center frequency. Typical values for the envelope detector values are $R_3=R_4=100$ kiloohms and $C_3=C_4=100$ pF (the RC time constant of each envelope detector is 10 μs.

Figure 5.4.5 shows the response behavior of these two resonant circuits. This same architecture of two resonant circuits, with one tuned to above and the other tuned to below the carrier frequency, is found in Armstrong's famous 1933 patent (US Patent 1,941,069) that is the basis for the claim he "invented FM" (Armstrong uses the word *non-reactive* for resonant). Armstrong's patent does indeed pre-date Travis's paper, but Travis observed that his inspiration actually came from the English inventor Henry Joseph Round (a long-time competitor of Armstrong's, mentioned in the opening of section 3.2), who had discussed the double-tuned idea *years* before Armstrong's patent. As Travis wrote (note 17): "In the development of the system here described the form taken by the discriminator [another word for *FM detector*] is fairly well crystallized. A *differential* [my emphasis added, and you'll soon see why] rectifier is used, following closely a disclosure by Round [Round's 1927 US Patent 1,642,173 is then cited]."

Because of the orientation of the diodes in Figure 5.4.4 it is clear that the envelope detector capacitors C_3 and C_4 charge with the polarities shown. When diode *D*1 is forward biased, C_3 receives a pulse of charge (increasing the voltage drop across C_3), and when diode *D*2 is forward biased, C_4 receives a pulse of charge (increasing the voltage drop across C_4). These pulses of charge arrive at the IF frequency, which is faster than one pulse each 0.1 µs.

C_3 and C_4 therefore quickly charge to the maximum values of e_1 and e_2, respectively, in a time interval much less than the period of even the highest frequency of the modulation (15 kHz).[19]

As a glance at Figure 5.4.5 shows, for frequencies $\omega < \omega_{IF}$ the response of the lower channel is greater than is the response of the upper channel, and so while both C_3 and C_4 will receive charge, the voltage drop across C_4 will be greater than the voltage drop across C_3. Now, the plus terminal of C_4 (point B) is connected to ground (zero volts), and so the minus terminal of C_4 (point C) is *less* than zero (that is, the voltage of point C in Figure 5.4.4 is negative). Then, since the voltage drop across C_3 is less than that of C_4, as we move through C_3 from point C up to point A, the voltage with respect to ground increases *but does not reach zero*.

In short, when $\omega < \omega_{IF}$, the voltage at point A is *negative* with respect to ground. If you repeat this line of reasoning for the case of

19 This minimum interval is greater than 70 µs, and so C_3 and C_4 receive a large number of charging pulses in that time interval, which means C_3 and C_4 are essentially fully charged to the *maximum* of the voltages (e_1 and e_2) driving their respective diodes. And yet, the 10 µs time constant of the envelope detectors is sufficiently less than 70 µs that the envelope detectors can easily follow the amplitude changes of a signal varying at the modulation's audio rate.

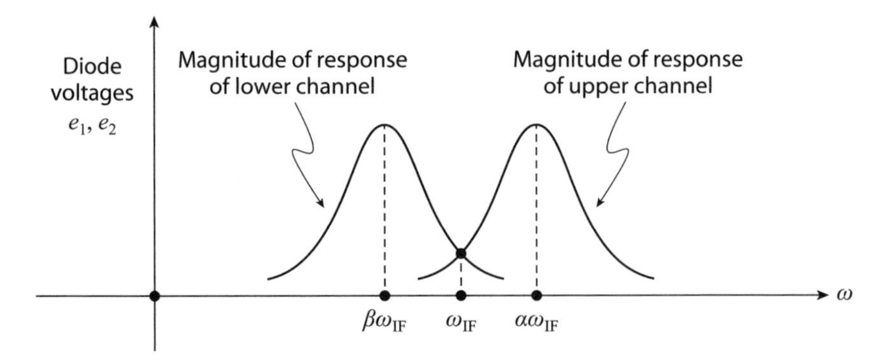

FIGURE 5.4.5. The magnitudes of the frequency responses of the upper and lower channels in the Travis detector. (The word *response* refers to the diode input voltages e_1 and e_2 in Figure 5.4.4.) The resonant frequencies of the upper and lower channels are $\alpha\omega_{IF}$ and $\beta\omega_{IF}$, respectively, with $\alpha > 1$ and $\beta < 1$. Selecting the values for α and β is what is meant by "adjusting" the Travis detector. A *proper* adjustment means, as explained in the text, that the two response curves cross (as shown) at $\omega = \omega_{IF}$.

$\omega > \omega_{IF}$, the conclusion is that the voltage at point A is *positive* with respect to point B (ground). When $\omega = \omega_{IF}$, the voltage drops across C_3 and C_4 are equal (but in opposite direction and so cancel) and the voltage at point A, *if the detector is properly adjusted*, is zero.

The output of a Travis detector at point A, with respect to ground, is the *difference* of the voltage drops across C_3 and C_4, and what makes this useful is that, for significant deviations of ω from ω_{IF}, this difference (or *differential*) voltage is very nearly *linearly* proportional to the deviation. Travis showed this in his paper by graphically plotting the difference voltage, as measured directly from the response curves of the upper and lower channels, as a function of frequency. Let me now show you (and Hardy, too, who no doubt would be appalled by the use of graph paper to establish a mathematical result) an *analytical* way to see this, with the aid of a computer to do all the messy math.

Concentrate your attention on the equivalent circuit of the L, r, and C_1 upper-channel loop, shown in Figure 5.4.6, where v is the IF signal induced in the upper half of the transformer secondary. Ignoring the presence of diode $D1$, a loop current i results, with

$$i = \frac{v}{r + j\omega L + \dfrac{1}{j\omega C_1}} = \frac{j\omega C_1}{1 - \omega^2 L C_1 + j\omega r C_1} v.$$

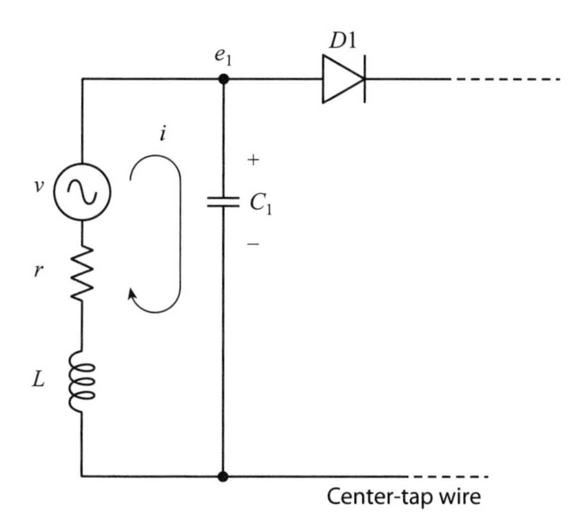

FIGURE 5.4.6. The resonant front end of the upper channel.

The loop is *resonant* (the inductive and capacitive impedances in the loop cancel, and so the loop current is *maximum* at $\frac{v}{r}$) at frequency $\frac{1}{\sqrt{LC_1}}$, which we'll choose to be *greater* than ω_{IF}. That is,

$$\frac{1}{\sqrt{LC_1}} = \alpha\omega_{IF}, \alpha > 1.$$

The loop current i charges C_1 to voltage e_1 with respect to the center tap, and this voltage is said to *drive* diode $D1$. When the diode conducts, charge is transferred to C_3, and after it receives a number of such charging pulses (see note 19 again), the voltage drop across C_3 is essentially the *maximum* or *peak* value of e_1. (Think of pumping up a bicycle tire with multiple strokes of a hand-operated pump, with each new burst of air like a burst of charge. C_3 is the "tire.")

The value of e_1 is

$$e_1 = i\frac{1}{j\omega C_1} = v\frac{1}{1-\omega^2 LC_1 + j\omega rC_1},$$

or as

$$LC_1 = \frac{1}{\alpha^2\omega_{IF}^2},$$

we have

$$e_1 = v\frac{1}{1-\omega^2\dfrac{1}{\alpha^2\omega_{IF}^2}+j\omega rC_1} = v\frac{1}{1-\dfrac{1}{\alpha^2}\left(\dfrac{\omega}{\omega_{IF}}\right)^2+j\omega rC_1}.$$

Since

$$C_1 = \frac{1}{\alpha^2\omega_{IF}^2 L},$$

then

$$\omega rC_1 = \frac{\omega r}{\alpha^2\omega_{IF}^2 L} = \frac{1}{\alpha^2\omega_{IF}^2\dfrac{L}{\omega r}} = \frac{1}{\alpha^2\omega_{IF}^2\dfrac{\omega L}{\omega^2 r}} = \frac{1}{\alpha^2\left(\dfrac{\omega_{IF}}{\omega}\right)^2\dfrac{\omega L}{r}}$$

$$= \frac{1}{\alpha^2}\left(\frac{\omega}{\omega_{IF}}\right)^2\frac{1}{\left(\dfrac{\omega L}{r}\right)}.$$

Writing $Q = \dfrac{\omega L}{r}$, as we did in section 2.6, we arrive at, for the upper channel, that the voltage drop across C_3 is the maximum of the absolute value[20] of e_1, given by

$$|e_1|_{\max} = v\left|\frac{1}{1-\dfrac{1}{\alpha^2}\left(\dfrac{\omega}{\omega_{IF}}\right)^2+j\dfrac{1}{\alpha^2}\left(\dfrac{\omega}{\omega_{IF}}\right)^2\dfrac{1}{Q^2}}\right|, \ \alpha > 1.$$

20 The presence of an imaginary component of e_1 (and e_2) simply reflects the fact e_1 is not in phase with v. Phase measures relative displacement in time, which need not concern us here, because the output voltage of the Travis detector varies so slowly compared with the IF frequencies. The absolute value takes into account the imaginary component of e_1 and e_2 when their *magnitudes* are computed.

If we repeat this entire argument for the lower channel, with the only difference being to define its resonant frequency as *less* than ω_{IF} (that is, as $\beta\omega_{IF}$ with $\beta<1$), then we'll arrive at the voltage drop across C_4 as the maximum of the absolute value of e_2, given by

$$|e_2|_{\max}=v\left|\frac{1}{1-\frac{1}{\beta^2}\left(\frac{\omega}{\omega_{IF}}\right)^2+j\frac{1}{\beta^2}\left(\frac{\omega}{\omega_{IF}}\right)^2\frac{1}{Q^2}}\right|,\ \beta<1.$$

So, finally, the Travis detector output (that is, the voltage at point A in Figure 5.4.4) with respect to point B (ground), is $|e_1|_{max}-|e_2|_{max}$. The obvious question now is, What are α and β? We want their values to be such that the detector output is both zero when $\omega=\omega_{IF}$ *and* varies with frequency as nearly linearly as we can make it *over the entire IF bandwidth of essentially* 200 kHz *(centered on* 10.7 MHz*)*. Suppose, then, we choose β so as to put the resonant frequency of the lower channel at the lower end of the IF frequency interval (10.6 MHz), and so $\beta(10.7)=10.6$. Thus, $\beta=\frac{106}{107}=0.9906\cdots$. For a given value of β, the value of α in the upper channel that gives a detector output of zero at $\omega=\omega_{IF}$ can be shown[21] to be

$$\alpha=\frac{1}{\sqrt{2-\frac{1}{\beta^2}-\frac{1}{Q^2}}}.$$

Is the output variation with frequency linear? The easy way to answer that is to simply see what happens. Using $\beta=\frac{106}{107}$ and the associated computed value of α for two typical values of transformer Q, the output of the Travis detector is shown in Figure 5.4.7. As those plots show, the detector output *does* vary linearly over the entire IF amplifier bandwidth (normalized frequency in the interval 0.99 to 1.01), and *is* indeed zero at $\omega=\omega_{IF}$. The output of the detector increases with

21 This is an interesting enough result, requiring only high school algebra (see Challenge Problem 5.5), that I think even Hardy would find it a good example for illustrating the power of math in technology for (perhaps) skeptical students.

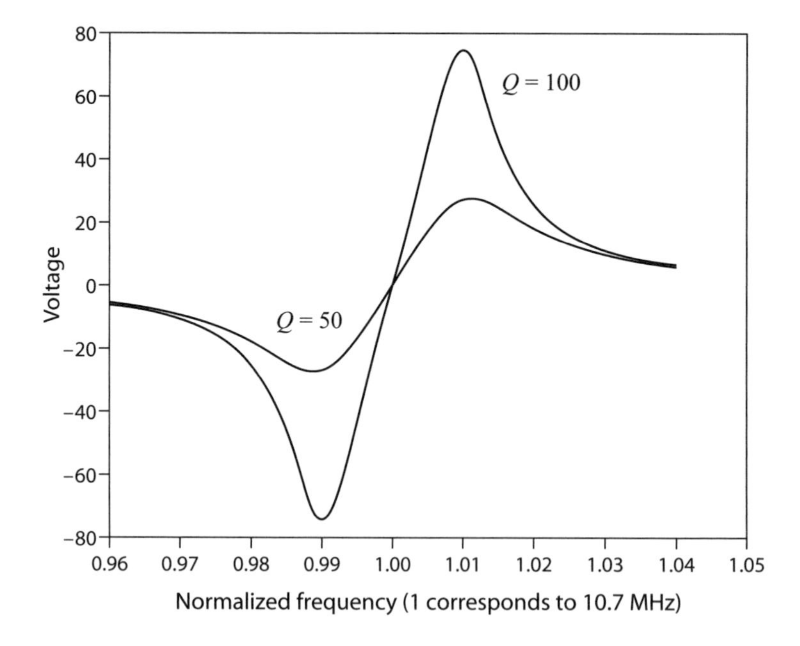

FIGURE 5.4.7. The Travis detector output versus frequency.

increasing Q, and our circuit would clearly work in a *wideband* FM radio application. The output is an AM signal ready for additional audio amplification (if needed) and then to be input to a loudspeaker.

The Travis detector was developed while Travis was an engineer at the RCA License Laboratory (1931–1935); his paper appeared the year he left RCA to join the Philco Radio and Television Corporation. One of his former colleagues at the License Laboratory was the electrical engineer Dudley Foster (1900–1990), who had joined RCA in 1934 and so was familiar with Travis's work, long before it appeared in print. The same year Travis left RCA, the electrical engineer Stuart Seeley (1901–1978) joined RCA, where he met Foster. The two then took Travis's circuit and, while retaining much of its form, developed a different way to generate the input voltages to the envelope detector diodes in the upper and lower channels. The result was the famous *Foster-Seeley detector* (a description of which was published in the March 1937 issue of the *Proceedings of the IRE*), used in many FM radio receivers until the 1960s. Like Travis, Foster and Seeley did not present their circuit as a contribution to FM radio practice but rather as a means of automatically controlling the frequency in tuning the local oscillator in an AM receiver, the same problem that had concerned Travis. Some years later, Seeley teamed up with the physicist Jack Avins (1911–1993) in developing yet another FM detector (the *ratio detector*), which was described in the June 1947 issue of the *RCA Review*. The ratio detector, unlike the Travis or the Foster-Seeley detectors, *was* presented as a contribution to FM radio practice. Both the ratio detector and the Foster-Seeley detector bear a great deal of resemblance to the Travis detector.

One final loose end to deal with is the issue of the *limiter* in Figure 5.4.1 (just before the FM detector) that until now we have imagined isn't there. For the Travis and Foster-Seeley detectors it *is* there (but, as it turns out, it actually *isn't* there for the ratio detector), and here's why. FM radio, just like AM radio, is subject to external phenomena that can affect the amplitude of the transmitted signal during propagation. These external amplitude variations are completely unrelated to the intentional modulation introduced at the transmitter, but when the FM detector receives an FM signal that is supposed to be of constant amplitude but has extraneous *amplitude* modulation present, such modulation passes right through the detector and appears in its output on top of the valid AM generated by the detector. The result is a *noisy* output. Unlike AM radio, however, FM radio can deal with this.

AM radio can't handle such external amplitude variations because it has no way to distinguish between what is valid AM and what is extraneous AM (one person's "noise" is another person's hot saxophone!). FM radio, however, "knows" that *any* AM in the input to the FM detector is "bad," and if it can be removed, the signal to the loudspeaker will be clearer. So, the job of the limiter is to remove any AM from the signal it receives from the IF amplifier. The limiter played a big role in Armstrong's 1933 FM patent, but in fact it can be traced back to years earlier, to the RCA electrical engineer Clarence Hansell (1898–1967), who used limiters in FM circuits in 1927.

How does a limiter work? The fundamental idea is simple, using the phenomenon of *amplifier saturation* that was discussed in Challenge Problem 3.5. If you did that problem, you found that *three* identical amplifiers in series, each with gain 5 and a saturation voltage of 10 volts, form a composite amplifier with a threshold of just 0.08 volt.

Now, suppose the input to this low-threshold composite amplifier is a sinusoid with an amplitude that varies (at an audio frequency rate, a rate *far* less than the IF frequency) from (just to pick some numbers) 0.7 volt to 7 volts, at any frequency in the IF amplifier bandpass. The output will be a *clipped* sinusoid with maximum amplitude at the saturation voltage of the composite amplifier, 10 volts. The AM of the input has been completely removed, but because of the clipping, additional frequencies have now been introduced, as the output, while still periodic, is no longer a single tone. The *fundamental* frequency of the clipped output *is*, however, equal to that of the input sinusoid, and that frequency can be recovered with a bandpass filter centered on the IF amplifier's 10.7 MHz frequency. The output of that bandpass filter is now a sinusoid of fixed amplitude (see Challenge Problem 5.6) at the frequency of the composite amplifier input. The bandpass filter output, *now with no AM*, is the input to the FM detector. The combination of the low-threshold composite amplifier with the bandpass filter forms the limiter in the FM receiver of Figure 5.4.1.

5.5 Bilotti's Phase Quadrature Detector

The theme of this book has been highly historical. That makes sense, I think, given my stated goal (some readers might actually call it a conceit!) of delivering all I have to say directly to a resurrected G. H. Hardy. Still, I would like to end the book on a "modern" note, in a continuation of our discussion of FM detectors.

You'll recall that earlier I mentioned that the Foster-Seeley detector was quite popular up to the 1960s. It may well have occurred to you then to wonder why the Foster-Seeley circuit seemed to fall out of favor after that period.

In fact, all the detectors I've mentioned so far (Travis, Foster-Seeley, ratio) ran into difficulty with the emergence of integrated circuit (IC) technology. Those early detector circuits were *discrete*-component circuits, and all were built around a center-tapped transformer, a relatively bulky gadget that does not easily lend itself to fabrication in the small, compact IC packages ("chips") that became ever more popular in the 1970s and later. So, we end this chapter with a discussion of the

Bilotti[22] phase quadrature FM detector (shown in Figure 5.5.1), which is probably the most popular FM detector in use today.[23]

The two capacitors, resistor, and inductor are a phase-shifting network (that introduces a frequency-dependent phase shift between the FM input $v_1(t)$ and the network signal $v_2(t)$), and a mixer (multiplier) and a low-pass filter that form a phase detector. The output of the LPF is a dc voltage linearly proportional to the instantaneous frequency of the FM input. The LPF output is, as in the Travis detector, positive when the input frequency is greater than f_{IF}, negative when the input frequency is less than f_{IF}, and zero when the input frequency is equal to f_{IF}. The circuit gets its name from the fact that at the IF frequency the signals v_1 and v_2 are 90° out of phase (are in *quadrature*). The detector has the following elegant mathematical description, one that I think Hardy would have admired.

The impedance Z of the parallel combination of C_2, R, and L is given by (where, to keep the algebra uncluttered, we'll use $s=j\omega$)

$$\frac{1}{Z}=\frac{1}{\frac{1}{sC_2}}+\frac{1}{R}+\frac{1}{sL},$$

because, you'll recall, for parallel circuits the *reciprocals* of the individual impedances add. With a little algebra we have

$$Z=\frac{sRL}{s^2RC_2L+sL+R}.$$

22 After the Argentine electrical/mechanical engineer Alberto Bilotti (1925–2012), who invented the phase quadrature detector in 1967.

23 Integrated circuits can easily fabricate resistors, capacitors, and transistors in their solid-state crystal structures. However, components that involve magnetic fields, like transformers and inductors, present complications. Their rapidly expanding and collapsing magnetic fields can induce large (damaging) voltages in the surrounding chip structure, and that has motivated electrical engineers to search for ways to accomplish electronic tasks using *only* resistors, capacitors, and transistors. The Bilotti detector is such a circuit, using no transformer; the inductor in Figure 5.5.1 is *simulated*, using, for example, the clever circuit discussed in Challenge Problem 3.6. Using only resistors, capacitors, and two differential amplifiers (which themselves use only resistors, capacitors, and transistors), the simulated inductor has no associated fluctuating magnetic field.

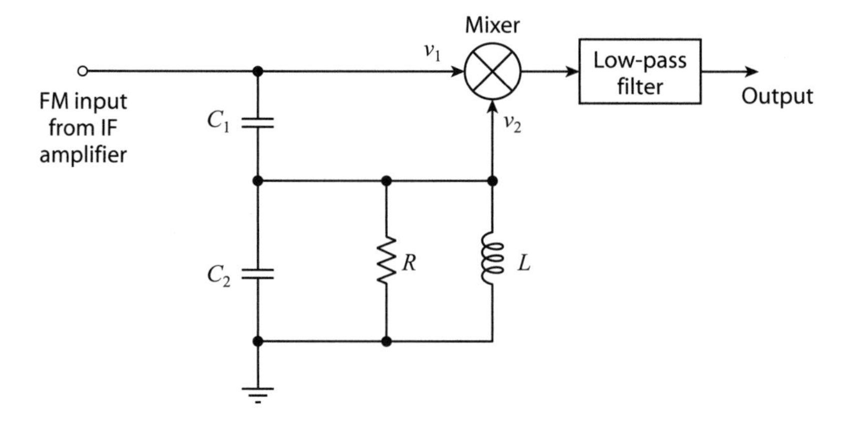

FIGURE 5.5.1. Bilotti's phase quadrature FM detector.

Next, we see by inspection that

$$V_2(\omega) = V_1(\omega)\frac{Z}{Z + \frac{1}{sC_1}}.$$

Inserting the expression that we just derived for Z, a bit of routine if slightly messy algebra (but not nearly as messy as it would be without using the s notation!) we arrive at

$$\frac{V_2(\omega)}{V_1(\omega)} = \frac{s^2 LC_1}{s^2(C_1 + C_2)L + s\frac{L}{R} + 1}.$$

Now, we replace s with $j\omega$ (and s^2 with $-\omega^2$) to get

$$\frac{V_2(\omega)}{V_1(\omega)} = \frac{\omega^2 LC_1}{\omega^2(C_1 + C_2)L - 1 - j\omega\frac{L}{R}}.$$

Our next step is to define

$$\omega_{IF} = \frac{1}{\sqrt{(C_1 + C_2)L}},$$

which "tunes" the Bilotti circuit to the IF amplifier's center frequency (the frequency to which the heterodyned FM carrier is shifted by the local oscillator at the front end of the FM receiver). With that, we have

$$\frac{V_2(\omega)}{V_1(\omega)}=\frac{\omega^2 LC_1}{\omega^2\frac{1}{\omega_{IF}^2}-1-j\frac{1}{\frac{R}{\omega L}}}.$$

We now define[24]

$$\hat{Q}=\frac{R}{\omega L},$$

and after yet a bit more algebra, we have

$$\frac{V_2(\omega)}{V_1(\omega)}=\frac{j\hat{Q}\frac{C_1}{C_1+C_2}\left(\frac{\omega}{\omega_{IF}}\right)}{1+j\hat{Q}\frac{\omega_{IF}}{\omega}\left[\frac{(\omega+\omega_{IF})(\omega-\omega_{IF})}{\omega_{IF}^2}\right]}.$$

So far, the analysis has been exact, but now we make a small approximation. For an FM signal in the IF passband, the ratio $\frac{\omega}{\omega_{IF}}$ varies from about 0.99 to 1.01 (10.6 to 10.8 MHz, centered on 10.7 MHz). So, we introduce very little error in writing

$$\frac{\omega}{\omega_{IF}}\approx\frac{\omega_{IF}}{\omega}\approx 1,$$

and

$$\omega+\omega_{IF}\approx 2\omega_{IF}.$$

24 $\hat{Q}$ is *not* the Q in our earlier discussion of the Travis detector. *That* Q (of the secondary coils in a center-tapped transformer) was $\omega\frac{L}{r}$, where r is the resistance of each coil. For $\hat{Q}$, the resistor R in the Bilotti circuit is separate and distinct, indeed, *totally independent* of the inductor L. The fact that $\hat{Q}$ looks similar to $\frac{1}{Q}$ is purely accidental, and in an attempt to emphasize that, I have put a "hat" on Q.

This results in

$$\frac{V_2(\omega)}{V_1(\omega)} = \frac{j\hat{Q}\frac{C_1}{C_1+C_2}}{1+j2\hat{Q}\left(\frac{\omega}{\omega_{IF}}-1\right)}.$$

Notice that when $\omega = \omega_{IF}$ we have

$$\frac{V_2(\omega)}{V_1(\omega)} = j\hat{Q}\frac{C_1}{C_1+C_2},$$

and so (because of the j) $v_1(t)$ and $v_2(t)$ are 90° out of phase at the IF frequency. More generally, the phase difference between $v_1(t)$ and $v_2(t)$, as a function of frequency, is

$$\theta(\omega) = 90^\circ - \tan^{-1}\left\{2\hat{Q}\left(\frac{\omega}{\omega_{IF}}-1\right)\right\}.$$

So, if we write the FM input signal to the Bilotti circuit as

$$v_1(t) = \sin(\omega t + \alpha),$$

where α is some arbitrary phase, then the other input to the mixer (multiplier) is

$$v_2(t) = \frac{\hat{Q}\frac{C_1}{C_1+C_2}}{\sqrt{1+4\hat{Q}^2\left(\frac{\omega}{\omega_{IF}}-1\right)^2}}\sin\{\omega t + \alpha - \theta(\omega)\}$$

$$= \frac{\hat{Q}\frac{C_1}{C_1+C_2}}{\sqrt{1+4\hat{Q}^2\left(\frac{\omega}{\omega_{IF}}-1\right)^2}}\sin\left[\omega t + \alpha - 90^\circ + \tan^{-1}\left\{2\hat{Q}\left(\frac{\omega}{\omega_{IF}}-1\right)\right\}\right].$$

Thus, the output $v_1(t)v_2(t)$ of the mixer (multiplier) is given by

$$v_1(t)v_2(t)=\frac{\hat{Q}\frac{C_1}{C_1+C_2}}{\sqrt{1+4\hat{Q}^2\left(\frac{\omega}{\omega_{IF}}-1\right)^2}}$$

$$\sin(\omega t+\alpha)\sin\left[\omega t+\alpha-90^\circ+\tan^{-1}\left\{2\hat{Q}\left(\frac{\omega}{\omega_{IF}}-1\right)\right\}\right].$$

This admittedly looks pretty awful, but you'll soon be amazed at how, with just a bit more algebra, this expression reduces to a very elementary form.

Recalling the trigonometric identity

$$\sin(A)\sin(B)=\frac{1}{2}\{\cos(A-B)-\cos(A+B)\}$$

we see that the $A+B$ term has a frequency of $2\omega\approx 2\omega_{IF}$, which will be suppressed by the low-pass filter. So, the output of the LPF is the dc voltage (notice that the arbitrary α disappears)

$$\frac{1}{2}\frac{\hat{Q}\frac{C_1}{C_1+C_2}}{\sqrt{1+4\hat{Q}^2\left(\frac{\omega}{\omega_{IF}}-1\right)^2}}\cos\left[90^\circ-\tan^{-1}\left\{2\hat{Q}\left(\frac{\omega}{\omega_{IF}}-1\right)\right\}\right]$$

$$=\frac{1}{2}\frac{\hat{Q}\frac{C_1}{C_1+C_2}}{\sqrt{1+4\hat{Q}^2\left(\frac{\omega}{\omega_{IF}}-1\right)^2}}\cos\left[90^\circ+\tan^{-1}\left\{2\hat{Q}\left(1-\frac{\omega}{\omega_{IF}}\right)\right\}\right].$$

Recalling another trigonometric identity,

$$\cos(A+B)=\cos(A)\cos(B)-\sin(A)\sin(B),$$

we have

$$\cos\left[90^\circ + \tan^{-1}\left\{2\hat{Q}\left(1-\frac{\omega}{\omega_{IF}}\right)\right\}\right] = -\sin\left[\tan^{-1}\left\{2\hat{Q}\left(1-\frac{\omega}{\omega_{IF}}\right)\right\}\right]$$
$$= -\frac{2\hat{Q}\left(1-\frac{\omega}{\omega_{IF}}\right)}{\sqrt{1+4\hat{Q}^2\left(\frac{\omega}{\omega_{IF}}-1\right)^2}}.$$

Thus, at last,

$$\text{LPF output} = \left[-\frac{1}{2}\hat{Q}\frac{C_1}{C_1+C_2}\right]\left[\frac{2\hat{Q}\left(1-\frac{\omega}{\omega_{IF}}\right)}{1+\left\{2\hat{Q}\left(1-\frac{\omega}{\omega_{IF}}\right)\right\}^2}\right].$$

The expression in this box for the LPF output is the product of a *gain factor* (in the first set of square brackets) and a *frequency-dependent factor* (in the second set of square brackets). We can explore the nature of the frequency-dependent factor by simply defining

$$x = 2\hat{Q}\left(1-\frac{\omega}{\omega_{IF}}\right)$$

and then plotting the function $\frac{x}{1+x^2}$ (as is done in Figure 5.5.2) as x varies around $x=0$ (which corresponds to $\omega=\omega_{IF}$).

Figure 5.5.2 shows that the output of the LPF varies linearly (to the eye, at least) over, roughly, the interval $-\frac{1}{2}<x<\frac{1}{2}$. At the extreme edges of that interval, the magnitude of x is (to the eye, at least) about $0.4=\frac{2}{5}$. So, if we wish that interval of x to correspond to the frequency interval $0.99\omega_{IF}<\omega<1.01\omega_{IF}$, the same interval we used in Figure 5.4.7 for the Travis FM detector, then we have

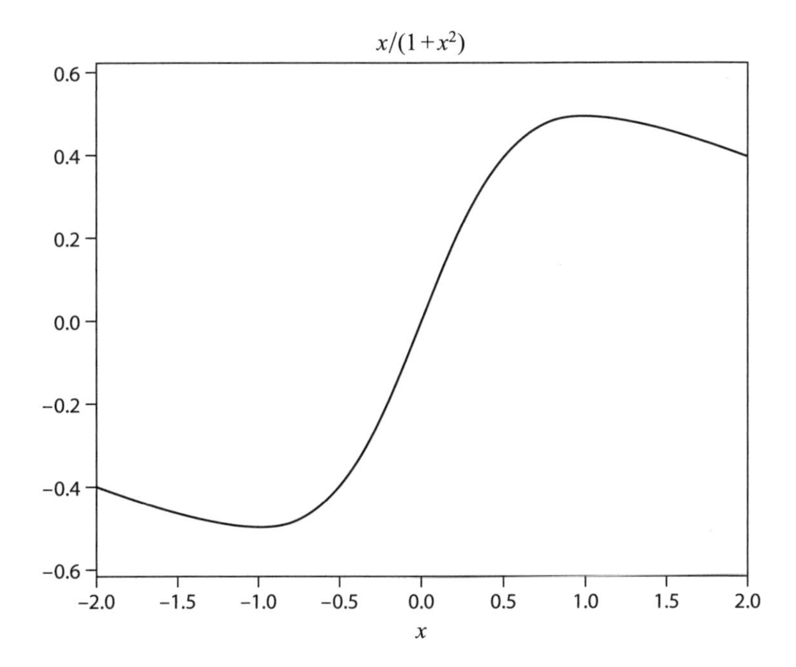

FIGURE 5.5.2. Locating where the LPF output varies *linearly* with frequency.

$$2\hat{Q}(1-0.99)=\frac{2}{5}=2\hat{Q}\frac{1}{100}=\frac{\hat{Q}}{50};$$

that is, $\hat{Q}=20$, a result we'll use momentarily to calculate the value of the gain factor.

Now, you'll recall that $R=\hat{Q}\omega L$, and so, at the IF frequency of FM radio we have

$$R=(20)\,(2\pi\times 10.7\times 10^{6})\,L=1.3446\times 10^{9}L.$$

Thus,

$$L=\frac{R}{1.3446}\times 10^{-9}\text{ H}.$$

So, suppose $R=13{,}446$ ohms (1.3446×10^{4}), a value easily fabricated on an IC chip. Then, $L=10^{-5}$ H, that is, $L=10\ \mu$H. Instead of trying to put an actual coil of wire in an IC chip, we'll instead simply simulate it with the circuit of Challenge Problem 3.6. We use the value of L to compute the values of C_1 and C_2. From earlier in this analysis, you'll recall that we defined

$$\omega_{IF} = \frac{1}{\sqrt{L(C_1 + C_2)}}.$$

Thus,

$$C_1 + C_2 = \frac{1}{\omega_{IF}^2 L} = \frac{1}{(2\pi \times 10.7 \times 10^6)^2 \times 10^{-5}} = 2210^{-12} = 22 \text{ pF}.$$

Finally, suppose we set $C_1 = C_2$ (each is 11 pF. This gives us a gain factor of

$$-\frac{1}{2}\hat{Q}\frac{C_1}{C_1 + C_2} = -\frac{1}{2}(20)\frac{1}{2} = -5.$$

Challenge Problem 5.1: Establish the Hilbert transform pair $\frac{1}{1+t^2} \leftrightarrow \frac{t}{1+t^2}$ that we used in section 5.2. *Hint*: Recall Challenge Problem 1.13, where you derived the Fourier transform pairs $\frac{1}{1+jt} \leftrightarrow 2\pi e^{\omega} u(-\omega)$ and $\frac{1}{1-jt} \leftrightarrow 2\pi e^{-\omega} u(\omega)$. Further, recall from section 4.3 that if $m(t)$ has Hilbert transform $\bar{m}(t)$, then $\bar{M}(\omega) = -j\ \mathrm{sgn}(\omega) M(\omega)$. So, suppose you define $m(t) = \frac{2}{1+t^2} = \frac{1}{1+jt} + \frac{1}{1-jt}$. You can then immediately write $\bar{M}(\omega)$, and from that . . . (you complete this argument).

Challenge Problem 5.2: In the text we gave a "conservation of energy proof" of the Bessel identity $1 = J_0^2(M) + 2\sum_{n=1}^{\infty} J_n^2(M)$, based on the observation that the power of an FM signal is constant, independent of the modulation index M. See if you can develop a purely mathematical derivation of this identity.

Challenge Problem 5.3: Suppose we relax the Carson rule for the FM bandwidth to require the transmitted sideband power to be only at least 90% of the total sideband power (in the text we derived the "traditional" Carson rule using 98%). Does this modification "save" much bandwidth? What if we change the requirement to transmission of at least 99% of the total sideband power? Does that greatly increase the bandwidth?

Challenge Problem 5.4: In commercial wideband FM radio, the IF frequency is 10.7 MHz. Where do you think this value comes from? *Hint*: Remembering that the FM band is 88 to 108 MHz (carrier frequencies run from 88.1 to 107.9 MHz), repeat the image-rejection calculations for AM radio in section 3.2.

Challenge Problem 5.5: Derive the result stated in the text for the relationship between the resonant frequencies in the upper and lower channels of a perfectly adjusted (that means the output is zero when $\omega = \omega_{IF}$) Travis detector:

$$\alpha = \frac{1}{\sqrt{2 - \frac{1}{\beta^2} - \frac{1}{Q^2}}}.$$

Hint: Start with $|e_1|_{max} = |e_2|_{max}$ at $\frac{\omega}{\omega_{IF}} = 1$. A trivial solution is $\alpha = \beta$, which is *not* of physical interest. At some point in your analysis you will have to explicitly use the fact that $\alpha \neq \beta$.

Challenge Problem 5.6: Imagine a sinusoidal signal with instantaneous frequency in the FM/IF passband that has been "clipped" to an arbitrarily low level by a limiter, to eliminate any AM that may be present in the signal. This clipped signal is then run through a bandpass filter centered on the FM/IF passband. See if you can develop a Fourier series argument to support the claim in the text that the instantaneous frequency of the preclipped signal can be recovered from the filter output. *Hint*: Remember that the rate at which the instantaneous frequency changes in broadcast FM is much lower than the instantaneous frequency itself.

Chapter 6

American AM Broadcast Radio

A Historical Postscript

> There are three cardinal ideals in Radio—Motherhood, Blessed Regularity, and Personal Daintiness.[1]

Starting in the early 1920s, radio graduated from being a plaything of amateur tinkerers, electrical engineers, and physicists and became big business. There was no lack of entrepreneurs who served up the tantalizing possibility for newcomers to break into the fascinating new technology. In the early days of broadcast radio, home-correspondence programs and residential vocational schools used the romantic image of radio to attract students. One of the largest schools was the National Radio Institute, which ran ads in the pulp-fiction magazines likely to be read by young men. A different approach was the use of a 1921 publicity photo (see Figure PS2) of Mary Texanna Loomis (1880–1960), who operated the Loomis College of Radio Engineering in Washington, DC, from 1920 to the mid-1930s.[2] The success of NRI and of Loomis's school—which were just two of numerous such operations in America in the 1920s, '30s, '40s, and even later—was indicative of the popular view of radio as a possible route to obtaining skills to qualify for a "glamorous" job.

1 From the essay "Thix" by the American humorist James Thurber (1894–1961), which is included in his 1948 book *The Beast in Me and Other Animals*. The title "Thix" is a childish pronunciation of *six*, the age of a child character in a parody of the blood-and-thunder radio dramas of the 1930s and '40s, which often had youngsters engaging in violent confrontations with criminals. The last two of Thurber's "ideals" are references to the products sold by many of the sponsors of radio dramas.

2 Figure PS2 is from a glass plate negative at the Library of Congress. From the hardware (an ac motor spinning a dc generator) visible on the lower shelf beneath a variac, the equipment rack appears to be a power supply that could have been the source of both the high-voltage dc and the large filament currents required by the then-available vacuum tubes. The hardware visible on the workbench in the background might explain what appear to be oil-stained overalls (but which might in fact just be artifacts caused by damage to the glass plate negative emulsion), but the Victorian-era shoes remain a mystery. Loomis was a distant cousin of Mahlon Loomis (see note 13 in the appendix).

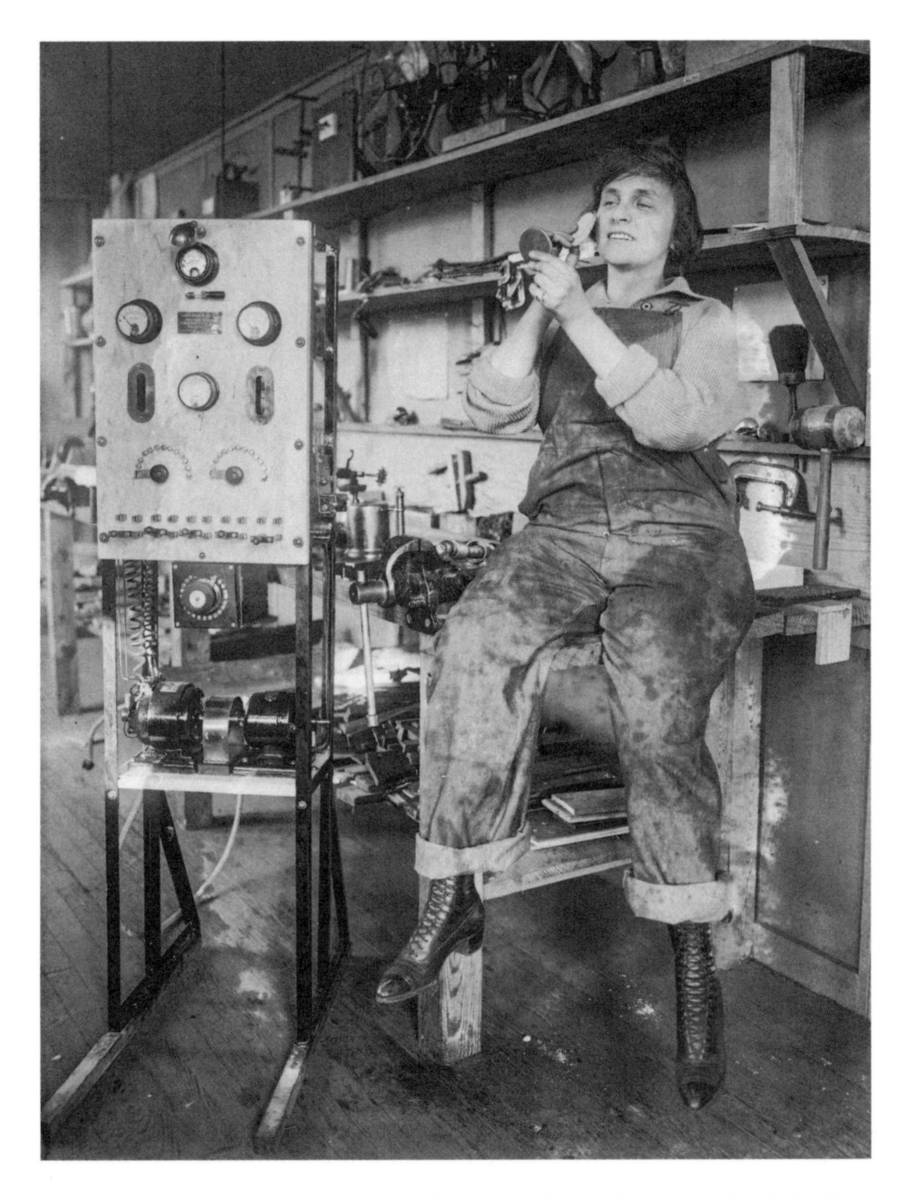

FIGURE PS2. The original caption read: "Feminine despite her occupation, Miss Mary Texanna Loomis, the only woman in the world to own and run a radio school, stops in the midst of a message from, say, Mars, to powder her nose."

There was another side to radio, however, that many people found not so attractive. The imagery of Figure PS3 (from a 1923 issue of the English humor magazine *Punch*) shows that the all-too-common modern scene of groups of people (think parents and their children) sitting around a restaurant table, saying nothing to one another all the while their noses are stuck in their smartphones, is actually nothing new. The fascination of the new "wireless" technology was already widespread more than a century ago, with headphones and primitive radios acting as great-grandpa's and great-grandma's "smartphones"!

The eventual widespread availability of AM radio receivers, in homes at all levels of economic status, caused major cultural shocks in entertainment, politics, and religion. Each was, individually, a big shock, but that they occurred almost simultaneously further magnified the impact of each on society.[3] When the BBC broadcast Mozart's *Magic Flute* live from the Royal Opera House in Covent Garden, in January 1924, it was received by the UK radio audience to great acclaim. But what exceeded that feat, by a wide margin (although that was not immediately appreciated) was the broadcast by 2LO that same month of the first BBC radio drama.

Titled *A Comedy of Danger*, it was the story of a rescue from a coal mine accident. Taking place totally in the dark, the story relied not only on speech and sound effects but also on the listener's imagination. It, too, was received with great enthusiasm by the UK audience. Ironically, when the author of that drama was in America a few months later, he found that radio executives there were of a radically different mind. Instead, they

> rejected the whole idea [of radio stories]. That sort of thing might be possible in England, they explained, where broadcasting was a monopoly[4] and a few crackpot highbrows . . . could

3 The 1930 US Census asked each head of household if their home included a radio set. Ten years later it was clear that radios were indeed popping up everywhere: in 1940 thirty million households had at least one. These households cut across the economic spectrum: in 1942 it was estimated that there were 10 million homes with a radio but no car, 13 million homes with a radio but no telephone, and 6 million homes with a radio but no bathtub. You might not be able to drive anywhere or call anybody or even just take a bath, but in 1942 you almost surely could listen to *The Lone Ranger* (since 1933).

4 Broadcast radio in the UK was, until the 1970s, a state-controlled monopoly, with broadcasting costs covered by listener-paid fees, an approach initially considered but

FIGURE PS3. Anticipating the future. (*Punch* Cartoon Library/TopFoto).

> impose what they liked on a suffering public. But the American set-up was different: it was competitive, so it had to be popular, and it stood to reason that plays you couldn't see could never be popular.[5]

Never have the "experts" been so wrong!

Within the next four years American radio dramas appeared on nationwide networks, starting with *Real Folks* on NBC in 1928, and then *Amos 'n' Andy* a year later (it had been on Chicago's local radio station WGN as *Sam 'n' Henry* since January 1926). These shows were merely the initial ripple of what would become a tidal wave. During the following years radio gave American listeners *Just Plain Bill* (the tale of a small-town barber with a mortgaged shop); *The Romance of Helen Trent*" (during 7,222 broadcasts this program showed that "just because a woman is 35 doesn't mean romance is over"); *Ma Perkins* (a kindly amateur philosopher who owned a small-town lumber yard); *John's Other Wife* (despite the naughty title, the "other wife" was John's proper secretary); *Pepper Young's Family* (set in middle America, the characters experienced love, hatred, and finally went crazy); *Our Gal Sunday* (which explored the question, Can a coal miner's daughter find happiness married to England's richest lord?); *Young Dr. Malone* (featuring a physician who had many adventures between operations, such as surviving being shot down over Germany in the Second World War, emerging victorious from a murder trial, and being saved by a blood transfusion from a fatal illness); *When a Girl Marries* (what happens when young lovers from "opposite sides of the tracks" marry—and it wasn't always pretty); *Backstage Wife* (the tale of the troubles a stenographer from the Iowa sticks has after she marries a Broadway matinee idol who is relentlessly pursued by every woman under the age of 85 that he encounters); *Young Widder*

eventually discarded in the US. For a discussion on whether American radio should be public or private, see Mary S. Mander, "The Public Debate about Broadcasting in the Twenties: An Interpretive History," *Journal of Broadcasting*, Spring 1984, pp. 167–185.

5 From an essay by the English novelist and dramatist Richard Hughes (1900–1976), "The Birth of the Radio Drama," *Atlantic Monthly*, December 1957, pp. 145–148. For an amusing parody of what was on American radio before the invention of radio drama, see H. Le B. Bercovici, "Station B-U-N-K," *American Mercury*, February 1929, pp. 233–240. That essay tells us that B-U-N-K's competitors were stations J-U-N-K and B-O-S-H.

Brown (widowed in her early 30s, the still devastatingly beautiful Ellen Brown spends the next 20 years being pursued by every bachelor in Simpsonville)—and on and on went the list.

These shows (in 1938 there were at least 50 of them on the air, uttering well over a million words each week) were *serials*, that is, continuously evolving 15-minute, five-day-a-week presentations from 10:30 in the morning to 6 o'clock in the evening. Sponsored mostly by the manufacturers of soaps and cleansing agents (in 1936 the top radio advertiser, *by far*, was Procter & Gamble, with the makers of breakfast foods and laxatives trailing behind), these programs became known as "soap operas," "dishpan dramas," and "washboard weepers." Specifically targeting the millions of stay-at-home women of the 1920s through the early 1950s, the soaps were tales of perpetually troubled people wallowing in melodrama, and they were both immensely popular and hugely profitable. In 1940 one-third of the total advertising income of NBC and CBS, combined, was due to the soaps.

The development of television, however, and the changing post–Second World War economic forces that encouraged the mass departure of working-age women from the house and into the labor force combined to spell doom for the radio soap opera. When "Ma Perkins" said her final goodbyes on Friday, November 25, 1960, after 7,065 broadcasts, it was the end of the road for radio serials after more than three decades of fabulous success.[6]

Along with the adult soaps, broadcast radio also introduced a markedly different sort of dramatic program: the crime-and-horror show. As one writer said of those programs:

> Come five o'clock each weekday afternoon, millions of American children drop whatever they are doing and rush to the nearest radio set. Here, with feverish eyes and cocked ears, they listen for that first earsplitting sound which indicates that the Children's Hour is at hand. This introductory signal may be the wail of a

6 For more on the radio soaps, see Whitefield Cook, "Be Sure to Listen In!" *American Mercury*, March 1940, pp. 314–319; Merrill Denison, "Soap Opera," *Harper's*, April 1940, pp. 498–505; and Max Wylie, "Washboard Weepers," *Harper's*, November 1942, pp. 633–638. In the late 1940s James Thurber wrote a five-part series of long, informative essays about the soaps, full of low-key humor, for the *New Yorker* magazine. You can find those essays reprinted (under the collective title "Soapland") in Thurber's 1948 book (see note 1).

> police siren, the rattle of a machine gun, the explosion of a hand grenade, the shriek of a dying woman, the bark of a gangster's pistol, or the groan of a soul in purgatory. Whatever it is, the implication is the same. Radio has resumed its daily task of cultivating our children's morals—with blood-and-thunder effects.[7]

The "children's" program that probably best illustrates what Gibson had in mind is *Gangbusters*, first heard on CBS in January 1936. The opening of each broadcast certainly was something to hear: tires screeching, a policeman's whistle, the shattering of a glass window, the wail of a siren—all as the background to a voice yelling "Calling the police! Calling the G-men! Calling all Americans to war on the underworld!" So raucous was this energetic opening that it gave birth to the still common use of the phrase "coming on like gangbusters!" as a description of anything with a strong start. Despite all the possible objections to the appropriateness of such a program for children, *Gangbusters* had a huge, loyal, enthusiastic audience, and it stayed on the air for more than 20 years (until 1957).

Other dramatic shows that didn't quite fit into the mold of simply fighting violent gangsters included the spooky *The Shadow* ("Who knows what evil lurks in the hearts of men? The Shadow knows!" followed by creepy laughter); *Inner Sanctum Mysteries* (a show that opened with the sound of a turning doorknob, then a *slowly* creaking door swinging on rusty hinges, followed by a voice inviting you to pass "through the gory portal" into that night's outlandish tale, one designed to simply scare the daylights out of you,[8] which it almost always did from 1941 to 1952); *The Strange Dr. Weird* (featuring additional creepy tales that always ended with the narrator saying: "Oh, you have to leave now—too bad! But perhaps you'll drop in on me again soon. I'm always home. Just look for the house on the other side of the cemetery—the house of Dr. Weird!")—which suggested listeners keep a hypo handy to get through "emotional emergencies";

7 Worthington Gibson, "Radio Horror: For Children Only," *American Mercury*, July 1938, pp. 294–296.

8 As one observer of early radio wrote: "[N]othing could be creepier than human voices stealing through space, preferably late on a stormy night, with a story of the supernatural . . . particularly when you are listening alone." From Roy S. Durstine, "We're On the Air," *Scribner's Magazine*, May 1928, pp. 623–631.

other more sophisticated versions of *Inner Sanctum* such as *Escape*, *The Whistler*, *The Mysterious Traveler*, and the murderously intense *Lights Out* (described as the ultimate horror show,[9] it was intentionally given a late-night air-time slot to keep it away from children).[10]

It's almost a guaranteed bet that Hardy would never have listened to any of these horror shows, and *certainly not* to the soaps (which "enjoyed" the occasional label of being "American rubbish" when rebroadcast abroad). Other kinds of broadcasts, however, gained the attention of nearly everyone as politicians "discovered" radio early on: on June 21, 1923, Warren Harding became the first president of the United States to be heard on the radio. A few months later, in November, former President Woodrow Wilson followed in Harding's footsteps. Harding's successor, Calvin Coolidge, had his 1925 inaugural address broadcast coast to coast over a network of 27 stations, and then, two years later, his February 1927 address to a joint session of Congress was transmitted to an audience of 20 million over a network of 42 stations stretching from Portland, Maine, to San Francisco, California. That speech was the first international political broadcast as well, as it was sent by shortwave radio to London, where the BBC rebroadcast it over 2LO to the whole of the UK, to Paris, and to South Africa. President Franklin Roosevelt and Germany's Adolf Hitler brought political radio to its peak, both before and during the Second World War, with Roosevelt's Fireside Chats and Hitler's loud, melodramatic (occasionally even unhinged) rants.[11]

As fast as politicians embraced radio, they did not outpace the "radio priests," the so-called *Bible-thumpers* (also known as *God-pumpers*) of the ether, who were the ancestors of today's television

9 In his 1975 book *This Was Radio*, the radio actor Joseph Julian (1911–1982) explained how the plot of one episode of *Lights Out* involved a man being turned inside out. This astonishingly gross event was a challenge met by a sound-effects man who *slowly* peeled a tight rubber glove off one hand while a colleague, *just as slowly*, crushed a strawberry box (to simulate breaking bones). As another example, to simulate the electrocution of a character, a sizzling pan of frying bacon did the trick! For a medium so heavily dependent on imagination, such creativity was routine.

10 For more on the dramas of old-time radio, see John Dunning's 1976 book *Tune in Yesterday*, Prentice-Hall, 1976, and the invaluable book by Frank Buxton and Bill Owen, *The Big Broadcast 1920–1950* (2d ed.), Scarecrow Press, 1997. Episodes of many of the programs I mention here can now be listened to on the Web.

11 The political theater of early radio is still, of course, with us today on television and on a multitude of internet platforms.

evangelists. The start of the twentieth century was the age of Elmer Gantry–type evangelists (fundamentalist preachers) who held wildly popular tent revivals attended by vast crowds. It is estimated that the best known of these masters of religious fervor, Billy Sunday (1862–1935), spoke directly, face-to-face, to a total of perhaps 100 million people over the span of his entire career (as early as 1929 he also had his own radio show, *The Back Home Hour*). That's an impressive number, to be sure, but it could easily be equaled in a *single month* of Sunday-morning radio broadcasts. The multiplicative force of radio made the preachers who came after Billy Sunday into Hollywood-style celebrities—even cult figures—to millions of listeners during the Great Depression of the 1930s.

The radio age of religion began just two months to the day after KDKA/ Pittsburgh broadcast the 1920 Harding-Cox presidential election results, when that same station broadcast the January 2, 1921, sermon of the pastor of Pittsburgh's Calvary Episcopal Church. The joining of radio and religion quickly blossomed from that simple beginning into what became known as the "Invisible Church," or the "Electric Church," or the "Electric Pulpit." These phrases describe what one writer called "the promise of GE & Jesus walking hand in hand to make radio [a] rousing commercial success."[12] It also made the radio preachers into influential forces to be reckoned with, whether with a regional or a national reach. Of the former, the Pentecostal evangelist Aimee Semple McPherson (1890–1944) is noteworthy, as in 1924 she started her very own radio station, KFSG, in Los Angeles (the call letters stood for **K**all **F**our**S**quare **G**ospel).

12 Dave Berkman, "Long Before Falwell: Early Radio and Religion—As Reported by the Nation's Periodical Press," *Journal of Popular Culture*, Spring 1988, pp. 1–11. As Berkman observed, radio and religion were a natural combination, as "both were built on words and music." Jerry Falwell Sr. (1933–2007) was a conservative Baptist pastor and televangelist. At his peak he had enormous influence, but his was only the most recent in a long line of such electronic pulpits, stretching back to the first days of radio. There was *National Vespers* (1927), *The Catholic Hour* (1929), *The Mormon Tabernacle Choir* (1929), *The CBS Church of the Air* (1931), *The Gospel Hour* (1936), *The Lutheran Hour* (1938), *Radio Bible Class* (1940), *The Voice of Prophecy* (1941), *The Hour of Faith* (1942), *The Back to God Hour* (1948), and *The Hour of Decision* (1950), featuring Billy Graham (1918–2018), spiritual advisor to multiple US presidents. In 1951, at the start of the decade in which radio drama would give way to television, the Baptist preacher Charles Fuller (1887–1968) had his *Old Fashioned Revival Hour* carried by ABC on a total of 650 (!) stations.

An example of a radio priest with *national* stature was Catholic Father Charles Coughlin (1891–1979), who, in the 1930s, became sufficiently prominent to attract the (negative) attention of the president of the United States.[13] The Church tolerated Coughlin's anti-Semitic broadcasts because he also attacked communism, the arch foe of Catholicism. Coughlin attracted millions of dollars in donations from faithful listeners all across the country, but when he went from being an ardent supporter of Roosevelt in the early 1930s to being an equally ardent attacker in the late 1930s, he found himself labeled a fascist. By 1942 Church superiors finally suppressed him. Coughlin died a forgotten man, but at his height the radio priest's broadcast words were, to millions of devoted listeners, second only to those of the pope.[14]

Even before 1942, however, there were keen observers who took a decidedly pragmatic view of religious broadcasting. One was E. B. White (1899–1983)—he achieved literary fame as the author of the children's classic *Charlotte's Web*—who wrote a regular column in *Harper's Magazine*. In his April 1939 essay, "Sabbath Morn," White declared: "One of the chief pretenders to the throne of God is radio. . . . [A]fter all, the Church merely holds out the remote promise of salvation; the radio tells you if it's going to rain tomorrow." Eternal bliss (maybe, perhaps) in the (it's hoped) distant future or certain knowledge of rain for the crops tomorrow? The choice was a clear and easy one for many of the farmers listening to the Sunday morning broadcast in middle America.

Even when such programs were at the height of popularity, much more was happening in broadcast radio in the 1920s, '30s, and '40s than simply soap opera, crime shows, church services, adventure and Western programs, and political talk. There was sports, starting with the July 2, 1921, broadcast of the world heavyweight championship boxing match between the American Jack Dempsey and the French

13 Marshall W. Fishwick, "Father Coughlin Time: The Radio and Redemption," *Journal of Popular Culture*, Fall 1988, pp. 33–47.

14 The Church began operating its own station, Vatican Radio HVJ, in February 1931. As Marconi said (speaking in Italian) at the inaugural ceremonies: "I have the highest honour of announcing that in only a matter of seconds the Supreme Pontiff, Pope Pius XI, will inaugurate the Radio Station of Vatican City State. The electric radio waves will carry to all the world his words of peace and blessing." (https://www.vaticannews.va/en/pope/news/2019-02/pope-pius-xi-vatican-radio-anniversary.htm).

light-heavyweight champion Georges Carpentier (300,000 listened to what was billed as "The Scientific Sensation of the Century"). There was news and gossip (*Walter Winchell*—"Good evening, Mr. and Mrs. America and all the ships at sea. Let's go to press!" accompanied by the frantic, staccato chatter of a telegraph key and a rapid-fire delivery at almost 200 words a minute); gossip shows (*Your Hollywood Reporter*); comedy (*The Life of Riley*); and game shows (*Hit the Jackpot*). In 1926 the US Department of Agriculture started its Radio Service to broadcast educational programming to farmers. That same year saw the start of the US Bureau of Home Economics's *Housekeepers' Chat* program, starring "Aunt Sammy" (so named because the star of the program was supposed to be the wife of Uncle Sam). These broadcasts, which continued until 1944, carried important information to America's homemakers on topics that included nutrition, sanitation, child care, and emergency plumbing repairs. It was an enormously popular production. When the meal plans that had been broadcast were brought out in printed form (*Aunt Sammy's Radio Recipes*) the Government Printing Office received more than a million orders.

Radio had become so popular by the early 1930s that it quickly linked up with another form of mass entertainment—the movies. Starting in 1932, Paramount Pictures and the Philco Radio Corporation became partners in a yearly series of films called *The Big Broadcast of* (fill in the year), featuring well-known radio personalities. You could listen to them during the day and then go to the local theater and watch them at night. "Technical" features of radio, well known to anyone who owned a radio, could also find their way into a movie. For example, in the 1939 film *Another Thin Man* (one of a series of movies featuring an amateur detective), the hero is about to be shot by gangsters. To cover up the sounds of their guns, they turn on a radio but have to wait for the tubes to warm up. The delay is just sufficient to allow help to arrive to save the hero. Movies weren't consistent on this issue, however, as was illustrated two years earlier in the 1937 film *The Black Legion*: when Humphrey Bogart's character turned on his home radio, it *instantly* burst into sound.

As anyone who listens (almost certainly while driving rather than at home, where television and video games provide the thrills) to AM radio today knows, the glory days of radio as entertainment are history. One of the characters in George Lucas's nostalgic tribute to radio, the 1995 film *Radioland Murders*, says "Radio will never die. It would be like killing the imagination"—but he was wrong. Today's mostly insipid radio content of loud rock music, highly partisan (often viciously nasty) call-in talk shows, a seemingly endless parade of slick talkers selling "miracle" vitamin drugs, or promising instant relief from the IRS chasing after you for unpaid taxes, or making claims they can cancel your ironclad, bulletproof time-share lease on a crummy apartment on a nowhere island that you would travel to only if you lost your mind—all of that is a pale ghost of the long-ago radio shows. As one writer so perfectly put it, "Radio was a dream and now it's a jukebox. It's as if planes stopped flying and sat on the runway showing travelogues."[15]

Old-time radio mystery, crime, and adventure shows did have one particularly interesting, long-lasting effect as the result of a tendency of their creators to use classical music for their themes. A cynic might say that was because such music was in the public domain, and thus free, but so what? The music was good! Who, after all, even today, can listen to the "Overture" from Rossini's *William Tell* and not instantly think of *The Lone Ranger* thundering out of "those thrilling days of yesteryear" on his fiery horse Silver with his "faithful Indian companion, Tonto"? And that was just the start.

The Green Hornet flew out of radio speakers on the frantic violin notes of Rimsky-Korsakov's "The Flight of the Bumblebee" (used by Woody Allen as the opening music to his sentimental homage to Second World War–era radio, the 1987 film *Radio Days*). *The Shadow* and *I Love a Mystery* crept into darkened bedrooms of kids hiding under blankets on the eerie sounds of Saint-Saëns's "Omphale's Spinning Wheel" and Sibelius's "Valse Triste," respectively. The wonderfully

15 From Garrison Keillor's hilarious 1991 novel of early radio, *WLT, A Radio Romance*. Despite its dust-jacket pitch of being "slightly steamy . . . full of romance, intrigue, tough business, and loose living," the book offers up a pretty good description of the realities of early AM broadcast radio in America. The call letters WLT stand for **W**ith **L**ettuce and **T**omato, a joke based on the fictional station being operated out of a sandwich shop. WLT had its very own fictional soap: *Avis Burnett, Small Town Librarian*, the story of "a woman who sacrifices her own happiness in the service of others." She does that by suggesting good books to read to the women of the town, all the while fending off the approaches of numerous men who simply cannot resist the lure of an attractive, unattached librarian. Keillor was the guiding spirit behind the long-running *A Prairie Home Companion*, which was a nostalgic reminder of how good radio once was.

horrific *Escape* used Mussorgsky's "Night on Bald Mountain," and the adventure program *Arabesque* introduced its young fans to Rimsky-Korsakov's *Scheherazade*.

The FBI in Peace and War did its thing to the notes of the "March" from Prokofiev's *The Love for Three Oranges*, while *The Count of Monte Cristo* used "Cortège de Bacchus" from Delibes's ballet *Sylvia*. *The Story of Mary Marlin* played Debussy's "Clair de Lune"; *When a Girl Marries* was accompanied by Drigo's "Serenade"; *Girl Alone* used the "Cécile Waltz" by McKee; *Kitty Keene* floated into homes on the notes of "None but the Lonely Heart" by Tchaikovsky; *A Tale of Today* opened with the "Coronation March" from Meyerbeer's opera *The Prophet*; and Brahms's "Lullaby" heralded the arrival of each new episode of *Hilltop House* (the story of a matron at an orphanage who must choose between love and raising other women's children).

This aspect of old-time radio carried over to the early television programs; when *The Lone Ranger* moved from radio to TV it took the *William Tell* "Overture" with it.[16] The children who listened to this music in the 1930s and '40s did not (in most cases) become the serial murderers that Worthington Gibson (see note 7) feared. Rather, they grew up to become the middle-class consumers who fueled the 1950s craze for hi-fi sets and long-playing recordings of the classical music they had first heard on the radio.[17]

Television killed old-time radio, but it wasn't a total victory. A price *was* paid, as illustrated by the final paragraph of Joseph Julian's book (see note 9):

> Trying to analyze the reasons for the broad, universal appeal of radio drama I find it expressed best by a little seven-year-old

16 Born in 1940, I was old enough, before the "'end" of old-time radio, to listen to my share of its dramatic presentations. In addition to fondly recalling *Yours Truly, Johnny Dollar* (the insurance investigator with the "action-packed expense account!") and remembering how long I laughed every time the closet full of junk at 79 Wistful Vista fell out on the floor—for what seemed *forever*—on *Fibber McGee and Molly*, one piece of beautiful music I first heard on the radio always instantly brings cigarettes (!) to my mind. That's because Philip Morris used Ferde Grofé's "On the Trail" (from his *Grand Canyon Suite*) in its radio commercials.

17 My father was one of those kids, and I still have vivid memories of *Prince Igor* vibrating our little house as the gloomy Russian opera slowly rumbled, much like a monstrous molasses tidal wave, out of dad's custom-built, king-size, acoustically perfect speaker enclosures, powered by a Heathkit vacuum tube amplifier of gargantuan wattage. The neighbors trembled, and who could blame them?

boy who . . . was asked which he liked better, plays on the radio or plays on television.

"On the radio," he said.

"Why?" he was asked.

He thought for a moment, then replied, "Because I can see the pictures better."

This perfectly illustrates why early radio became known as "the theatre of the mind."

This chapter has been all about AM broadcast radio, but in parallel with the AM radio that was so visible in the everyday world of the 1920s, '30s, and '40s, an intense behind-the scenes debate was going on concerning FM broadcast radio—or what Armstrong wanted to see *develop* into FM broadcasting. What he ran into, however, according to the "traditional" story, was the huge inertia of already-established AM radio business concerns that were not interested in profoundly altering the status quo. Armstrong's 1954 suicide is imagined to have been, at least in part, prompted by his deep emotional (and monetary) investment in FM—investments that were seemingly frustrated by those who liked things just as they were. The result was years of legal wrangling, reminiscent of Armstrong's battles with De Forest. After his death, Armstrong's legal efforts did, finally, pay off (his estate received millions of dollars in judgments), and of course every automobile made in the US today has a radio that receives both AM *and* FM (and the audio signal in television is FM in nature).

But, ultimately, whether we discuss AM or FM or SSB or any other type of modulation, in the end it all began with Maxwell. In 1878, less than a year before his death, Maxwell wrote a poem that reads astonishingly like a description of all the old radio broadcasts (and the ones from last night, too) that, even as you read this, are on their way to the next galaxy and beyond.[18] Titled "A Paradoxical Ode," at one point Maxwell's poem says:

18 Earth is, today, a blazing, lit-up beacon in the universe, with countless sources of radiation produced by radio and television transmitters, civilian and military radars, cellphone satellite links, and numerous other modern electronic gadgets. In particular, the earliest radio shows are on the surface of an electromagnetic sphere now 200 light-years in diameter, a sphere expanding at the speed of light. If there is intelligence "out there," looking for signs of intelligent life elsewhere in the cosmos, its first knowledge of

Till in the twilight of the Gods,
When earth and sun are frozen clods,
When, all its energy degraded,
Matter to ether shall have faded;
We, that is, all the work we've done,
As waves in the ether, shall forever run
In ever widening spheres through heavens beyond the sun.

Maxwell's poem[19] may well have been the inspiration for a British physicist who, decades later, wrote[20] of *his* vision of the end of the world:

It would seem that the universe will finally become
a ball of radiation. . . . The longest waves are Hertzian waves
of the kind used in broadcasting. . . . [T]his ball of radio waves
will . . . go on expanding . . . forever. Perhaps then I may describe
the end of the physical world as—one stupendous broadcast.

As the spreading waves of Eddington's "broadcast" heralding "the end of the physical world" expand into the universe, they will relentlessly grow weaker. That image reminds me of how Woody Allen narrates the end of his wonderful movie *Radio Days*: with the passing of time, old-time radio voices grow ever more faint to those who heard them "live, on the air," until at the end only memories remain.

But that is just the end of a particular era. I vividly recall the thrill of hearing, after the end of old-time radio (but still more than half a century ago), the first voice radio message from the Moon, and, I predict, there are readers of this book who will hear the first voice radio transmission from the surface of Mars, a transmission not from alien space invaders but from humans who are alive on Earth right now.

Won't that be something!

---

humans will come from (probably) *Amos 'n' Andy*. Now *that* is, indeed, a sobering, even disquieting, thought.

19 You can find the complete poem in *The Life of James Clerk Maxwell* by Lewis Campbell and William Garnett (first published in 1882 and available today in reprint), p. 650.

20 Arthur Eddington, *New Pathways in Science*, Macmillan, 1935, p. 71.

A Final Author's Note to the Reader

I opened this book with a **Note to the Reader**, and I'll close it with another. When I was in high school in the mid-1950s I didn't have a car and didn't get one—a 1952 Pontiac with twin mufflers and an awesome straight-eight engine—until 1962, when I started graduate school. When the final sad day came, the day the ancient Pontiac made its last trip (to the junkyard), I remember the radio was playing Don McLean's "American Pie," in which he "drove his Chevy to the levee, but the levee was dry." It was a fitting last song for my first "wheels." In high school I depended on my pals for wheels, during which we listened on the car radio to all the music hits of the day. Today I drive a Jeep bristling with electronics that would have been science fiction in the '50s, including not just AM and FM but satellite radio, too. As I roar down the highway and tune SiriusXM to the "Golden '50s" and hear the Chordettes's 1954 "Mr. Sandman"; or Bill Haley's 1955 "Rock Around the Clock"; or Jim Lowe's "The Green Door" or LaVern Baker's "Jim Dandy" or Gogi Grant's "The Wayward Wind," all from 1956; or Don Rondo's 1957 "White Silver Sands" or Buddy Holly's "That'll Be the Day" of the same year; or The Big Bopper's 1958 "Chantilly Lace"; or (best of all) Marty Robbins's 1959 "Big Iron," my car instantly acts like Doc Brown's "flux capacitor" has been installed, and like Marty McFly in *Back to the Future*, I am suddenly in the wonderful time machine I mentioned in my opening **Note** and back in the past.

No matter when *you* were in high school, even if just a few years ago, I'll bet you have had the same experience. R. V. Jones (note 13 in the preface) was right—radio truly *is* magic.

Appendix

Maxwell's Theory, the Poynting Vector, and a Simple Radio Transmitting Antenna

> Ten thousand years from now—there can be little doubt that the most significant event of the nineteenth century will be judged as Maxwell's discovery of the laws of electrodynamics. The American Civil War will pale into provincial insignificance in comparison.
>
> —Richard Feynman, 1965 Nobel Prize laureate in Physics, in his *Lectures on Physics*

Before Maxwell published his revolutionary *Treatise on Electricity and Magnetism* in 1873 (the equal in its impact on the world of science to Newton's *Principia*) everything known of the physics of electricity and magnetism was expressible in the following equations (using the modern notation—*not* used by Maxwell—of vector calculus): With $\boldsymbol{E}$, $\boldsymbol{B}$, and $\boldsymbol{J}$ denoting the vector quantities (all vectors are in boldface) of the electric field, the magnetic field, and the electric current density, respectively, all at the same arbitrary point in space, with ρ denoting the scalar quantity of net electric charge density at that point, and with ∇ denoting the vector differential operator (in rectangular coordinates),

$$\nabla = \frac{\partial}{\partial x}\boldsymbol{i} + \frac{\partial}{\partial y}\boldsymbol{j} + \frac{\partial}{\partial z}\boldsymbol{k},$$

where $\boldsymbol{i}, \boldsymbol{j}$, and $\boldsymbol{k}$ are unit vectors in the x, y, and z space directions, respectively, we have[1]

1 The discussion in this appendix assumes you know what it means to talk of (and how to compute) the *scalar dot product* and the *vector cross product* of two vectors.

$$\text{(A1.1)} \qquad \nabla \cdot \boldsymbol{E} = \rho / \epsilon,$$

$$\text{(A1.2)} \qquad \nabla \cdot \boldsymbol{B} = 0,$$

$$\text{(A1.3)} \qquad \nabla \times \boldsymbol{B} = \mu \boldsymbol{J},$$

$$\text{(A1.4)} \qquad \nabla \times \boldsymbol{E} = -\frac{\partial \boldsymbol{B}}{\partial t},$$

$$\text{(A1.5)} \qquad \nabla \cdot \boldsymbol{J} = -\frac{\partial \rho}{\partial t},$$

where μ and ε are constants (called the *magnetic permeability* and the *electric permittivity*, respectively, of the matter in which our physical quantities exist), constants that can be measured in the lab via some very simple (simple, at least, in *concept*) experiments. In MKS units (meters, kilograms, seconds) the numerical values of μ and ε in a vacuum are $\mu = 4\pi \times 10^{-7}$ and $\varepsilon = 8.85 \times 10^{-12}$. In air, these values don't change by much.

These five equations have the following physical meaning. (A1.1) says the divergence of the electric field is directly proportional to the net charge density. We imagine electric field lines (imagery due to the English experimenter Michael Faraday (1791–1867)) starting from positive electric charge and ending on negative electric charge. That is, ρ is a *source/sink* for $\boldsymbol{E}$. (A1.2) says magnetic field lines *always* have zero divergence *everywhere* (that is, there is no such thing as "magnetic charge"). Therefore, magnetic field lines *always* form closed loops, with no beginning or end. (A1.3) says an electric current creates a magnetic field, that is, $\boldsymbol{J}$ is a source for $\boldsymbol{B}$. (A1.4) says a *time-varying* magnetic field creates an electric field (a time-varying $\boldsymbol{B}$ is a source for $\boldsymbol{E}$). And finally, (A1.5) says any change in the net charge density at a point requires a current (the minus sign means a *reduction* in net charge, which implies a *positive* current flowing *away* from the point). These statements were known, in essence,

---

Furthermore, the assumption is that you know if ∇ is one of those vectors and $\boldsymbol{F}$ is any other vector, then $\nabla \cdot \boldsymbol{F}$ is the *divergence* of $\boldsymbol{F}$ (a scalar), and $\nabla \times \boldsymbol{F}$ is the *curl* of $\boldsymbol{F}$ (a vector). This would all have been "old hat" for Hardy.

through the work of Faraday, the French physicist André-Marie Ampere (1775–1836), and the Danish experimenter Hans Christian Oersted (1777–1851).

When Maxwell published his *Treatise*, he added an extra touch of his own to these equations, based (amazingly) on absolutely no experimental evidence at all. This new term, called the *displacement current*, is what gives life to radio. How Maxwell arrived at his displacement current is an issue that historians of science have speculated about for decades, but the following is how it might be explained to a resurrected Hardy.

We start by noticing that (A1.3) and (A1.5) are not entirely consistent. We see this by taking the divergence of (A1.3) to write

$$\nabla \bullet (\nabla \times \boldsymbol{B}) = \nabla \bullet (\mu \boldsymbol{J}) = \mu(\nabla \bullet \boldsymbol{J}).$$

Now, it is easy to confirm, by direct calculation, that for *any* vector the divergence of the curl is identically zero. That is, $\nabla \bullet (\nabla \times \boldsymbol{B}) = 0$, and so (A1.3) says $\nabla \bullet \boldsymbol{J} = 0$, *always*. But (A1.5) says that's not so! Let's see if we can avoid this seeming paradox by imagining that (A1.3) is slightly changed to read

$$\nabla \times \boldsymbol{B} = \mu(\boldsymbol{J} + \boldsymbol{D}), \tag{A1.6}$$

where $\boldsymbol{D}$ is some vector that we add to $\boldsymbol{J}$ to make (A1.3) and (A1.5) consistent. So, what is $\boldsymbol{D}$?

Well, what we now have is

$$\nabla \bullet (\nabla \times \boldsymbol{B}) = \nabla \bullet \mu(\boldsymbol{J} + \boldsymbol{D}) = \mu \nabla \bullet (\boldsymbol{J} + \boldsymbol{D}) = 0,$$

and so

$$\nabla \bullet \boldsymbol{J} = -\frac{\partial \rho}{\partial t} = -\nabla \bullet \boldsymbol{D},$$

or

$$\nabla \bullet \boldsymbol{D} = \frac{\partial \rho}{\partial t}.$$

From (A1.1) we have $\rho = \varepsilon \nabla \bullet \boldsymbol{E}$, and so

$$\nabla \bullet \boldsymbol{D} = \frac{\partial}{\partial t}(\varepsilon \nabla \bullet \boldsymbol{E}),$$

or, because the time derivative and the space derivatives of the ∇ operator commute, we have

$$\nabla \bullet \boldsymbol{D} = \nabla \bullet \frac{\partial}{\partial t}(\varepsilon \boldsymbol{E}),$$

from which we conclude the $\boldsymbol{D}$ in (A1.6),

$$\boldsymbol{D} = \varepsilon \frac{\partial \boldsymbol{E}}{\partial t},$$

is Maxwell's *displacement current density*[2] term necessary to make (A1.3) and (A1.5) consistent.

So, since 1873 the following have been what are called *Maxwell's electromagnetic field equations in space* (where $\rho = 0$ and $\boldsymbol{J} = 0$):

$$\nabla \bullet \boldsymbol{E} = 0,$$

$$\nabla \bullet \boldsymbol{B} = 0,$$

$$\nabla \times \boldsymbol{E} = -\frac{\partial \boldsymbol{B}}{\partial t},$$

$$\nabla \times \boldsymbol{B} = \mu \epsilon \frac{\partial \boldsymbol{E}}{\partial t}.$$

Now, we add one final touch to these equations. Let's define a scaled version of the magnetic field, called $\boldsymbol{H}$, such that $\boldsymbol{B} = \mu \boldsymbol{H}$. We then arrive at

(A1.7) $$\nabla \bullet \boldsymbol{E} = 0,$$

(A1.8) $$\nabla \bullet \boldsymbol{H} = 0,$$

2 Why this term is called a *current density* will be explained in the box at the end of the appendix. The word *displacement* comes from Maxwell's mechanical visualization of the term. Today that mechanical view is of only historical interest.

(A1.9)
$$\nabla \times \boldsymbol{E} = -\mu \frac{\partial \boldsymbol{H}}{\partial t},$$

(A1.10)
$$\nabla \times \boldsymbol{H} = \epsilon \frac{\partial \boldsymbol{E}}{\partial t}.$$

This form of the field equations displays a nice symmetry, and our next step is to solve them for $\boldsymbol{E}$ and $\boldsymbol{H}$.

We start by taking the curl of the first curl equation, (A1.9), to write

$$\nabla \times \nabla \times \boldsymbol{E} = -\mu \frac{\partial}{\partial t}(\nabla \times \boldsymbol{H}) = -\mu\epsilon \frac{\partial^2 \boldsymbol{E}}{\partial t^2}.$$

Next, it is easy (if just a bit messy) to show by direct expansion that if $\boldsymbol{A}$, $\boldsymbol{B}$, and $\boldsymbol{C}$ are *any* vectors, it then follows that

$$\boldsymbol{A} \times (\boldsymbol{B} \times \boldsymbol{C}) = \boldsymbol{B}(\boldsymbol{A} \bullet \boldsymbol{C}) - \boldsymbol{C}(\boldsymbol{A} \bullet \boldsymbol{B}),$$

an identity remembered by all physics and electrical engineering students as the "BAC-CAB" rule. So, associating ∇ with both $\boldsymbol{A}$ and $\boldsymbol{B}$, and $\boldsymbol{E}$ with $\boldsymbol{C}$, we have

$$\nabla \times \nabla \times \boldsymbol{E} = \nabla(\nabla \bullet \boldsymbol{E}) - (\nabla \bullet \nabla)\boldsymbol{E} = -(\nabla \bullet \nabla)\boldsymbol{E},$$

where we've used (A1.7) and replaced $\boldsymbol{C}(\boldsymbol{A} \bullet \boldsymbol{B})$ with $(\boldsymbol{A} \bullet \boldsymbol{B})\boldsymbol{C}$. The question that arises now is, What does $(\nabla \bullet \nabla)\boldsymbol{E}$ mean? It has to be a vector, of course, because the other terms are vectors,[3] but what is $(\nabla \bullet \nabla)\boldsymbol{E}$, *specifically*?

Writing $\nabla \bullet \nabla$ out in detail, we have

$$\nabla \bullet \nabla = \left(\frac{\partial}{\partial x}\boldsymbol{i} + \frac{\partial}{\partial y}\boldsymbol{j} + \frac{\partial}{\partial z}\boldsymbol{k}\right) \bullet \left(\frac{\partial}{\partial x}\boldsymbol{i} + \frac{\partial}{\partial y}\boldsymbol{j} + \frac{\partial}{\partial z}\boldsymbol{k}\right)$$
$$= \frac{\partial}{\partial x}\left(\frac{\partial}{\partial x}\right) + \frac{\partial}{\partial y}\left(\frac{\partial}{\partial y}\right) + \frac{\partial}{\partial z}\left(\frac{\partial}{\partial z}\right) = \frac{\partial^2}{\partial x^2} + \frac{\partial^2}{\partial y^2} + \frac{\partial^2}{\partial z^2},$$

3 The first term on the right, $\nabla(\nabla \bullet \boldsymbol{E})$, immediately disappeared from any future concern because of (A1.7), but formally it is a vector ($\nabla \bullet \boldsymbol{E}$ is a scalar field, and ∇ operating on a scalar field yields a vector field called the *gradient*). This would, of course, have all been known to Hardy.

which is a scalar operator, written as ∇^2. Note carefully, this operator is *not* in boldface. So, we've arrived at

$$-\nabla^2 \boldsymbol{E} = -\mu\epsilon \frac{\partial^2 \boldsymbol{E}}{\partial t^2},$$

or

$$\nabla^2 \boldsymbol{E} = \mu\epsilon \frac{\partial^2 \boldsymbol{E}}{\partial t^2},$$

which is the well-known *wave equation* (second-order space derivatives on the left, and second-order time derivatives on the right), recognizable to all who have studied differential equations. Mathematicians had, in fact, been solving wave equations for vibrating strings a century or more before Maxwell. The equation says $\boldsymbol{E}$ propagates through space at speed $\frac{1}{\sqrt{\mu\varepsilon}}$. Repeating this entire procedure to isolate $\boldsymbol{H}$ gives the same result, that $\boldsymbol{H}$, too, propagates through space at speed $\frac{1}{\sqrt{\mu\varepsilon}}$.

When Maxwell inserted the MKS values of μ and ε for space, he calculated the MKS propagation speed to be

$$\frac{1}{\sqrt{(4\pi \times 10^{-7})(8.85 \times 10^{-12})}} = 3 \times 10^8 \text{ m/s (meters per second)}$$

which he immediately recognized as being quite close to the known speed of light, c (as determined by other, independent experimental setups). That was enough for Maxwell to make the leap to concluding that light is electromagnetic in nature.

We'll return to the wave equation, and its solutions, in more detail later in the appendix. But first, let's derive one more result contained in the field equations, a result central to radio. Maxwell died young, at age 48, but perhaps if he had lived, he would have made the next great deduction from the field equations. Instead, in virtually simultaneous and independent discoveries, two Englishmen did so, in 1884, five years after Maxwell's death: John Henry Poynting (1852–1914), a professor of physics who was Third Wrangler in the 1876 Tripos, and Oliver Heaviside (1850–1925), who was an unemployed, self-taught eccentric genius. Their discovery extended the idea of propagating *fields* to that of propagating *energy* through what appears to be empty space.[4] Modern textbook discussions are invariably based on Heaviside's analysis, which follows.

As the discussion between Professors Tweedle and Twombly in the preface showed, the energy stored in a capacitor C charged to V volts is $W_e = \frac{1}{2}CV^2$. A common physical form for a capacitor is that of two parallel plates, each with area A, separated by a gap of width d. Since the voltage difference between the plates is V, the electric field in the gap is $E = \frac{V}{d}$ (and so $V = Ed$). Some freshman physics (which we skip here) shows that $C = \varepsilon\frac{A}{d}$, and so the total electrical energy in the electric field is

$$W_e = \frac{1}{2}\left(\varepsilon\frac{A}{d}\right)(Ed)^2 = \frac{1}{2}\varepsilon(Ad)E^2.$$

That is,

$$\frac{W_e}{Ad} = \frac{1}{2}\varepsilon E^2.$$

Now, Ad is the volume of the gap, and so $\frac{W_e}{Ad}$ can be thought of as the *energy density* in the gap due to the electric field. In the same way, a study of the stored magnetic field energy (in, say, the magnetic

4 The Victorians thought space *not* to be empty but, rather, to be filled with an invisible something called the *aether* (or *ether*). Just like waves at the beach need water to "wave in," Victorian science thought electromagnetic radiation needed something to "wave in," too, and so the ether was thought to be an inescapable necessity. That there was no observable effect of this "substance" and that no experiment was ever devised that could detect it *were indeed* cause for a bit of concern, but there seemed to be no way to do without it. The negative result of the famous 1887 Michelson-Morley experiment (to detect the Earth's motion through the ether) convinced physicists that the ether simply doesn't exist. Energy stored in and propagating through a vacuum, therefore still remains a rather mysterious business even for twenty-first-century science.

field surrounding a current-carrying inductor) says that the magnetic field energy density is $\frac{1}{2}\mu H^2$. So, writing $E^2 = \boldsymbol{E} \bullet \boldsymbol{E}$ and $H^2 = \boldsymbol{H} \bullet \boldsymbol{H}$, Heaviside wrote the total electromagnetic field energy density (energy per unit volume) as

$$W = \frac{1}{2}\mu \boldsymbol{H} \bullet \boldsymbol{H} + \frac{1}{2}\varepsilon \boldsymbol{E} \bullet \boldsymbol{E}.$$

The net rate at which energy flows out of a region of space is $-\frac{\partial W}{\partial t}$, where the minus sign makes this a positive quantity if W decreases. If we define $\boldsymbol{P}$ as a power vector that describes the rate of the flow of energy per unit area across a boundary surface enclosing a region of space, then this positive quantity is simply the divergence of $\boldsymbol{P}$. That is,

$$\nabla \bullet \boldsymbol{P} = -\frac{\partial W}{\partial t} = -\frac{1}{2}\mu\left\{\boldsymbol{H} \bullet \frac{\partial \boldsymbol{H}}{\partial t} + \frac{\partial \boldsymbol{H}}{\partial t} \bullet \boldsymbol{H}\right\} - \frac{1}{2}\varepsilon\left\{\boldsymbol{E} \bullet \frac{\partial \boldsymbol{E}}{\partial t} + \frac{\partial \boldsymbol{E}}{\partial t} \bullet \boldsymbol{E}\right\},$$

or

$$\nabla \bullet \boldsymbol{P} = -\mu \boldsymbol{H} \bullet \frac{\partial \boldsymbol{H}}{\partial t} - \varepsilon \boldsymbol{E} \bullet \frac{\partial \boldsymbol{E}}{\partial t}.$$

From (A1.9) we have

$$\frac{\partial \boldsymbol{H}}{\partial t} = -\frac{1}{\mu}\nabla \times \boldsymbol{E},$$

and from (A1.10) we have

$$\frac{\partial \boldsymbol{E}}{\partial t} = \frac{1}{\epsilon}\nabla \times \boldsymbol{H},$$

and so

$$\nabla \bullet \boldsymbol{P} = \boldsymbol{H} \bullet (\nabla \times \boldsymbol{E}) - \boldsymbol{E} \bullet (\nabla \times \boldsymbol{H}).$$

A fundamental identity from vector calculus tells us that for *any* two vectors $\boldsymbol{A}$ and $\boldsymbol{C}$ we have $\nabla \bullet (\boldsymbol{A} \times \boldsymbol{C}) = \boldsymbol{C} \bullet (\nabla \times \boldsymbol{A}) - \boldsymbol{A} \bullet (\nabla \times \boldsymbol{C})$, and so, associating $\boldsymbol{C}$ with $\boldsymbol{H}$, and $\boldsymbol{A}$ with $\boldsymbol{E}$, Heaviside wrote

$$\nabla \bullet \mathbf{P} = \nabla \bullet (\boldsymbol{E} \times \boldsymbol{H})$$

and made the obvious deduction that

$$\boldsymbol{P} = \boldsymbol{E} \times \boldsymbol{H}.$$

$\boldsymbol{P}$ is called the *Poynting vector* (because Poynting's more complicated analysis appeared in print before Heaviside's), which points[5] in the direction in which energy is transported through space by the traveling $\boldsymbol{E}$ and $\boldsymbol{H}$ fields. This direction is, by virtue of how a vector cross product works, perpendicular to the plane in which $\boldsymbol{E}$ and $\boldsymbol{H}$ lie. Such is the case, for example, for the electromagnetic waves generated by the transmitting antenna of a radio station.[6]

To summarize what we have learned so far, from a mathematical description[7] of the electric and magnetic fields $\boldsymbol{E}$ and $\boldsymbol{H}$ we have found that $\boldsymbol{E}$ and $\boldsymbol{H}$ each obey a traveling wave equation. These equations also tell us that $\boldsymbol{E}$ and $\boldsymbol{H}$ transport energy through space at the speed of light, in a direction normal to the plane in which $\boldsymbol{E}$ and $\boldsymbol{H}$ lie. To examine these conclusions in more detail, consider the following special case. Suppose $\boldsymbol{E}$ lies totally along the z-direction (the direction of the $\boldsymbol{E}$ vector is called the *polarization* of the wave), and in addition, imagine that $\boldsymbol{E}$ travels in the x-direction at speed c (the speed of

5 The name is, of course, purely a historical coincidence, but can you think of a better name, considering what $\boldsymbol{P}$ represents?

6 There is, I should tell you, some ambiguity in $\boldsymbol{P}$, as more generally, $\boldsymbol{P} = \boldsymbol{E} \times \boldsymbol{H} + \boldsymbol{G}$, where $\boldsymbol{G}$ is *any* vector with zero divergence. Such a vector represents energy flowing endlessly along a closed-loop path, which strikes many analysts as quite odd. $\boldsymbol{G}$ is therefore usually taken to be zero, but there is some lively debate, even today, about doing that. See my book *Oliver Heaviside*, The Johns Hopkins University Press, 2001, pp. 130–131.

7 The partial differential equations for $\boldsymbol{E}$ and $\boldsymbol{H}$ of (A1.7) through (A1.10). They are *differential* equations because the values of the electric and magnetic fields at every point in space, for every instant of time, can be related to the values of the fields at nearby points in space and time. They are *partial* differential equations because there are multiple independent variables (time and at least one space variable).

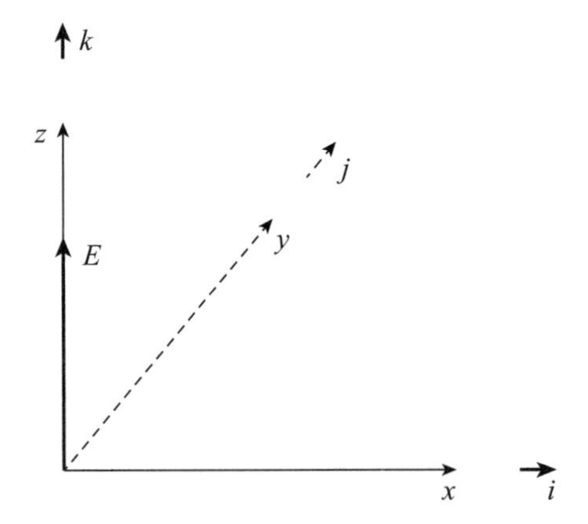

FIGURE A1.1. A particular traveling electric field.

light), as shown in Figure A1.1. Further, suppose that $\boldsymbol{E}$ is a *sinusoidal* oscillation at radian frequency ω. We can then write

$$\boldsymbol{E}(x,t) = E_0 \sin\left\{\omega\left(\frac{x}{c} - t\right)\right\}\boldsymbol{k}, \tag{A1.11}$$

where E_0 is the peak amplitude of the electric field vector (note, carefully, that the positive y-axis is *into* the plane of the paper, which is why it's drawn as a dashed line).

The claim is that (A1.11) satisfies a wave equation, and that's easy to check, as follows. Since

$$\nabla^2 = \frac{\partial^2}{\partial x^2} + \frac{\partial^2}{\partial y^2} + \frac{\partial^2}{\partial z^2},$$

then, as $\boldsymbol{E}$ has no y or z dependency, the wave equation for this example reduces to

$$\frac{\partial^2 \boldsymbol{E}}{\partial x^2} = \mu\varepsilon \frac{\partial^2 \boldsymbol{E}}{\partial t^2} = \frac{1}{c^2}\frac{\partial^2 \boldsymbol{E}}{\partial t^2}.$$

Since

$$\frac{\partial^2 \boldsymbol{E}}{\partial x^2} = -E_0 \frac{\omega^2}{c^2} \sin\left\{\omega\left(\frac{x}{c} - t\right)\right\}\boldsymbol{k},$$

and since

$$\frac{1}{c^2}\frac{\partial^2 \boldsymbol{E}}{\partial t^2} = -E_0\frac{\omega^2}{c^2}\sin\left\{\omega\left(\frac{x}{c}-t\right)\right\}\boldsymbol{k},$$

we see that our assumed $\boldsymbol{E}$ *does* work.

Now, for the assumed $\boldsymbol{E}$, what is $\boldsymbol{H}$? From Maxwell we know $\nabla\times\boldsymbol{E} = -\mu\frac{\partial \boldsymbol{H}}{\partial t}$. We can calculate $\nabla\times\boldsymbol{E}$ by expanding the 3×3 determinant for the curl of a vector as follows:

$$\nabla\times\boldsymbol{E} = \begin{vmatrix} \boldsymbol{i} & \boldsymbol{j} & \boldsymbol{k} \\ \frac{\partial}{\partial x} & \frac{\partial}{\partial y} & \frac{\partial}{\partial z} \\ 0 & 0 & E_0\sin\left\{\omega\left(\frac{x}{c}-t\right)\right\} \end{vmatrix}$$

$$= -\boldsymbol{j}\frac{\omega}{c}E_0\cos\left\{\omega\left(\frac{x}{c}-t\right)\right\} = -\mu\frac{\partial \boldsymbol{H}}{\partial t},$$

and so

$$\frac{\partial \boldsymbol{H}}{\partial t} = \frac{\omega E_0}{\mu c}\cos\left\{\omega\left(\frac{x}{c}-t\right)\right\}\boldsymbol{j}.$$

Integrating with respect to t, we have

$$\boldsymbol{H} = \frac{\omega E_0}{\mu c}\sin\left\{\omega\left(\frac{x}{c}-t\right)\right\}\left(-\frac{1}{\omega}\right)\boldsymbol{j},$$

or

$$\boldsymbol{H} = -\frac{E_0}{\mu c}\sin\left\{\omega\left(\frac{x}{c}-t\right)\right\}\boldsymbol{j}, \tag{A1.12}$$

and we see that $\boldsymbol{E}$ and $\boldsymbol{H}$ are perpendicular to each other, with both lying in the yz-plane (which is normal to the x-axis). That is, the $\boldsymbol{E}$ and $\boldsymbol{H}$ traveling fields are *transverse* waves with the field vectors at right angles to the direction of propagation, unlike the *longitudinal* compression waves of sound in air, or pressure waves in rock, whose oscillations occur *along* the direction of propagation.

Finally, what can we say about the Poynting vector $\boldsymbol{P}=\boldsymbol{E}\times\boldsymbol{H}$? Expanding the 3×3 determinant for the vector cross product, we have

$$\boldsymbol{P}=\boldsymbol{E}\times\boldsymbol{H}=\begin{vmatrix} \boldsymbol{i} & \boldsymbol{j} & \boldsymbol{k} \\ 0 & 0 & E_0\sin\left\{\omega\left(\frac{x}{c}-t\right)\right\} \\ 0 & -\frac{E_0}{\mu c}\sin\left\{\omega\left(\frac{x}{c}-t\right)\right\} & 0 \end{vmatrix},$$

or

$$\boldsymbol{P}=\boldsymbol{E}\times\boldsymbol{H}=\frac{E_0{}^2}{\mu c}\sin^2\left\{\omega\left(\frac{x}{c}-t\right)\right\}\boldsymbol{i}, \tag{A1.13}$$

and since the sine *squared* is never negative, we see that $\boldsymbol{P}$ *everywhere and always* points in the direction of the positive x-axis, normal to the plane in which $\boldsymbol{E}$ and $\boldsymbol{H}$ lie.[8]

The $\boldsymbol{P}$ vector allows me to end this appendix with a physical description of how an elementary radio antenna radiates energy into space. This is where I have to hope that our resurrected Hardy would be so fascinated by the earlier mathematics of this appendix that he would sit still for the next several paragraphs as we "talk a little physics." Hardy came of age as a cultured Victorian, so I think it's reasonable to assume this would be the case. At the least, I can promise him (and you) that there will be little engineering jargon.

The fundamental physics boils down to that of controlling some of the electrons in the wire we'll call the antenna. The electrons I'm referring to are the *valence* electrons (recall Professor Twombly's comments in the preface), that is, those in the outermost orbit around the nucleus of the antenna's atoms. These electrons are the ones that

8 Here's a little for-fun exercise for you. It's always a good idea to make sure the derived results from a calculation are dimensionally correct. So, since $\boldsymbol{P}$ is a power vector describing the rate of flow of energy per unit area, the quantity $\frac{E_0{}^2}{\mu c}$ should have the units of power/area. See if you can show that is, indeed, the case. You might object that I haven't told you the units of μ but only its numerical value in the MKS system. In fact, you *do* have enough information from the appendix and the preface to do this. If you get stuck, take a look at the box at the end of this appendix.

participate in chemical reactions because they are so weakly bound to an atom that it doesn't take much energy to get them to move from atom to atom. Because of this movement they are also known as *conduction electrons*, as their movement constitutes an electrical current. Recall that, *in space*, we took the electric current density vector $\boldsymbol{J}$ in Maxwell's equations as zero, but our antenna works only because, *in it*, $\boldsymbol{J} \neq 0$. For a metal, there is a positive constant σ, called the *conductivity*, such that $\boldsymbol{J} = \sigma \boldsymbol{E}$, a statement that is a version of Ohm's law. It is important to realize that this "law" is *not* a fundamental law of physics, as are Maxwell's field equations, but, rather, is only approximately true (but it is a very good approximation in a metal like copper).[9]

One common misconception about the motion of the conduction electrons in a wire is that they move at nearly the speed of light. In fact, it is easy to show that, even for quite large currents, the moving electrons that *are* the current are *barely* moving! And yet, electrical *signals* in wire *do* move pretty fast (not at the speed of light but certainly at speeds measured in miles per second). To see how this is *not* a paradox, imagine a yardstick on a tabletop, with one end resting against an upright domino. When you gently tap the other end of the yardstick with a finger, the domino 3 ft distant seems to fall over instantly, and yet the individual atoms in the yardstick hardly moved.

To relate this example to the electrons in an antenna, let n be the conduction electron density (see note 9), and so if q is the electron charge (see note 9), then $\rho = nq$ is the conduction charge density (coulombs per cubic meter, C/m^2). Suppose the antenna is a simple copper wire, with a uniform cross-sectional area of A square meters (m^2). If s denotes the speed of the conduction electrons (in meters per second, m/s) when the antenna current is i amperes (an ampere

9 The valence electrons move because, in an electric field $\boldsymbol{E}$, they experience a force $\boldsymbol{F} = q\boldsymbol{E}$, where q is the electron charge of -1.6×10^{-19} C (coulombs) (the coulomb is the MKS unit of charge), which is enough to break the electron free. There are a *lot* of available valence electrons in a metal. For example, in copper their density is $n = 8.43 \times 10^{28}$ per cubic meter. Because $q < 0$, the force $\boldsymbol{F}$ is *opposite* the direction of $\boldsymbol{E}$: electrons move in the direction *against* the electric field. Since negative charge moving in one direction can be thought of as positive charge moving in the opposite direction, the conduction electric current is equivalent to a positive current in the direction of $\boldsymbol{E}$, which is just what $\boldsymbol{J} = \sigma \boldsymbol{E}$ with $\sigma > 0$ says.

is the charge of one coulomb flowing past a point on the wire each second) then, dimensionally, we can write

$$i=\rho sA\left(\frac{\textit{coulombs}}{\text{cubic meter}}\bullet\frac{\textit{meters}}{\textit{second}}\bullet\text{square meters}=\frac{\textit{coulombs}}{\textit{second}}=\text{amperes}\right).$$

That is,

$$s=\frac{i}{\rho A}=\frac{i}{nqA}$$

is the conduction electron speed when the antenna current is i. For example, suppose $A=0.0006\ \text{m}^2$ (a square about one inch on a side, the size of a heavy-duty electrical cable), and $i=100$ A.[10] Then, in copper,

$$s=\frac{100}{8.43\times10^{28}\times1.6\times10^{-19}\times0.0006}\ \text{m/s},$$

which is about 1 m/day. That's pretty slow. Snails move faster than that!

Maxwell's equations predict that *accelerated* charges will radiate energy, and in just a moment I'll show how, using ***P***, you can see this without having to solve the field equations in detail. We can accelerate the conduction electrons in a wire by connecting an oscillating voltage generator at the center of the wire (as in Figure A1.2). This will create an oscillating electric field in the wire, which will drive the conduction electrons back and forth along the wire. The wire, *in total*, is always electrically neutral, that is, has zero net charge; the oscillating electric field changes the *local distribution* of the conduction electron charge along the length of the wire.

10 For reference, a modern single-family home is typically wired for 200 A. With a voltage of 120 volts, this is 24,000 W (watts) of power. In comparison, the maximum power allowed in the US for an AM radio broadcast station is 50,000 W. The one exception was WLW/Cincinnati, which from 1934 to 1939 was allowed by the FCC to operate at 500,000 W (at 700 kHz). WLW's signal could be heard across the entirety of North America (and for that reason was dubbed "The Nation's Station"), a capability thought in those anxious years leading up to the Second World War to be in the interest of national security.

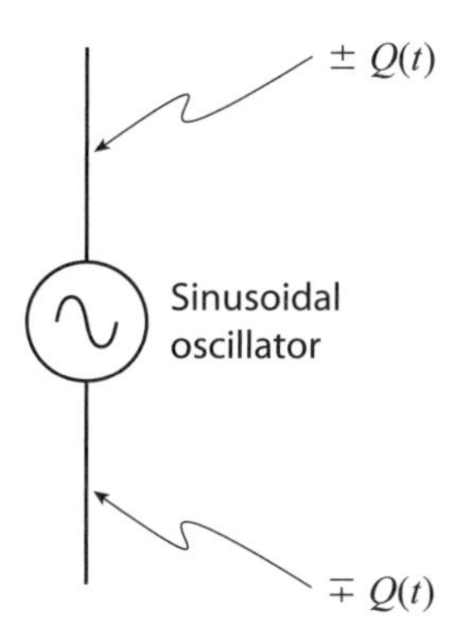

FIGURE A1.2. A center-driven dipole antenna, so called because the upper and lower halves of the antenna form a *dipole* (two time-varying charges, slightly displaced, of equal magnitude at every instant of time but of opposite sign, $\pm Q(t)$ and $\mp Q(t)$). At every instant of time the total electrical charge of the antenna is zero. It was with just such an antenna that the German physicist Heinrich Hertz (1857–1894) experimentally generated the electromagnetic waves predicted by Maxwell's equations. Hertz's spark-gap transmitter (1887–1888) worked in the frequency band 50 to 500 MHz, the same general region where FM radio operates today.

In the antenna $\boldsymbol{J} \neq 0$, while outside the antenna $\boldsymbol{J} = 0$, and so to rigorously understand the launching of radio waves into space by the antenna we have to solve Maxwell's equations taking into account the transition as the waves move across the antenna surface into space. This is, however, more than we need to do to gain a "plausible feeling" for how wave launching works, using a beautiful argument due to the English physicist J. J. Thompson (1856–1940), who was Second Wrangler in the 1880 Tripos and winner of the 1906 Nobel Prize in Physics for his 1897 discovery of the electron. Thompson's argument dates from 1903, but it is based on the seminal concept of the field, due to Faraday.

So far in this appendix I haven't revealed much about just what a *field* is; it's easy to write $\boldsymbol{E}$ or $\boldsymbol{H}$ in equations but, *physically*, what actually *is* a field? Well, the honest answer is, today, still pretty much just what Faraday put forth two centuries ago. Faraday was no mathematician, but his physical insight was stupendous, and his imagery is still (I think) the best we have. Given a point electric charge (which is the modern view of the charged particles called electrons and protons), Faraday thought of the space around a lone charge as being permeated with a three-dimensional, spherically symmetrical field of

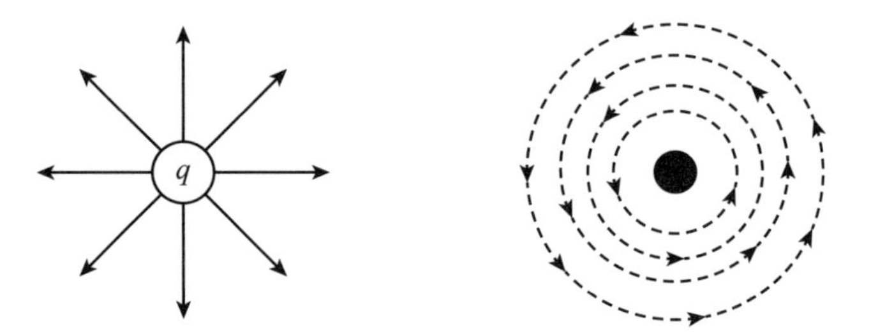

FIGURE A1.3. Faraday's electric field lines for a stationary positive charge q (radially symmetrical in three dimensions) and Oersted's circular magnetic field lines for a wire (black circle) carrying a positive current out of the paper toward you. The sense of the magnetic field generated by the current is given by the *right-hand rule*: with your right hand curled about the wire with your thumb pointing in the direction of the current, your fingers point in the direction of the field lines.

radial electric lines of force (directed away from a positive charge and toward a negative charge). Faraday also visualized a field of magnetic lines of force around a permanent magnet, beginning on the north pole and terminating on the south pole, an image almost certainly motivated by how iron filings line up on a piece of paper placed over the magnet. These lines of force were mechanically interpreted by Faraday as stresses in the ether (see note 4). Oersted showed (in 1820) that circular magnetic field lines also exist around a wire when there is a current in the wire.[11] This was a discovery that Einstein thought so important he referenced it in his address at the 7th Berlin Radio Exhibition, mentioned in the preface. The fields of Faraday and Oersted are shown in Figure A1.3.

Now, Thompson's idea was to imagine the electric field lines of a charge to be continuous at all times, and so as that charge is made to move back and forth along the antenna the field lines become *kinked*—think of a snapped rope—as shown in Figure A1.4. (That figure shows a *positive* charge moving upward, and so the electron motion is actually downward.) From the right-hand rule (see the

11 Oersted determined the structure of the magnetic field around a current-carrying wire with the simple technique of holding a compass near the wire and observing the deflection of the needle. It was in 1831 that Faraday made *his* momentous discovery of the reverse of Oersted's effect (a current, that is, moving electric charges, generates a magnetic field), which is Maxwell equation (A1.3). What Faraday discovered is called *electromagnetic induction* (a *time-varying* magnetic field generates a current in a conductor), which is Maxwell equation (A1.4).

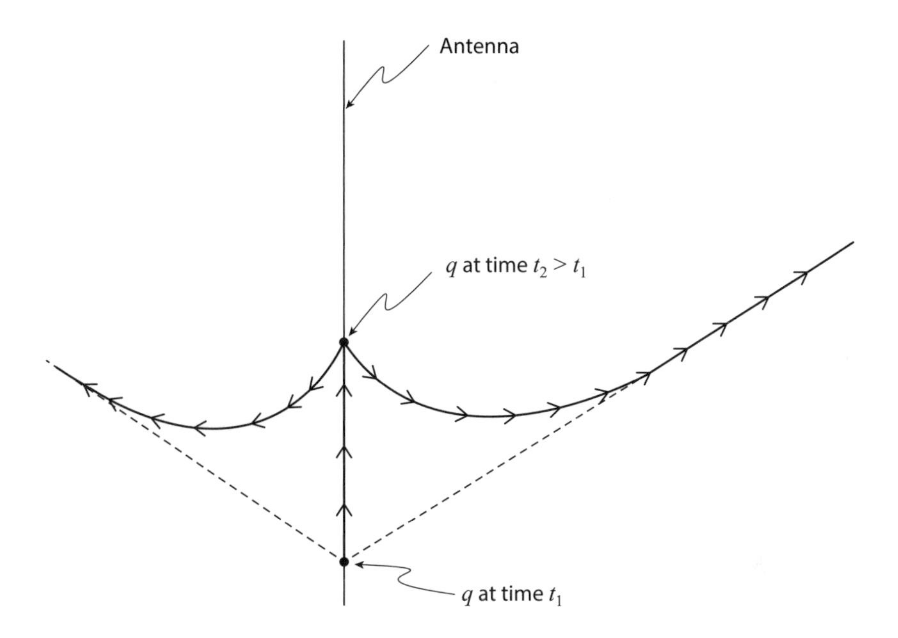

FIGURE A1.4. Suddenly moving a positive charge creates kinks in the continuous electric field lines, which propagate outward from the antenna. Note, *carefully*, that the two kinks shown, both in the plane of the paper, are in fact reproduced in three dimensions *all around* the antenna.

caption for Figure A1.3) we see that at the kink on the right, the magnetic field lines are pointing *into* the plane of the paper, while at the kink on the left the magnetic field lines are pointing *out of* the plane of the paper. From Figure A1.4 you can also see that at both kinks the electric field has a component pointing downward *in* the plane of the paper. Thus, at both kinks the Poynting vector $\boldsymbol{P} = \boldsymbol{E} \times \boldsymbol{H}$ points away from the antenna.[12]

When the current in the antenna reverses direction then so, too, will the direction of $\boldsymbol{H}$, but so, too, will the direction of the $\boldsymbol{E}$-field kink, and so $\boldsymbol{P}$ will *not* reverse: $\boldsymbol{P}$ will *always* have a component pointing away from the antenna, and this component is the *radiation field*. A more detailed study of Maxwell's equations for a radio antenna powered by an oscillating energy source shows that there is also an alternating component of $\boldsymbol{P}$, which points inward and outward. This component of $\boldsymbol{P}$ decays rapidly with distance from the antenna, at

12 The vector cross product obeys a right-hand rule. The direction of $\boldsymbol{E} \times \boldsymbol{H}$ is the direction a right-hand-threaded screw would advance if $\boldsymbol{E}$ was rotated into $\boldsymbol{H}$.

least as fast as the fourth power of distance; it represents an alternating energy flow into and out of the antenna and is called the *near field* or the *induction field* ***P***. That energy therefore remains coupled to the antenna and is unimportant in the operation of true radio. The radiation ***P***, however, decays much more slowly with distance (as the square of the distance), never to return. It is energy that has decoupled from the antenna.[13]

There are other components of ***P*** that *encircle* the antenna, but to "see" them is really beyond the capability of Thompson's elementary kink imagery. Rather, a detailed mathematical analysis of Maxwell's equations is required for truly understanding those components, but since it is the *radiation field* that makes radio possible, we really don't need to pursue those fields.[14]

To finish our simple discussion of antennas, we can relate the frequency at which efficient energy radiation occurs to the physical length of the antenna. In radio, length is most naturally expressed in units of wavelength. If λ and ω denote, respectively, the wavelength and the frequency (radians per second (rad/s), with $\omega = 2\pi f$, with f in hertz), then $\lambda f = c$, the speed of light. Since $c = 3 \times 10^8$ m/s, then at 1 MHz (right in the middle of the AM rf band) one wavelength is 300 m. Now, imagine a vertical antenna, as shown in Figure A1.5.

13 These decay rates for the induction and radiation ***P***'s have a profound mathematical implication. No matter what the superadvanced state of technology may be at some far-future date, there will always be a *finite* distance beyond which the induction field will be undetectable, whereas the radiation field will remain detectable at *any* distance. That's because, independent of distance, a receiving antenna subtending a *fixed* solid angle in space will intercept a fixed amount of energy per unit time from the radiation ***P*** but an ever-decreasing amount of energy per unit time from the induction ***P*** with increasing distance from the transmitting antenna (refer to note 18 in chapter 6, concerning *Amos 'n' Andy* and how extraterrestrials may come to view humans). You *can* transmit with the induction field, but that isn't radio: *radio* uses the radiation field. Induction-field voice communication was actually achieved as early as 1892 (Thomas W. Hoffer, "Nathan B. Stubblefield and His Wireless Telephone," *Journal of Broadcasting*, Summer 1971, pp. 317–329), and *maybe* even earlier, in 1866 (Otis B. Young, "The Real Beginning of Radio: The Neglected Story of Mahlon Loomis," *Saturday Review*, March 7, 1964, pp. 48–50). Stubblefield (1859–1928), like Loomis (1826–1886), didn't understand the distinction between induction and radiation.

14 This entire discussion depends on the creation of kinks in the electric field lines of *accelerated* charges, which occurs only because the antenna current is *time varying*. An antenna with a constant current in it has no electric field kinks and does *not* radiate rf energy.

An oscillating signal source is at the base of the antenna, with the instantaneous maximum antenna current at the source (with peak value I_0), with the current then continually decreasing in amplitude up the antenna until the current becomes zero at the top. Such an antenna is the dipole antenna of Figure A1.2 with the lower arm replaced by the Earth.

To understand the steady decrease in the current up the antenna, realize that the antenna is capacitively coupled to all nearby objects (e.g., the Earth). If we think of the complete antenna as made up of numerous dipoles, each of short length and all in series, then each of these short dipoles can be thought of as half of a capacitor (the Earth is the other half), in analogy with the two halves of a parallel-plate capacitor. These capacitances (shown by the dashed lines in Figure A1.5) provide paths to ground for the antenna current. The current then returns through the Earth to the signal source at the base of the antenna.

A (much) more analytical study[15] of antennas than we are doing here shows that this current distribution can be described, very well, by

$$I(z) = I_0 \sin\left\{\frac{2\pi}{\lambda}(h - z)\right\}\sin(2\pi ft)$$

where h is the height of the antenna. This gives $I(h) = 0$ at the top of the antenna at every instant of time and says $I(0) = I_0 \sin\left\{\frac{2\pi}{\lambda}h)\right\}\sin(2\pi ft)$ at the bottom of the antenna. To have $I(0) = I_0 \sin(2\pi ft)$, we require $\frac{2\pi}{\lambda}h = \frac{\pi}{2}$, or $h = \frac{1}{4}\lambda$, resulting in an antenna called *a grounded quarter-wavelength monopole*. For an AM radio frequency of 1 MHz, $\frac{1}{4}\lambda$ is 75 m(that is, 246 ft). The next time you drive by your local AM radio station's transmitter site, you'll notice the antenna height *is* in that range (since the AM radio band is from 540 to 1,600 kHz, a quarter wavelength varies from 456 ft (at 540 kHz) to 154 ft (at 1,600 kHz)).

15 Entire books have been written on antennas, and a good one (at the math level of this book) is the classic by John Kraus, *Antennas*, first published by McGraw Hill in 1950 and since then in many editions.

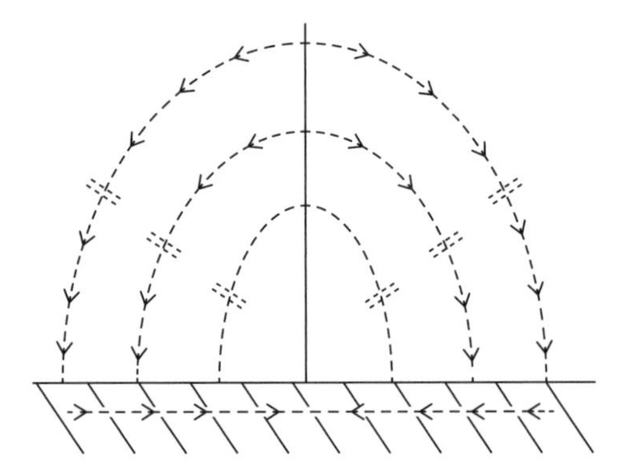

FIGURE A1.5. A vertical antenna with distributive capacitive coupling of its current to the Earth.

Answer to Note 8

If our calculations haven't gone off the rails somewhere, the expression for P in (A1.13), namely, $\frac{E_0^2}{\mu c}$ should have the units of *power per unit area.* We can show this is the case, as follows. First, we know $\mu\epsilon = \frac{1}{c^2}$, and so we know $\mu = \frac{1}{\epsilon c^2}$, which says $\frac{E_0^2}{\mu c} = \frac{E_0^2}{\frac{1}{\epsilon c^2}c} = \frac{E_0^2}{\frac{1}{\varepsilon c}} = \varepsilon c E_0^2$. The MKS units of cE_0^2 are $\frac{\text{meters}}{\text{second}} \bullet \left(\frac{\text{volts}}{\text{meter}}\right)\text{squared} = \frac{\text{volts squared}}{\text{meters} \bullet \text{second}}$. We can find the units of ε by recalling the capacitance of a parallel-plate capacitor is $\varepsilon\frac{A}{d}$, where A is the plate area (units of square meters), and d is the gap spacing (units of meters). So, capacitance has the units of $(\textit{units of } \varepsilon) \bullet \frac{\text{square meters}}{\textit{meters}} = (\textit{units of } \varepsilon) \bullet \text{meters}$. From the discussion between Professors Twombly and Tweedle in the preface, we know the equation for a capacitor is $i = C\frac{dv}{dt}$, which says the units of capacitance are those of $\frac{i}{\frac{dv}{dt}}$, or $\frac{\text{amperes}}{\frac{\text{volts}}{\text{second}}} = \frac{\text{amperes} \bullet \text{second}}{\text{volts}}$.

Thus, $(units\ of\ \varepsilon) \bullet \text{meters} = \frac{\text{amperes} \bullet \text{second}}{\text{volts}}$, or units of $\varepsilon = \frac{\text{amperes} \bullet \text{second}}{\text{meters} \bullet \text{volts}}$. Thus, combining this expression with the units of cE_0^2, we have the units of P as $\left(\frac{\text{amperes} \bullet \text{second}}{\text{meters} \bullet \text{volts}}\right) \bullet \left(\frac{\text{volts squared}}{\text{meters} \bullet \text{second}}\right) = \frac{\text{amperes} \bullet \text{volts}}{\text{square meters}}$. Since amperes • volts are the units of power, the units of P are *power per unit area*. With the units of ε in hand, we can now understand why Maxwell's seminal contribution to the field equations, the displacement current, is a *current density*. That's because $\varepsilon \frac{\partial \mathbf{E}}{\partial t}$ has units of $\frac{\text{amperes} \bullet \text{second}}{\text{meters} \bullet \text{volts}} \bullet \frac{\text{volts / meter}}{\text{second}} = \frac{\text{amperes}}{\text{square meters}}$.

This appendix opened on a somewhat poetic note from physicist Richard Feynman, so let's close it in a similar fashion. In a 1928 magazine advertisement by radio manufacturer Atwater Kent (see note 19 in chapter 3), a company that sold high-end home radio receivers in the 1920s and '30s, these words were "borrowed" (that is, were unattributed) from an 1850 poem by Henry Wadsworth Longfellow:

In the elder days of Art,
Builders wrought with greatest care
Each minute and unseen part;
For the Gods are everywhere.

Atwater Kent meant these words to be taken as a pronouncement by the company of the rightful pride it had in its excellent craftsmanship. But they also apply equally well to the creator of our universe (in whatever form that may be) that brought into existence the beautiful, delicately constructed electromagnetic field equations discovered by Maxwell, the equations that make radio possible.[16]

16 In his *Apology*, Hardy named some people he considered to be "'real" mathematicians. Along with Einstein (see note 14 in the preface) he included (and rightfully so) Maxwell. Hardy evidently didn't hold the development of radio technology against Maxwell! So, perhaps, Hardy wasn't quite the rigid, pure theoretician I may have led you to believe.

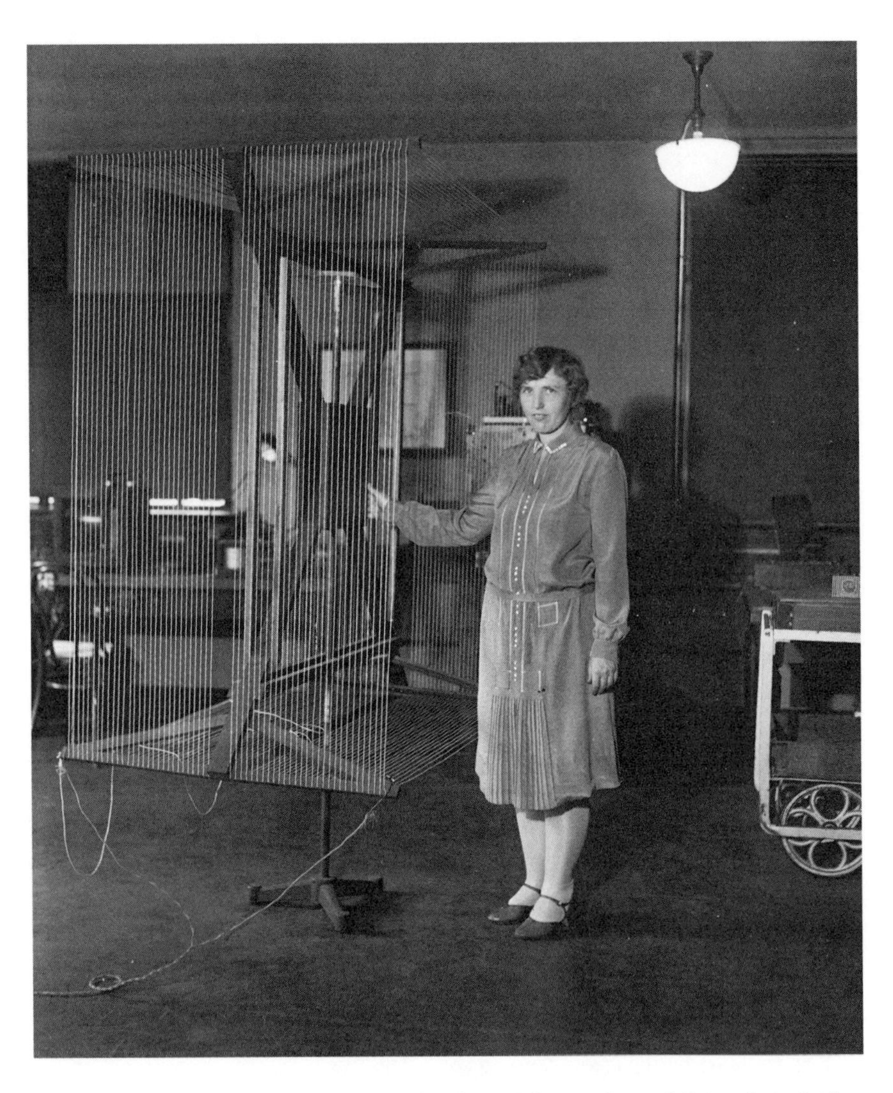

FIGURE A1.6. Mary Texanna Loomis posing in 1928 for another publicity photo (at her radio school—see chapter 6) with a "birdcage" antenna that is just a bit more complicated than the quarter-wavelength monopole.

Solutions, Partial Answers, and More Hints to Most of the End-of-Chapter Challenge Problems in *The Mathematical Radio*

CP1.1: To confirm the claim made for Figure 1.5.3, we do the following. Since the series L and R is in parallel with the series C and R, we can write by inspection that Z is the product of the two series impedances divided by their sum.

That is, $Z = \dfrac{(j\omega L + R)\left(R + \dfrac{1}{j\omega C}\right)}{2R + j\omega L + \dfrac{1}{j\omega C}}$, or after a bit of algebra,

$$Z = \frac{2R\left(R^2 + \frac{L}{C}\right) + R\left(\omega L - \frac{1}{\omega C}\right)^2 + j2R^2\left(\omega L - \frac{1}{\omega C}\right) - j\left(\omega L - \frac{1}{\omega C}\right)\left(R^2 + \frac{L}{C}\right)}{4R^2 + \left(\omega L - \frac{1}{\omega C}\right)^2}.$$

Now, choose $R = \sqrt{\dfrac{L}{C}}$ (and so $\dfrac{L}{C} = R^2$). The imaginary part of Z thus vanishes, with the result $Z = \dfrac{2R(R^2 + R^2) + R\left(\omega L - \dfrac{1}{\omega C}\right)^2}{4R^2 + \left(\omega L - \dfrac{1}{\omega C}\right)^2}$

$$= \frac{4R^3 + R\left(\omega L - \frac{1}{\omega C}\right)^2}{4R^2 + \left(\omega L - \frac{1}{\omega C}\right)^2} = R\frac{4R^2 + \left(\omega L - \frac{1}{\omega C}\right)^2}{4R^2 + \left(\omega L - \frac{1}{\omega C}\right)^2} = R = \sqrt{\frac{L}{C}} \text{ for all } \omega.$$

Looking next at Figure CP1.1, and writing Z_1 as the impedance of R_1 in parallel with C_1, and Z_2 as the impedance of R_2 in parallel with C_2, we have $\dfrac{E}{V} = \dfrac{Z_2}{Z_1 + Z_2}$. Now, as the impedance of an R in parallel with a C is given by $Z = \dfrac{R\dfrac{1}{j\omega C}}{R + \dfrac{1}{j\omega C}} = \dfrac{R}{1 + j\omega RC}$,

then $\dfrac{E}{V} = \dfrac{\dfrac{R_2}{1 + j\omega R_2 C_2}}{\dfrac{R_1}{1 + j\omega R_1 C_1} + \dfrac{R_2}{1 + j\omega R_2 C_2}}$. So, if $R_1C_1 = R_2C_2$, then

$\dfrac{E}{V} = \dfrac{R_2}{R_1 + R_2} = k.$

CP1.2: In Van der Pol's equation $\dfrac{d^2u}{dt^2} - \varepsilon(1 - u^2)\dfrac{du}{dt} + u = 0$ we define $u = \dfrac{dy}{dt}$. Then, $\dfrac{du}{dt} = \dfrac{d^2y}{dt^2}$, and $\dfrac{d^2u}{dt^2} = \dfrac{d^3y}{dt^3}$, and so Van der Pol's equation becomes $\dfrac{d^3y}{dt^3} - \varepsilon\left[1 - \left(\dfrac{dy}{dt}\right)^2\right]\dfrac{d^2y}{dt^2} + \dfrac{dy}{dt} = 0$; that is,

$\dfrac{d^3y}{dt^3} - \varepsilon\dfrac{d^2y}{dt^2} + \varepsilon\left(\dfrac{dy}{dt}\right)^2\dfrac{d^2y}{dt^2} + \dfrac{dy}{dt} = 0$. Then, integrating term by term, we get $\dfrac{d^3y}{dt^3} - \varepsilon\dfrac{dy}{dt} + \varepsilon\dfrac{1}{3}\left(\dfrac{dy}{dt}\right)^3 + y = C$, where C an arbitrary constant, which is Van der Pol's equation if $C = 0$.

CP1.3: The meter, designed for a sinusoidal input, will measure the peak value F_M of $f(t)$—we'll discuss peak detectors later in this book and how they occur in radio—and display $\dfrac{F_M}{\sqrt{2}}$. The actual rms value of the sawtooth $f(t)$, however, is $\sqrt{\dfrac{1}{T}\int_0^T f^2(t)\,dt} = \sqrt{\dfrac{1}{T}\int_0^T \left(F_M\dfrac{t}{T}\right)^2 dt} = F_M\sqrt{\dfrac{1}{T^3}\int_0^T t^2\,dt}$

$= F_M\sqrt{\dfrac{1}{T^3}\left(\dfrac{1}{3}t^3\right)\Big|_0^T} = F_M\sqrt{\dfrac{1}{T^3}\left(\dfrac{T^3}{3}\right)} = \dfrac{F_M}{\sqrt{3}} < \dfrac{F_M}{\sqrt{2}}.$

So, the meter reads high, creating an error of

$$\frac{\dfrac{F_M}{\sqrt{2}} - \dfrac{F_M}{\sqrt{3}}}{\dfrac{F_M}{\sqrt{3}}} \times 100\% = \left(\sqrt{\frac{3}{2}} - 1\right) \times 100\% \approx 22\%.$$

CP1.4: $Z_i = j\omega L + \dfrac{\dfrac{1}{j\omega C}\left[j\omega L + Z(\omega)\right]}{\dfrac{1}{j\omega C} + j\omega L + Z(\omega)} = j\omega L + \dfrac{j\omega L + Z(\omega)}{1 - \omega^2 LC + j\omega CZ(\omega)}$

$$= \frac{j\omega L - j\omega^3 L^2 C - \omega^2 LCZ(\omega) + j\omega L + Z(\omega)}{1 - \omega^2 LC + j\omega CZ(\omega)}$$

$$= \frac{Z(\omega) - \omega^2 LCZ(\omega) + j2\omega L - j\omega^3 L^2 C}{1 - \omega^2 LC + j\omega CZ(\omega)}$$

$= \dfrac{Z(\omega)\left[1 - \omega^2 LC\right] + j\omega L\left[2 - \omega^2 LC\right]}{1 - \omega^2 LC + j\omega CZ(\omega)}$. Now, suppose $1 = \omega^2 LC$. Then, $Z_i = \dfrac{j\omega L}{j\omega CZ(\omega)} = \dfrac{L/C}{Z(\omega)}$. If we repeat this analysis for the modified circuit, we have

$$Z_i = \frac{1}{j\omega C} + \frac{\left(\dfrac{1}{j\omega C} + Z(\omega)\right)j\omega L}{\dfrac{1}{j\omega C} + Z(\omega) + j\omega L} = \frac{1}{j\omega C} + \frac{\left(\dfrac{L}{C} + j\omega LZ(\omega)\right)j\omega C}{1 - \omega^2 LC + j\omega CZ(\omega)}$$

$= \dfrac{1}{j\omega C} + \dfrac{j\omega L - \omega^2 LCZ(\omega)}{1 - \omega^2 LC + j\omega CZ(\omega)}$. Now, suppose $1 = \omega^2 LC$. Then,

$$Z_i = \frac{1}{j\omega C} + \frac{j\omega L - Z(\omega)}{j\omega CZ(\omega)} = \frac{1}{j\omega C} + \frac{L/C}{Z(\omega)} - \frac{1}{j\omega C} = \frac{L/C}{Z(\omega)}.$$

CP1.5: To "see" these two limiting cases, recall that as $\omega \to 0$ inductors short (approach zero impedance) and capacitors open (approach infinite impedance), while as $\omega \to \infty$ inductors open and capacitors short.

CP1.7: We have $A = V\dfrac{R}{R + \dfrac{1}{j\omega C}}$ and $B = V\dfrac{R}{R + j\omega L}$.

So, $\dfrac{A}{B} = \dfrac{\dfrac{R}{R + \dfrac{1}{j\omega C}}}{\dfrac{R}{R + j\omega L}} = \dfrac{R + j\omega L}{R + \dfrac{1}{j\omega C}} \; \dfrac{-\omega^2 LC + j\omega RC}{1 + j\omega RC}$

$$=\frac{[-\omega^2LC+j\omega RC][1-j\omega RC]}{1+(\omega RC)^2}=\frac{-\omega^2LC+\omega^2(RC)^2+j\omega RC+j\omega^3RLC^2}{1+(\omega RC)^2}.$$

Now, suppose $R=\sqrt{\frac{L}{C}}$, that is, suppose $L=R^2$. We define $\omega_0^2=\frac{1}{LC}$; that is, $R^2C^2=LC=\frac{1}{\omega_0^2}$. Then,

$$\frac{A}{B}=\frac{-\omega^2\frac{1}{\omega_0^2}+\omega^2\frac{1}{\omega_0^2}+j\omega\frac{1}{\omega_0}+j\omega^3RCLC}{1+\left(\omega\frac{1}{\omega_0}\right)^2}$$

$$=\frac{j\left(\frac{\omega}{\omega_0}\right)+j\omega^3\left(\frac{1}{\omega_0}\right)\left(\frac{1}{\omega_0^2}\right)}{1+\left(\frac{\omega}{\omega_0}\right)^2}=j\frac{\left(\frac{\omega}{\omega_0}\right)+\left(\frac{\omega}{\omega_0}\right)^3}{1+\left(\frac{\omega}{\omega_0}\right)^2}$$

$$=j\left(\frac{\omega}{\omega_0}\right)\frac{1+\left(\frac{\omega}{\omega_0}\right)^2}{1+\left(\frac{\omega}{\omega_0}\right)^2}=j\left(\frac{\omega}{\omega_0}\right).$$

So, for all ω, from dc to infinity, the phase difference between A and B is (because of the j) a constant 90°, but the amplitudes of the two signals A and B do vary with ω. The magnitudes of A and B are equal only at the single frequency $\omega=\omega_0$.

CP1.8: This problem is solved in section 4.2.

CP1.9: The sampling theorem assumes $\varphi(t)$ is *continuous* at the location of the impulse. $\delta(t)$ occurs at $t=0$, but $\frac{1}{t}$ is discontinuous at $t=0$, and so the sampling theorem is not applicable. Since $\delta(t)$ is even, and $\frac{1}{t}$ is odd, then we immediately conclude that $\int_{-\infty}^{\infty}\frac{1}{t}\delta(t)dt=0$, because the integrand is odd. For a different way to arrive at this conclusion, refer to note 11 in chapter 4.

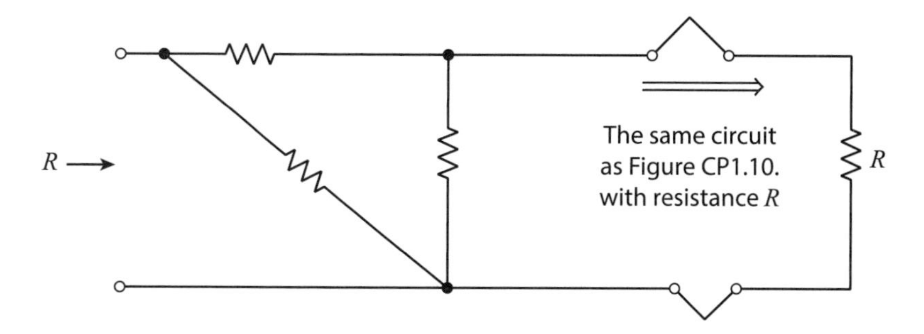

FIGURE SCP1.10. The finite equivalent of an infinite circuit.

CP1.10: Notice that Figure CP1.10 can be redrawn as shown in Figure SCP1.10, because the original circuit is *infinitely* long (what makes this problem "look hard" is what, in fact, makes it doable): The input resistance R is the result of a 1 ohm resistor in parallel with a 1 ohm resistor that is itself in series with the parallel combination of a 1 ohm resistor and a resistor of value R. Mathematically, we write $R = 1 \| (1 + [1 \| R])$. Now, as $1 + [1 \| R] = 1 + \frac{R}{1+R} = \frac{1+2R}{1+R}$, we have $R = \frac{\frac{1+2R}{1+R}}{1 + \frac{1+2R}{1+R}} = \frac{1+2R}{1+R+!+2R} = \frac{1+2R}{2+3R}$, or $2R + 3R^2 = 1 + 2R$, or $3R^2 = 1$, and so $R = \frac{1}{\sqrt{3}}$.

CP1.11: $A\cos(\omega_c t) + B\sin(\omega_c t) = A\frac{e^{j\omega_c t} + e^{-j\omega_c t}}{2} + B\frac{e^{j\omega_c t} - e^{-j\omega_c t}}{2j}$

$$= \left(\frac{1}{2}A + \frac{1}{2j}B\right)e^{j\omega_c t} + \left(\frac{1}{2}A - \frac{1}{2j}B\right)e^{-j\omega_c t} = \frac{A - jB}{2}e^{j\omega_c t} + \frac{A + jB}{2}e^{-j\omega_c t}$$

$$= \frac{1}{2}\sqrt{A^2 + B^2}e^{-j\tan^{-1}\left(\frac{B}{A}\right)}e^{j\omega_c t} + \frac{1}{2}\sqrt{A^2 + B^2}e^{j\tan^{-1}\left(\frac{B}{A}\right)}e^{-j\omega_c t}$$

$$= \sqrt{A^2 + B^2}\,\frac{e^{j\left\{\omega_c t - \tan^{-1}\left(\frac{B}{A}\right)\right\}}}{2} + \sqrt{A^2 + B^2}\,\frac{e^{-j\left\{\omega_c t - \tan^{-1}\left(\frac{B}{A}\right)\right\}}}{2}$$

$$= \sqrt{A^2 + B^2}\cos\left\{\omega_c t - \tan^{-1}\left(\frac{B}{A}\right)\right\}.$$

CP1.12: We recall the Fourier duality theorem derived (using the symbol interchange trick) at the end of section 1.7: the pair $g(t) \leftrightarrow G(\omega)$ establishes the associated pair $G(t) \leftrightarrow 2\pi g(-\omega)$. So, suppose $g(t) = e^{-t}u(t)$. Then,

$$G(\omega) = \int_{-\infty}^{\infty} g(t)e^{-j\omega t}dt = \int_{0}^{\infty} e^{-t}e^{-j\omega t}dt = \int_{0}^{\infty} e^{-(1+j\omega)t}dt = \frac{e^{-(1+j\omega)t}}{-(1+j\omega)}\Big|_{0}^{\infty}$$

$= \frac{1}{1+j\omega}$. That is, $e^{-t}u(t) \leftrightarrow \frac{1}{1+j\omega}$. So, by the duality theorem,

$\frac{1}{1+jt} \leftrightarrow 2\pi e^{\omega}u(-\omega)$. Next, we suppose $g(t) = e^{t}u(-t)$. Then,

$G(\omega) = \int_{-\infty}^{0} e^{t}e^{-j\omega t}dt = \int_{-\infty}^{0} e^{(1-j\omega)t}dt = \frac{e^{(1-j\omega)t}}{1-j\omega}\Big|_{-\infty}^{0} = \frac{1}{1-j\omega}$. That is,

$e^{t}u(-t) \leftrightarrow \frac{1}{1-j\omega}$. So, by the duality theorem, $\frac{1}{1-jt} \leftrightarrow 2\pi e^{-\omega}u(\omega)$.

CP2.1: Let $x(t)$ be the voltage at ①. Since the voltage at ③ is $v_0(t)$, then the voltage at ② must also be $v_0(t)$ to have a zero input to the differential amplifier. Summing currents at ① and ②, and remembering the current into the differential amplifier is zero, we have (we are now in the frequency domain) $\frac{V_i - X}{Z_i} + \frac{V_0 - X}{Z_3} + \frac{V_0 - X}{Z_2} = 0$ and $\frac{X - V_0}{Z_2} = \frac{V_0}{Z_4}$. After some not very difficult algebra, the transfer function is $\frac{v_0(\omega)}{V_i(\omega)} = \frac{Z_3 Z_4}{Z_1 Z_2 + Z_3(Z_1 + Z_2) + Z_3 Z_4}$.

CP2.2: Since the voltage at ① is v, then the voltage at ② must also be v to have a zero input to the differential amplifier. Summing currents at ① and ②, we have $i = \frac{v - v_0}{R_1}$ and $\frac{v_0 - v}{R_2} = C\frac{dv}{dt}$. Thus,

$iR_1 = v - v_0$, or $v = v_0 + iR_1$, and so $\frac{v_0 - (v_0 + iR_1)}{R_2} = C\frac{dv}{dt} = -i\frac{R_1}{R_2}$, and

so $-C\frac{R_2}{R_1}\frac{dv}{dt} = i = C_{eq}\frac{dv}{dt}$, where $C_{eq} = -C\frac{R_2}{R_1} < 0$, because the two resistors and C are all greater than 0.

CP2.3: (a) $Z = \dfrac{\dfrac{1}{j\omega C}(R + j\omega L)}{R + j\omega L + \dfrac{1}{j\omega C}} = \dfrac{R + j\omega L}{1 - \omega^2 LC + j\omega RC}$

$= R\dfrac{1 + j\dfrac{\omega L}{R}}{1 - \omega^2 LC + j\omega RC}$. Since $LC = \dfrac{1}{\omega_0^2}$, then

$$Z = R\frac{1 + j\dfrac{\omega}{\omega_0}\dfrac{\omega_0 L}{R}}{1 - \omega^2\dfrac{1}{\omega_0^2} + j\omega\dfrac{RCL}{L}} = R\frac{1 + j\left(\dfrac{\omega}{\omega_0}\right)Q}{1 - \omega^2\dfrac{1}{\omega_0^2} + j\omega\dfrac{LC}{\dfrac{L}{R}}}$$

$= R\dfrac{1 + j\left(\dfrac{\omega}{\omega_0}\right)Q}{1 - \left(\dfrac{\omega}{\omega_0}\right)^2 + j\omega\dfrac{1}{\omega_0^2}\dfrac{1}{\dfrac{L}{R}}} = \dfrac{1 + j\left(\dfrac{\omega}{\omega_0}\right)Q}{1 - \left(\dfrac{\omega}{\omega_0}\right)^2 + j\dfrac{\omega}{\omega_0}\dfrac{1}{\dfrac{\omega_0 L}{R}}}$, or at

last, $Z = \dfrac{1 + j\left(\dfrac{\omega}{\omega_0}\right)Q}{1 - \left(\dfrac{\omega}{\omega_0}\right)^2 + j\left(\dfrac{\omega}{\omega_0}\right)\dfrac{1}{Q}}$. We define $x = \dfrac{\omega}{\omega_0}$. Then,

$Z = R\dfrac{1 + jxQ}{1 - x^2 + j\dfrac{x}{Q}} = R\dfrac{(1 + jxQ)\left(1 - x^2 - j\dfrac{x}{Q}\right)}{(1 - x^2)^2 + \left(\dfrac{x}{Q}\right)^2}$. Thus, Z is real when

the numerator is real (when the imaginary part is zero). That is, Z is real when $xQ(1 - x^2) - \dfrac{x}{Q} = 0$, or $Q(1 - x^2) = \dfrac{1}{Q}$. This is easily solved to give $x = \dfrac{\omega}{\omega_0} = \sqrt{1 - \dfrac{1}{Q^2}}$. For part (b), we first calculate $|Z|^2$ and then set $\dfrac{d|Z|^2}{dx} = 0$. This leads to a quartic equation in x that is quadratic in x^2 and so easily solved for x.

CP2.4: Call the voltage at ① X, and the voltage at ② Y. Then, summing currents at ①, ②, and ③ we have, respectively,

$$\frac{V_i - X}{R} = \frac{X}{\frac{1}{j2\omega C}} + \frac{X - V_0}{R}, \frac{V_i - Y}{\frac{1}{j\omega C}} = \frac{Y}{\frac{1}{2}R} + \frac{Y - V_0}{\frac{1}{j\omega C}}, \frac{Y - V_0}{\frac{1}{j\omega C}} = \frac{V_0 - X}{R}.$$

After some algebra these equations give

$$\left|\frac{V_0}{V_i}\right| = \sqrt{\frac{\left[1 - \left(\frac{\omega}{\omega_0}\right)^2\right]^2}{\left[1 - \left(\frac{\omega}{\omega_0}\right)^2\right]^2 + 16\left(\frac{\omega}{\omega_0}\right)^2}} \text{ where } \omega_0 = \frac{1}{RC}.$$

CP2.5: $\cos^4(\omega_m t)\cos(\omega_c t) = \left(\frac{e^{j\omega_m t} + e^{-j\omega_m t}}{2}\right)^4 \cos(\omega_c t)$

$= \frac{1}{16}\left(\frac{e^{j\omega_m t} + e^{-j\omega_m t}}{2}\right)^4 \cos(\omega_c t) = \frac{1}{16}\left[e^{j\omega_m t}(1 + e^{-j2\omega_m t})\right]^4 \cos(\omega_c t)$

$= \frac{e^{j4\omega_m t}}{16}(1 + e^{-j2\omega_m t})^4 \cos(\omega_c t)$. From the binomial theorem we have $(1+a)^4 = \sum_{k=0}^{4}\binom{4}{k}a^k 1^{4-k} = \sum_{k=0}^{4}\binom{4}{k}a^k$, or writing out the sum, $(1+a)^4 = \binom{4}{0} + \binom{4}{1}a + \binom{4}{2}a^2 + \binom{4}{3}a^3 + \binom{4}{4}a^4 = 1 + 4a$ $+ 6a^2 + 4a^3 + a^4$. So, with $a = e^{-j2\omega_m t}$, we have $(1 + e^{-j2\omega_m t})^4$ $= 1 + 4e^{-j2\omega_m t} + 6e^{-j4\omega_m t} + 4e^{-j6\omega_m t} + e^{-j8\omega_m t}$, and so

$$\frac{e^{j4\omega_m t}}{16}(1 + e^{-j2\omega_m t})^4 \cos(\omega_c t)$$

$$= \frac{1}{16}(e^{j4\omega_m t} + 4e^{j2\omega_m t} + 6 + 4e^{-j2\omega_m t} + e^{-j4\omega_m t})\cos(\omega_c t)$$

$$\left\{\frac{6}{16} + \frac{1}{16}(e^{j4\omega_m t} + e^{-j4\omega_m t}) + \frac{4}{16}(e^{j2\omega_m t} + e^{-j2\omega_m t})\right\}\cos(\omega_c t).$$

So, remembering that $\cos(A)\cos(B)=\frac{1}{2}\{\cos(A-B)+\cos(A+B)\}$,

we have $\cos^4(\omega_m t)\cos(\omega_c t)=\frac{3}{8}\cos(\omega_c t)+\frac{1}{16}\left[\cos\{(\omega_c-4\omega_m)t\}\right.$

$\left.+\cos\{(\omega_c+4\omega_m)t\}\right]+\frac{1}{4}\left[\cos\{(\omega_c-2\omega_m)t\}+\cos\{(\omega_c+2\omega_m)t\}\right].$

As a partial check on the arithmetic, notice that at $t=0$ this result reduces to the undeniably true $1=1$. This doesn't prove we haven't made an error somewhere, but if we hadn't gotten that equality, that would indicate we went astray somewhere.

CP2.6: To find the time function that pairs with $M(\omega)G(\omega)$, we'll use the inverse Fourier transform to write that time function as

$$\frac{1}{2\pi}\int_{-\infty}^{\infty} M(\omega)G(\omega)e^{j\omega t}d\omega=\frac{1}{2\pi}\int_{-\infty}^{\infty} M(\omega)\left\{\int_{-\infty}^{\infty} g(u)e^{-ju\omega}du\right\}e^{j\omega t}d\omega$$

$$=\int_{-\infty}^{\infty} g(u)\left\{\frac{1}{2\pi}\int_{-\infty}^{\infty} M(\omega)e^{j(t-u)\omega}d\omega\right\}du.$$

Since $m(t)=\frac{1}{2\pi}\int_{-\infty}^{\infty} M(\omega)e^{j\omega t}d\omega$, we see that

$\frac{1}{2\pi}\int_{-\infty}^{\infty} M(\omega)e^{j(t-u)\omega}d\omega=m(t-u)$, and so the time function we are after is $\int_{-\infty}^{\infty} g(u)m(t-u)du=g(t)*m(t)$, as was to be shown.

For the second question, a direct evaluation of $g(t)*\delta(t)$ says

$\int_{-\infty}^{\infty} g(u)\delta(t-u)du=g(u)|_{u=t}=g(t)$, where the sampling theorem (see section 1.7) has been used. Alternatively, to use the transform pair $m(t)*g(t)\leftrightarrow M(\omega)G(\omega)$, let $m(t)=\delta(t)$, which says $M(\omega)=1$. Thus, $\delta(t)*g(t)\leftrightarrow G(\omega)$. But since $g(t)\leftrightarrow G(\omega)$, we immediately have $\delta(t)*g(t)=g(t)$, and we are done.

CP2.7: Call the voltage at ② X. The voltage at ① is zero. Summing at ② and ③, respectively, $\frac{V-X}{R_f}=\frac{X}{\frac{1}{j\omega C_2}}+\frac{X-V_f}{j\omega L}$ and

$\frac{X-V_f}{j\omega L}=\frac{V_f}{\frac{1}{j\omega C_1}}+\frac{V_f}{R_1}$. With some algebra these two equations reduce to $V_f = -V \dfrac{j\dfrac{1}{\omega L R_f}}{\left(j\omega C_2 + \dfrac{1}{j\omega L} + \dfrac{1}{R_f}\right)\left(j\omega C_1 + \dfrac{1}{j\omega L} + \dfrac{1}{R_1}\right) + \dfrac{1}{\omega^2 L^2}}$.

This looks like a huge mess, but if we think carefully about what we are attempting to calculate, it is actually not difficult at all. We know, from the problem statement that summing at ① says V_f is a *negative real* quantity $\left(-\dfrac{R_1}{R_2}\right)$ times V. So, the messy equation we just derived has to have that property, as well, and, in fact, it does—but only at a specific frequency (the frequency at which the Colpitts circuit oscillates). We see from the derived equation that V_f is $-V$ multiplied by a pretty complicated complex quantity, and we want to find the frequency that makes that complex quantity positive real. The standard approach is to multiply through denominator and numerator by the conjugate of the denominator, giving a positive real denominator and a new, complex numerator whose imaginary part will equal zero at the oscillation frequency. We could do all that, but here's a somewhat shorter route. If we calculate just the real part of the denominator (which is, of course, the real part of the denominator conjugate) and then multiply that real part by the numerator (which is purely imaginary), that will give us the complex portion of the new numerator (which is zero at the oscillation frequency). The real part of the denominator is $-\omega^2 C_1 C_2 + \dfrac{C_1 + C_2}{L} + \dfrac{1}{R_1 R_f}$, which when set equal to zero gives the oscillation frequency as $\omega = \sqrt{\dfrac{1}{L}\left(\dfrac{C_1 + C_2}{C_1 C_2}\right) + \dfrac{1}{R_1 R_f C_1 C_2}}$. As a final step, you should multiply the complex part of the denominator conjugate by the numerator and show that the result is positive at the oscillation frequency; that is, $V_f = -V\dfrac{\text{positive}}{\text{positive}}$, in agreement with $V_f = -V\dfrac{R_1}{R_2}$.

CP2.8: Summing at ① and ②, respectively, $\dfrac{V_1 - X}{\dfrac{1}{j\omega C}} = \dfrac{X}{R} + \dfrac{X - V_2}{\dfrac{1}{j\omega C}}$ and $\dfrac{V_1 - V_2}{R} + \dfrac{X - V_2}{\dfrac{1}{j\omega C}} = 0$. After a bit of careful algebra, these two equations reduce to $\dfrac{V_1}{V_2} = H = \dfrac{1-(\omega RC)^2 + j2\omega RC}{1-(\omega RC)^2 + j3\omega RC}$, or after normalization, $\dfrac{V_1}{V_2} = H = \dfrac{1-\left(\dfrac{\omega}{\omega_0}\right)^2 + j2\left(\dfrac{\omega}{\omega_0}\right)}{1-\left(\dfrac{\omega}{\omega_0}\right)^2 + j3\left(\dfrac{\omega}{\omega_0}\right)}$, from which the plots in the problem statement result.

CP3.1: The two plate currents, when $v_i = 0$, are each 2.25 mA. So, a total of 4.5 mA flows in R_k to give a 1.5 volt drop. Thus, $R_k = \dfrac{1.5}{4.5 \times 10^{-3}} = 333$ ohms. To have a 250 volt drop across each tube, from plate to cathode, the plate voltage must be 251.5 volts, and so with 2.25 mA in each R_L we have a 48.5 volt drop, which says $R_L = \dfrac{48.5}{2.25 \times 10^{-3}} = 21{,}556$ ohms. Using the subscripts 1 and 2 to denote the triodes V1 and V2, respectively, we have
(a) $v_{gk1} = v_i - (i_{p1} + i_{p2})R_k$, (b) $i_{p1} = (6 + 2.5v_{gk1}) \times 10^{-3}$,
(c) $v_{gk2} = -(i_{p1} + i_{p2})R_k$, (d) $i_{p2} = (6 + 2.5v_{gk2}) \times 10^{-3}$, and
(e) $v_{p2} = 300 - i_{p2}R_L$. After some algebra, these equations reduce to $v_{p2} = 300 - \dfrac{6{,}000R_L - 6.25v_iR_kR_L}{(1{,}000 + 2.5R_k)^2 - 6.25R_k^2}$, and so

$\dfrac{dv_{p2}}{dv_i} \approx \dfrac{6.25R_kR_L}{(1{,}000 + 2.5R_k)^2 - 6.25R_k^2}$, or after plugging in numbers,

$\dfrac{dv_{p2}}{dv_i} \approx 16.8.$

CP3.2: Calling the voltage at ① X, then summing at ① and ② we have, respectively, $\frac{V_i - X}{R_1} + \frac{V_0 - X}{\frac{1}{j\omega C_1}} = \frac{X}{R_2} + \frac{X}{\frac{1}{j\omega C_2}}$, and $\frac{X}{\frac{1}{j\omega C_2}} + \frac{V_0}{R_3} = 0$. After some algebra these two equations reduce to $\frac{V_0}{V_i} = \frac{j\omega R_3 R_2 C_2}{\omega^2 R_1 R_2 R_3 C_1 C_2 - R_1 - R_2 - j\omega R_1 R_2 (C_1 + C_2)}$. If we make the real part of the denominator vanish, then $\frac{V_0}{V_i}$ is purely real; that is, if $\omega^2 = \frac{R_1 + R_2}{R_1 R_2 R_3 C_1 C_2}$ (which varies as R_2 is varied), and at that frequency $\frac{V_0}{V_i} = -\frac{R_3 C_2}{R_1 (C_1 + C_2)}$, which is independent of R_2.

CP3.3: Filter #1 is centered at $\frac{1}{6}\omega_c$. Filter #2 is centered at $\frac{1}{2}\omega_c$. Filter #3 is centered at $\frac{2}{3}\omega_c$.

CP3.5: With N identical amplifiers, each with gain A and saturated output of S, an input of $\frac{S}{A^N}$ to the first amplifier will saturate the final amplifier. With $S = 10$ volts and $A = 5$, an input of $\frac{10}{5^N}$ will saturate the final amplifier. If the input to the first amplifier is to be 0.1 volt to achieve saturation, then $\frac{10}{5^N} \leq 0.1$, or $100 \leq 5^N$. So, $N = 3$, and the actual voltage that saturates the final (third) amplifier is $\frac{10}{125} = 0.08$ volt.

CP3.6: We start by writing $i = \frac{v - x}{R_1}$. Then, summing currents at the nodes *not* connected to a differential amplifier output (starting at the upper node with voltage v), we have $\frac{x - v}{R_2} = \frac{v - y}{R_3}$, and $C\frac{d}{dt}(v - y) = \frac{v}{R_4}$. Manipulation of these three equations leads to

$$v = \frac{R_1 R_3 R_4}{R_2} C \frac{di}{dt}, \text{ and so } L_{eq} = \frac{R_1 R_3 R_4}{R_2} C.$$

CP4.1: Calling the voltage at ① *and* at ② X (the input voltages at the differential inputs are equal), we sum currents at ① to write $\frac{V_i - X}{R_1} = \frac{X - V_0}{R}$ and at ② to write $\frac{V_i - X}{R} = \frac{X}{\frac{1}{j\omega C}}$. These two equations lead to $\frac{V_o}{V_i} = \frac{1 - j\omega RC}{1 + j\omega RC}$, and so $\left|\frac{V_o}{V_i}\right|^2 = \frac{1 + (\omega RC)^2}{1 + (\omega RC)^2} = 1$, the definition of an all-pass filter. Notice that the phase is *not* independent of frequency; $\varphi = -2\tan^{-1}(\omega RC)$.

CP4.2: Starting where the problem statement ends, we have the output of a Hilbert transformer due to the input $m(t-a)$ as $\frac{1}{\pi}\int_{-\infty}^{\infty} \frac{m(u-a)}{t-u} du = \frac{1}{\pi}\int_{-\infty}^{\infty} \frac{m(z)}{t-z-a} dz = \frac{1}{\pi}\int_{-\infty}^{\infty} \frac{m(u)}{t-u-a} du$; that is, the input $m(t-a)$ produces the output $\frac{1}{\pi}\int_{-\infty}^{\infty} \frac{m(u)}{t-u-a} du$, which is just what we said would be the output under the assumption of time invariance.

CP4.3: If we let $p \to 0$, as suggested in (6) (and so $e^{-px} \to 1$), we have (since the integrand is even)

$$\int_{-\infty}^{\infty} \frac{\sin(ax)}{x} dx = 2 \lim_{p \to 0} \tan^{-1}\left(\frac{a}{p}\right) = \begin{cases} 2\tan^{-1}(+\infty) = \pi \text{ if } a > 0 \\ 0 \\ 2\tan^{-1}(-\infty) = -\pi \text{ if } a < 0 \end{cases}.$$

CP4.4: (a) Since $\cos(\omega t) = \frac{e^{j\omega t} + e^{-j\omega t}}{2} = \frac{e^{j\omega t} + e^{j(-\omega t)}}{2}$ we have

$$\overline{\cos(\omega t)} = \frac{e^{j\left(\omega t - \frac{\pi}{2}\right)} + e^{j\left(-\omega t + \frac{\pi}{2}\right)}}{2} = \frac{e^{-j\frac{\pi}{2}} e^{j\omega t} + e^{j\frac{\pi}{2}} e^{-j\omega t}}{2} = \frac{-je^{j\omega t} + je^{-j\omega t}}{2}$$

$$= -j\frac{e^{j\omega t} - e^{-j\omega t}}{2} = -j\frac{2j\,\sin(\omega t)}{2} = \sin(\omega t).$$

(b) Let the constant be k. Then $\overline{k} = \frac{1}{\pi}\int_{-\infty}^{\infty} \frac{k}{t-u} du$. To avoid integrating through the singularity at $u = t$, we write

$$\overline{k} = \frac{k}{\pi} \lim_{\varepsilon \to 0, T \to \infty} \left[\int_{-T}^{t-\epsilon} \frac{du}{t-u} + \int_{t+\varepsilon}^{T} \frac{du}{t-u}\right].$$

Change variable to $s=t-u$ $(ds=-du)$. Then,

$$\overline{k}=\frac{k}{\pi}\lim_{\varepsilon\to 0,T\to\infty}\left[\int_{t+T}^{\epsilon}\frac{-ds}{s}+\int_{-\varepsilon}^{t-T}\frac{-ds}{s}\right]$$

$$=\frac{k}{\pi}\lim_{\varepsilon\to 0,T\to\infty}\left[\int_{\varepsilon}^{t+T}\frac{ds}{s}+\int_{t-T}^{-\epsilon}\frac{ds}{s}\right]$$

$$=\frac{k}{\pi}\lim_{\varepsilon\to 0,T\to\infty}\left[\{\ln(s)\}\big|_{\epsilon}^{t+T}+\{\ln(s)\}\big|_{t-T}^{-\varepsilon}\right]$$

$$=\frac{k}{\pi}\lim_{\varepsilon\to 0,T\to\infty}\left[\ln(t+T)-\ln(\varepsilon)+\ln(-\epsilon)-\ln(t-T)\right]$$

$$=\frac{k}{\pi}\lim_{\varepsilon\to 0,T\to\infty}\left[\ln\left(\frac{t+T}{\epsilon}\right)+\ln\left(\frac{-\varepsilon}{t-T}\right)\right]$$

$$=\frac{k}{\pi}\lim_{\varepsilon\to 0,T\to\infty}\left[\ln\left(\frac{t+T}{\epsilon}\right)+\ln\left(\frac{\varepsilon}{T-t}\right)\right]$$

$$=\frac{k}{\pi}\lim_{\varepsilon\to 0,T\to\infty}\left[\ln\left\{\left(\frac{t+T}{\epsilon}\right)\left(\frac{\varepsilon}{T-t}\right)\right\}\right].$$

Notice that the two epsilons cancel even before we take the limit, and we have, $\overline{k}=\frac{k}{\pi}\lim_{T\to\infty}\ln\left(\frac{T+t}{T-t}\right)=\frac{k}{\pi}\ln(1)=0.$

(c) From the start of section 4.3 we have the impulse response of a Hilbert transformer as $\frac{1}{\pi t}$. But the output of a Hilbert transformer is, by definition, the Hilbert transform of the input, and so if the input is $\delta(t)$, then the output is $\overline{\delta(t)}=\frac{1}{\pi t}$. Thus, $\overline{\overline{\delta(t)}}=-\delta(t)=\frac{\overline{1}}{\pi t}$, which says $\frac{\overline{1}}{t}=-\pi\delta(t)$.

CP4.5: The formal Hilbert transform integral for the unit pulse from $t=0$ to $t=1$ is $\frac{1}{\pi}\int_{-\infty}^{\infty}\frac{du}{t-u}$ with a singularity at $u=t$. There are three distinct situations to be considered in turn: (a) the singularity occurs *before* the start of the pulse $(u=t<0)$; (b) the singularity occurs *during* the pulse $(0<t<1)$; (c) the singularity occurs *after* the pulse $(u=t>1)$. For (a), the transform integral reduces to $\frac{1}{\pi}\int_{0}^{1}\frac{du}{t-u}=\frac{1}{\pi}\ln\left(\frac{t}{t-1}\right)$, $t<0$. For (b) we use the same limiting idea we used in Challenge Problem **4.4** and write the integral as $\lim_{\varepsilon\to 0}\left\{\int_0^{t-\varepsilon}\frac{du}{t-u}+\frac{1}{\pi}\int_{t+\varepsilon}^{1}\frac{du}{t-u}\right\}=\frac{1}{\pi}ln\left(\frac{t}{1-t}\right)$, $0<t<1$. For (c) the integral reduces to $\frac{1}{\pi}\int_0^1\frac{du}{t-u}=\frac{1}{\pi}\ln\left(\frac{t}{t-1}\right)$, $t>1$. We can write the integral as one expression: $\frac{1}{\pi}\ln\left|\frac{t}{t-1}\right|$, for all t.

CP5.1: $M(\omega)=2\pi e^{-\omega}u(\omega)+2\pi e^{\omega}u(-\omega)$, and so $\overline{M(\omega)}=-j2\pi e^{-\omega}u(\omega)+j2\pi e^{\omega}u(-\omega)$. Thus, $\overline{m(t)}=\frac{-j}{1-jt}+\frac{j}{1+jt}=\frac{2t}{1+t^2}$, and so, as we started with $m(t)=\frac{2}{1+t^2}$, we have $\frac{2}{1+t^2}\leftrightarrow\frac{2t}{1+t^2}$, or $\frac{1}{1+t^2}\leftrightarrow\frac{t}{1+t^2}$.

CP5.5: Following the hint, we have (at $\frac{\omega}{\omega_{IF}}=1$) $e_1=\frac{1}{1-\frac{1}{\alpha^2}+j\frac{1}{\alpha^2 Q}}$, and $e_2=\frac{1}{1-\frac{1}{\beta^2}+j\frac{1}{\beta^2 Q}}$, and so setting $|e_1|=|e_2|$ gives $\left(1-\frac{1}{\alpha^2}\right)^2+\left(1-\frac{1}{\alpha^2 Q}\right)^2=\left(1-\frac{1}{\beta^2}\right)^2+\left(1-\frac{1}{\beta^2 Q}\right)^2$.

With some algebra this reduces to

$\frac{1}{\beta^2}\left(2-\frac{1}{Q^2}\right)-\frac{1}{\beta^4}=\frac{1}{\alpha^2}\left(2-\frac{1}{Q^2}\right)-\frac{1}{\alpha^4}$. Let $x=\frac{1}{\beta^2}$ and $y=\frac{1}{\alpha^2}$. Then, with $c=2-\frac{1}{Q^2}$ we have (with a bit of algebra) that $cx-x^2=cy-y^2$, which becomes $c(x-y)=(x-y)(x+y)$. We can cancel the $x-y$ factors *if* we take $x\neq y$, that is, $\alpha\neq\beta$. Then, $c=x+y$, or $y=c-x$. So, $\frac{1}{\alpha^2}=2-\frac{1}{Q^2}-\frac{1}{\beta^2}$, and we have $\alpha=\frac{1}{\sqrt{2-\frac{1}{Q^2}-\frac{1}{\beta^2}}}$.

Acknowledgments

I have a *lot* of people to thank for helping me, in any number of ways, as I wrote this book. I recall once reading that the best a writer can hope for, on their own, is to be *good*. To be better than *good* a writer needs *great* editing. I believe I have been pretty lucky on that score. Let me start with the editorial spirits, past and present, at Princeton University Press. Susannah Shoemaker offered me a contract for the book. Her successor as the Press's math editor (Diana Gillooly) and Diana's assistant (Kiran Pandey) guided the book through its final acceptance by the Princeton Editorial Board. To that end, two anonymous reviewers provided the Press very helpful commentary that greatly improved the book. The skills of Mark Bellis (production editor), Barbara Liguori (copyeditor), and Dimitri Karetnikov (illustration editor) are on obvious display as you hold this book in your hands.

All the computer-generated plots were done with the aid of MATLAB, a product of The MathWorks, Natick, MA, who kindly provided me a license pro bono for its use.

I owe very special thanks to Andrew Simoson (mathematics professor at King University in Bristol, Tennessee) who graciously agreed to write the foreword to the book. The result is an erudite math-history essay, illustrated with his wonderful art. I knew, from reading his own books in which his drawings have appeared, that Andy is a talented artist, and I approached him about contributing a foreword in the secret hope he would do the same for this book. Andy didn't fail to deliver in grand style, and I hope you get as big a kick out of Mary Cartwright's cat as I do!

The photo illustrations in the book were enhanced by the efforts of three particular individuals. In the UK, Steven Archer, Sub-Librarian at Trinity College, Cambridge, provided the wonderful photograph of G. H. Hardy used in the preface; and Ioan Nedelcu, rights and licensing manager at TopFoto, Kent, threw me a life preserver as I floundered in the choppy copyright and intellectual property waters of reprinting *Punch* cartoons. In the US, Michael J. Ward, publisher of magazineart .org, provided me with a high-resolution photograph of the frontispiece art, a work I had searched for (in vain) *for years*!

If this book wasn't already dedicated to my wife of more than 60 years, it would be dedicated to the memory of John T. Frye (1910–1985), who authored 119 "Carl and Jerry" stories for *Popular Electronics* magazine (see note 20 in the preface). Frye was a 1930 graduate of Logansport High School, in Indiana, and his classmates printed this line beneath his graduation yearbook photograph: "'All roads lead to Rome,' and all the high school corridors seem to lead to John's chair." Those enigmatic words were a reference to the wheelchair in which, after an early childhood bout with polio, he lived the rest of his life. Physically confined, he let his imagination soar and earned a living writing electronics-oriented fiction. I believe his "Carl and Jerry" tales alone convinced more youngsters to become electrical and/or electronic engineers or physicists than did even the most sophisticated textbook.[1] Professors Tweedle and Twombly in the preface are clearly just Carl and Jerry grown up and transplanted from electrical engineering to the math department.

I spent the academic year 1999–2000 on sabbatical leave from the University of New Hampshire, Durham, as a visiting professor at the University of Virginia in Charlottesville. While there I taught mostly in the electrical engineering department, but the classroom experience that stands out in my memory is an experimental course on American broadcast radio history that I offered in the Technology, Culture, and Society Department. Chapter 6 of this book is almost totally based on that class, and I owe much to the TCS faculty (and students!) at UVA for taking a chance with "the new guy on campus."

I want to express HUGE thanks to my favorite writing spot, the Exeter, New Hampshire, coffee shop Me & Ollie's, on Water Street.

1 Years before "Carl and Jerry," Frye had already developed his storytelling technique of a dialogue between two characters with his *Mac's Radio Service Shop* tales. First appearing in a 1948 issue of *Radio News* magazine (and later in *Electronics World* magazine), the stories revolved around the owner of a radio-repair shop (a business, alas, that has gone the same way into oblivion as has that of the village smithies who once shod the townsfolk's horses) and his young, eager-to-learn assistant Barney (who, by his own admission, is "plenty dumb about radio"). Over the next 30 years Frye wrote dozens more stories of the encounters between Mac and Barney, in which Barney (and, of course, the reader) learned about radio and electronics (at least two of the stories discuss SSB radio). In a 1962 newspaper interview Frye is quoted as saying, about his body of published work, "This doesn't look like much, but there really isn't much." There are thousands of now-senior professionals in engineering and physics who would strongly disagree.

Shoved into one corner of the shop is a really scruffy, beat-up, ready-for-the-junkyard (but oh-so-comfy) stuffed leather chair that always seems to be available for me (maybe nobody else but I would sit in it!). But *I* sit in it for two hours, nearly every morning, drink coffee, and just write. And I am happy.

Finally, I asked friends from high school, college, and later life to read various parts of early drafts of the book. Keeping my fingers crossed that I haven't left anyone out, here they are (in no particular order): Judy Grabiner, Pat and John Patterson, Kaylyn and Kent Warner, David Wunsch, Trevor Lipscombe, Ron and Jamie Addison, John Molinder, Gail and Edmund Tomlinson, John Baker, and Michael Blasgen.

My sincere thanks to all.

Exeter, New Hampshire
February 2023

Illustration Credits

The illustrations "When Uncle Sam Wants to Talk to All the People" (artist Lynn Bogue Hunt (1877–1960)), "It's Great to be a Radio Maniac" (artist Frederick G. Cooper (1883–1962)), "The Wonders of Radio" (artist Norman Rockwell (1894–1978)), the National Radio Institute advertisement, and the photos of Mary Texanna Loomis are all in the public domain. The *Punch* illustrations "The Secret Hope" (artist Ernest Howard Shepard (1879–1976)) and "A Lost Art" (artist David Louis Ghilchik (1892–1972)) are reproduced here by permission of TopFoto, Edenbridge, Kent, UK. The photographs of Loomis are reproduced here with the permission of Shorpy American Historical Photo Archive. The frontispiece and the illustration "And Now—The Auto Radio-Phone" (artist Howard Vachel Brown (1878–1945)) are both in the public domain; photographs of both were taken by Michael Ward of magazineart.org and are reproduced here with his permission. The photograph of G. H. Hardy is reproduced here by kind permission of the Master and Fellows of Trinity College, Cambridge.

Index

Also by Paul J. Nahin

Oliver Heaviside (1988, 2002)
Time Machines (1993, 1999)
The Science of Radio (1996, 2001)
Time Travel (1997, 2011)
An Imaginary Tale (1998, 2007, 2010)
Duelling Idiots (2000, 2002)
When Least Is Best (2004, 2007, 2021)
Dr. Euler's Fabulous Formula (2006, 2011)
Chases and Escapes (2007, 2012)
Digital Dice (2008, 2013)
Mrs. Perkins's Electric Quilt (2009)
Number-Crunching (2011)
The Logician and the Engineer (2012, 2017)
Will You Be Alive Ten Years From Now? (2013)
Holy Sci-Fi! (2014)
Inside Interesting Integrals (2015, 2020)
In Praise of Simple Physics (2016, 2017)
Time Machine Tales (2017)
How to Fall Slower Than Gravity (2018)
Transients for Electrical Engineers (2019)
Hot Molecules, Cold Electrons (2020)
In Pursuit of Zeta-3 (2021)
The Probability Integral (2023)